CHEVRON-NOTCHED SPECIMENS: TESTING AND STRESS ANALYSIS

A symposium
sponsored by
ASTM Committee E-24
on Fracture Testing
Louisville, Ky., 21 April 1983

ASTM SPECIAL TECHNICAL PUBLICATION 855
J. H. Underwood, Army Armament R&D Center,
S. W. Freiman, National Bureau of Standards, and
F. I. Baratta, Army Materials and Mechanics
Research Center, editors

ASTM Publication Code Number (PCN)
04-855000-30

1916 Race Street, Philadelphia, Pa. 19103

Library of Congress Cataloging in Publication Data

Chevron-notched specimens, testing and stress analysis.

(ASTM special technical publication; 855)
Papers presented at the Symposium on Chevron-notched Specimens: Testing and Stress Analysis.
Includes bibliographies and index.
1. Notched bar testing—Congresses. 2. Strains and stresses—Congresses. I. Underwood, J. H. II. Freiman, S. W. III. Baratta, F. I. IV. ASTM Committee E-24 on Fracture Testing. V. Symposium on Chevron-notched Specimens: Testing and Stress Analysis (1983; Louisville, Ky.) VI. Series.
TA418.17.C48 1984 620.1′126 84-70336
ISBN 0-8031-0401-4

Library of Congress Catalog Card Number: 84-70336

NOTE

The Society is not responsible, as a body, for the statements and opinions advanced in this publication.

Printed in Baltimore, Md. (b)
November 1984

Foreword

This publication, *Chevron-Notched Specimens: Testing and Stress Analysis,* contains papers presented at the Symposium on Chevron-Notched Specimens: Testing and Stress Analysis which was held 21 April 1983 at Louisville, Kentucky. ASTM's Committee E-24 on Fracture Testing sponsored the symposium. J. H. Underwood, Army Armament R&D Center, S. W. Freiman, National Bureau of Standards, and F. I. Baratta, Army Materials and Mechanics Research Center, served as symposium chairmen and editors of this publication.

The symposium chairmen are pleased to credit D. P. Wilhem, Northrop Corp., for proposing and initiating this symposium.

Related ASTM Publication

Probabilistic Fracture Mechanics and Fatigue Methods: Applications for Structural Design and Maintenance, STP 798 (1983), 04-798000-30

Fracture Mechanics: Fourteenth Symposium, Volume I: Theory and Analysis; Volume II: Testing and Application, STP 791 (1983), 04-791000-30

Fracture Mechanics for Ceramics, Rocks, and Concrete, STP 745 (1981), 04-745000-30

Fractography and Materials Science, STP 733 (1981), 04-733000-30

A Note of Appreciation to Reviewers

The quality of the papers that appear in this publication reflects not only the obvious efforts of the authors but also the unheralded, though essential, work of the reviewers. On behalf of ASTM we acknowledge with appreciation their dedication to high professional standards and their sacrifice of time and effort.

ASTM Committee on Publications

ASTM Editorial Staff

Contents

SUMMARY

Introduction

The Symposium on Chevron-Notched Specimens: Testing and Stress Analysis was held at the Galt House, Louisville, Kentucky, 21 Apr 1983, as part of the Spring meetings of ASTM Committee E-24 on Fracture Testing. Chevron-notched testing and analysis has been a topic of considerable interest to ASTM Committee E-24. The work at NASA Lewis Research Center and Terra Tek Systems, which made up much of the initial chevron-notched work, has been presented often at E-24 subcommittee and task group meetings. Mr. David P. Wilhem, while chairman of ASTM Subcommittee E24.01 on Fracture Mechanics Test Methods, proposed this symposium to bring together the most up-to-date investigations on chevron-notched testing. The current focus is on cooperative, comparative test and analysis programs, and a proposed standard test method, coordinated by task groups of Subcommittee E24.01 and Subcommittee E24.07 on Fracture Mechanics of Brittle Materials.

The most important advantage in using chevron-notched specimens for fracture testing is that a precrack can be produced in a single load application, with the precrack self-initiating at the tip of the chevron. The sometimes difficult, and always time consuming, fatigue precracking operation can be eliminated. One important purpose of the work described in this publication, given the precracking and other differences in chevron-notched testing compared with existing tests, is to identify the conditions which will yield reproducible results. These conditions involve specimen material, specimen size and geometry, test procedures, and the stress analysis procedures used to evaluate results. Once consistent results are obtained, then detailed comparisons of test data obtained by chevron-notched techniques can be made with results from standard tests.

The papers in the volume are presented in three sections:

1. *Stress Analysis,* including primarily finite element stress analysis of several chevron-notched geometries, but also encompassing boundary integral, photoelastic, and analytical and experimental compliance methods of stress analysis.

2. *Test Method Development,* both experimental and analytical investigations of key concerns with chevron-notched testing, such as specimen size effects, different material behavior including metals and nonmetals, and various methods for measuring crack growth.

3. *Fracture Toughness Measurements,* with primary emphasis on chevron-notched measurement of fracture toughness of structural materials, including

aluminum alloys and a variety of hard/brittle materials such as oxides and carbides.

This publication is the first collection of information on chevron-notched testing, and it should provide a resource for the development and use of this type of specimen for fracture testing. The symposium chairmen/editors are pleased to acknowledge the help of the ASTM editorial staff listed herein and Committee E-24 staff manager, Matt Lieff. Each of us also acknowledges the support of his respective laboratory and support staff.

John H. Underwood
Army Armament Research and Development Center, Watervliet, N.Y. 12189; symposium cochairman and coeditor.

Stephen W. Freiman
National Bureau of Standards, Washington, D.C. 20234; symposium cochairman and coeditor.

Francis I. Baratta
Army Materials and Mechanics Research Center, Watertown, Mass. 02172; symposium cochairman and coeditor.

Stress Analysis

J. C. Newman, Jr.[1]

A Review of Chevron-Notched Fracture Specimens

REFERENCE: Newman, J. C., Jr., "**A Review of Chevron-Notched Fracture Specimens,**" *Chevron-Notched Specimens: Testing and Stress Analysis, ASTM STP 855,* J. H. Underwood, S. W. Freiman, and F. I. Baratta, Eds., American Society for Testing and Materials, Philadelphia, 1984, pp. 5–31.

ABSTRACT: This paper reviews the historical development of chevron-notched fracture specimens; it also compares stress-intensity factors and load line displacement solutions that have been proposed for some of these specimens. The review covers the original bend-bar configurations up to the present day short-rod and bar specimens. In particular, the results of a recent analytical round robin that was conducted by an ASTM Task Group on Chevron-Notched Specimens are presented.

In the round robin, three institutions calculated stress-intensity factors for either the chevron-notched round-rod or square-bar specimens. These analytical solutions were compared among themselves, and then among the various experimental solutions that have been proposed for these specimens. The experimental and analytical stress-intensity factor solutions that were obtained from the compliance method agreed within 3% for both specimens. An assessment of the consensus stress-intensity factor (compliance) solution for these specimens is made.

The stress-intensity factor solutions proposed for three- and four-point bend chevron-notched specimens are also reviewed. On the basis of this review, the bend-bar configurations need further experimental and analytical calibrations.

The chevron-notched rod, bar, and bend-bar specimens were developed to determine fracture toughness of brittle materials, materials that exhibit flat or nearly flat crack-growth resistance curves. The problems associated with using such specimens for materials that have a rising crack-growth resistance curve are reviewed.

KEY WORDS: fracture mechanics, stress-intensity factor, cracks, finite-element method, boundary-element method, crack-opening displacement, chevron-notched specimen

Nomenclature

- A Normalized stress-intensity factor defined by Barker
- a Crack length measured from either front face of bend bar or load line
- a_0 Initial crack length (to tip of chevron notch)
- a_1 Crack length measured to where chevron notch intersects specimen surface
- b Length of crack front
- B Thickness of bar specimen or diameter of rod specimen

[1]Senior scientist, NASA Langley Research Center, Hampton, Va. 23665.

C^*	Normalized compliance, EBV_L/P, for chevron-notched specimen
E	Young's modulus of elasticity
F	Normalized stress-intensity factor for straight-through crack specimen
F^*	Normalized stress-intensity factor for chevron-notched specimen
F_c^*	Normalized stress-intensity factor determined from compliance for chevron-notched specimen
F_m^*	Minimum normalized stress-intensity factor for chevron-notched specimen
H	Half of bar specimen height or radius of rod specimen
K	Stress-intensity factor (Mode I)
K_m	Minimum stress-intensity factor for chevron-notched specimen
K_{Ic}	Plane-strain fracture toughness (ASTM E 399)
K_{Icv}	Plane-strain fracture toughness from chevron-notched specimen
K_R	Crack-growth resistance
k	Shear-correction parameter in Bluhm's slice model
P	Applied load
P_{max}	Maximum test (failure) load
V_L	Load-point half-displacement
V_T	Half-displacement measured at top of specimen along load line
w	Specimen width
x,y,z	Cartesian coordinates
α	Crack-length-to-width (a/w) ratio
α_i	Crack-length-to-width (a_i/w) ratios defined in Fig. 2
ν	Poisson's ratio

Chevron-notched specimens (Fig. 1) are gaining widespread use for fracture toughness testing of ceramics, rocks, high-strength metals, and other brittle

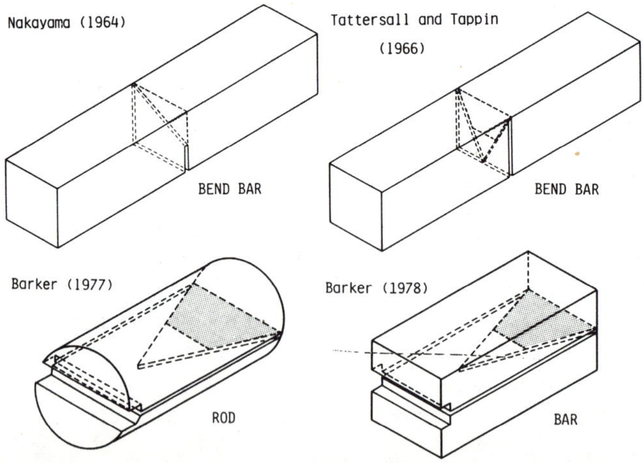

FIG. 1—*Various chevron-notched fracture specimen configurations.*

materials [1–7]. They are small (5 to 25-mm thick), simple, and inexpensive specimens for determining the plane-strain fracture toughness, denoted herein as K_{Icv}. Because they require no fatigue precracking, they are also well suited as quality control specimens. The unique features of a chevron-notched specimen, over conventional fracture toughness specimens, are: (1) the extremely high-stress concentration at the tip of the chevron notch, and (2) the stress-intensity factor passes through a minimum as the crack grows. Because of the high-stress concentration factor at the tip of the chevron notch, a crack initiates at a low applied load, so costly precracking of the specimen is not needed. From the minimum stress-intensity factor, the fracture toughness can be evaluated from the maximum test load. Therefore, a load-displacement record, as is currently required in the ASTM Test Method for Plane-Strain Fracture Toughness of Metallic Materials (E 399-83) is not needed. Because of these unique features, some of these specimens are being considered for standardization by the American Society for Testing and Materials (ASTM).

This paper reviews the historical development of chevron-notched fracture specimens. The paper also compares the stress-intensity factor and load-line displacement solutions that have been proposed for some of these specimens. The review is presented in four parts.

In the first part, the review covers the development of the original chevron-notched bend bars, the present day short-rod and bar specimens, and the early analyses for these specimens.

In the second part, the results of a recent "analytical" round robin conducted by the ASTM Task Group on Chevron-Notched Specimens are presented. Three institutions participated in the calculations of stress-intensity factors for either the chevron-notched round-rod or square-bar specimen. They used either three-dimensional finite-element or boundary-integral equation (boundary-element) methods. These analytical solutions were compared among themselves and among the various experimental solutions that have been determined for the rod and bar specimens. An assessment of the consensus stress-intensity factor (compliance) solution for these specimens is presented.

In the third part, some recent stress-intensity factor solutions, proposed for three- and four-point bend chevron-notched specimens, are reviewed.

In the last part, the applicability of chevron-notched specimens to materials that have a rising crack-growth resistance curve is discussed.

History of Chevron-Notched Specimens

In 1964, Nakayama [1,2] was the first to use a bend specimen with an unsymmetrical chevron notch. His specimen configuration is shown in Fig. 1. He used it to measure fracture energy of brittle, polycrystalline, refractory materials. All previous methods which had been developed for testing homogeneous materials were thought to be inadequate. This specimen is unique in that a crack initiates at the tip of the chevron notch at a low load, then propagates stably until catastrophic fracture. Because of the low load, the elastic stored energy in the test specimen and testing apparatus was small so that the fracture energy could be estimated from the area under a load-time history record.

Tattersall and Tappin [*3*] in 1966 proposed using a bend bar with a chevron notch symmetrical about the centerline of the specimen, as shown in Fig. 1. They used this specimen to measure the work of fracture on ceramics, metals, and other materials. The work of fracture was determined from the area under the load-displacement record divided by the area of the fracture surfaces.

In 1972, Pook [*4*] suggested using a chevron-notched bend bar to determine the plane-strain fracture toughness of metals. He stated that, "If the K_I against crack length characteristic is modified, by the introduction of suitably profiled side grooves, so that there is a minimum at $a/w \approx 0.5$, and the initial K_I is at least twice this minimum, it should be possible to omit the precracking stage, and obtain a reasonable estimate of K_{Ic} from the maximum load in a rising load test." Pook's "suitably profiled side grooves" is the present-day chevron notch. However, he considered only the analytical treatment needed to obtain stress-intensity factors as a function of crack length for various types of chevron notches. He did not study the experimental aspects of using a chevron-notched specimen to obtain K_{Ic}.

The nomenclature currently used for a straight-sided chevron notch in a rectangular cross section specimen is shown in Fig. 2. The specimen width, w, and crack length, a, are measured from the front face of the bend bar (or from the load line in the knife-edge-loaded specimen). The dimensions a_0 and a_1 are measured from the edge of the bend bar (or load line) to the vertex of the chevron and to where the chevron intersects the specimen surface, respectively. The specimen is of thickness B and the crack front is of length b.

Pook [*4*] used the stress-intensity factor solution for a three-point bend bar with a straight-through crack (STC) [*8*] and a side-groove correction proposed by Freed and Kraft [*9*] to obtain approximate solutions for various shape chevron notches (CN). The stress-intensity factors for a chevron-notched specimen, K_{CN}, was given by

$$K_{CN} = K_{STC}\left(\frac{B}{b}\right)^{1/2} \tag{1}$$

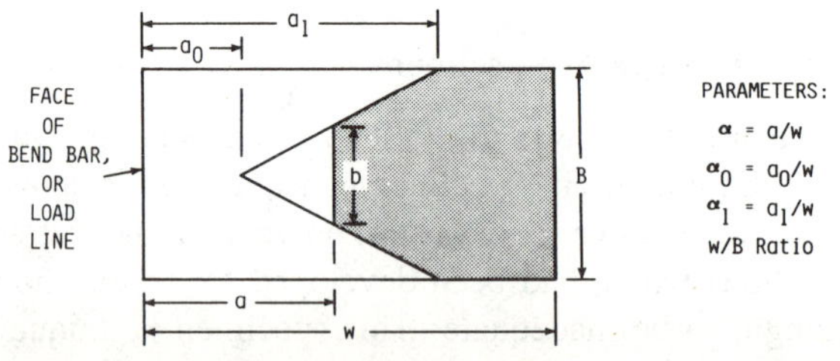

FIG. 2—*Chevron-notched fracture specimen nomenclature.*

where K_{STC} is the stress-intensity factor for a straight-through crack in a bar having the same overall dimensions. Figure 3 illustrates the unique stress-intensity factor solution for a chevron-notched specimen compared to a straight-through crack specimen. The dashed curve shows the normalized stress-intensity factors for the straight-through crack as a function of a/w. This curve is a monotonically increasing function with crack length. The solid curve shows the solution for the chevron-notched specimen. For $a = a_0$, the stress-intensity factor is very large, but it rapidly drops as the crack length increases. A minimum value is reached when the crack length is between a_0 and a_1. For $a \geqslant a_1$, the stress-intensity factors for the chevron-notched specimen and for the straight-through crack specimen are identical because the configurations are identical.

The analytical procedure used by Pook [*4*] to determine the stress-intensity factor as a function of crack length was an engineering approximation. At that time, no rigorous analysis had been conducted to verify the accuracy of Eq 1.

In 1975, Bluhm [*10*] made the first serious attempt to analyze the chevron-notched bend bars. The three-dimensional crack configuration was analyzed in an approximate "two-dimensional" fashion. The specimen was treated as a series of slices in the spanwise direction. Both beam bending and beam shear effects on the compliance of each slice were considered but the inter-slice shear stresses were neglected in the analysis. Then by a synthesis of the slice behavior, the total specimen compliance was determined. The slice model, however, introduced a "shear correction" parameter (k) which had to be evaluated from experimental compliance measurements. Experimental compliance measurements made on an "uncracked" chevron-notched bend bar ($\alpha_0 = 0$ and $\alpha_1 = 1$) were used to determine a value for the shear correction parameter for three- and four-

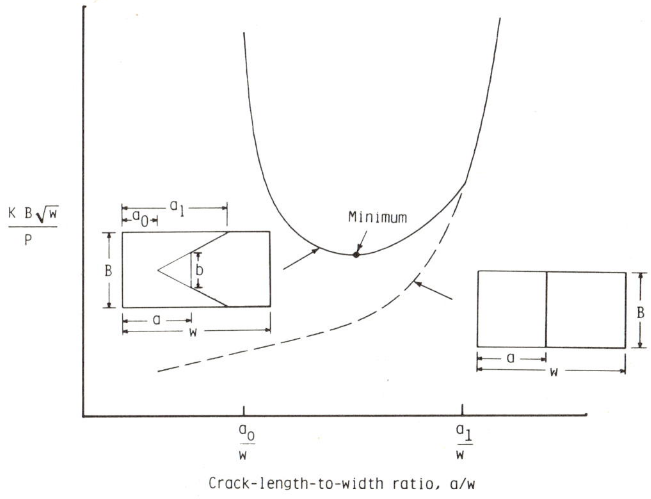

FIG. 3—*Comparison of normalized stress-intensity factors for chevron-notched and straight-through crack specimens.*

point bend specimens. Bluhm estimated that the slice model was capable of predicting the compliance of the cracked Tattersall-Tappin type specimen (see Fig. 1) to within 3%. Bluhm did not, however, calculate stress-intensity factors from the compliance equations. Later, Munz et al [7] did use Bluhm's slice model to calculate stress-intensity factors for various chevron-notched bar specimens.

In the following, the concept proposed by Pook [4] to determine the K_{Ic}-value for brittle materials using chevron-notched specimens will be illustrated. Figure 4 shows stress-intensity factor, K, plotted against crack length. The solid line beginning at a_0 and leveling off at K_{Ic} is the "ideal" crack-growth resistance curve for a brittle material. The dashed curves show the "crack-driving force" curves for various values of applied load on a chevron-notched specimen. Because of the extremely large K-value at $a = a_0$, a small value of load, like P_1, is enough to initiate a crack at the vertex of the chevron. At load P_1, the crack grows until the crack-drive value is equal to K_{Ic}, that is, the intersection point between the dashed curve and horizontal line at point A. Further increases in load are required to extend the crack to point B and C. When the maximum load, P_{max}, is reached the crack-drive curve is tangent to the K_{Ic} line at point D. Thus, the K-value at failure is equal to K_{Ic}. The tangent point also corresponds to the minimum value of stress-intensity factor on the crack-drive curve (denoted with a solid symbol). Therefore, K_{Icv} is calculated by

$$K_{Icv} = \frac{P_{max}}{B\sqrt{w}} F_m^* \qquad (2)$$

where P_{max} is the maximum failure load and F_m^* is the minimum value of the normalized stress-intensity factor. Because F_m^* is a predetermined value for the

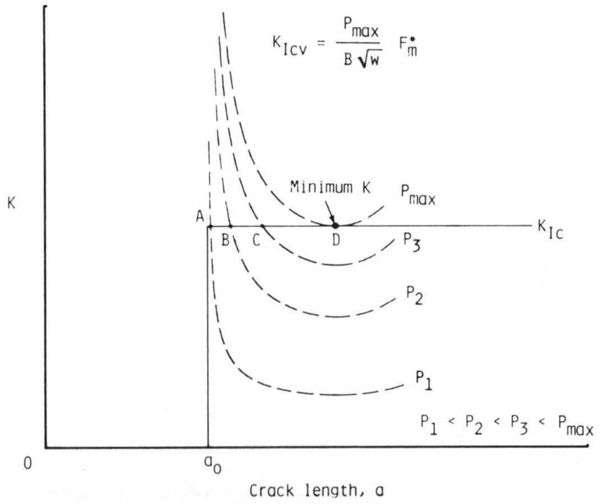

FIG. 4—*Fracture of "brittle" material using a chevron-notched specimen.*

particular chevron-notched configuration, it is necessary only to measure the maximum load to calculate K_{Icv}.

This maximum load test procedure can be only applied to brittle materials with flat or nearly flat crack-growth resistance curves. Many engineering materials, however, have a rising crack-growth resistance curve. The problems associated with using chevron-notched specimens for these materials will be discussed later.

Chevron-Notched Rod and Bar Specimens

Although the bend bars were the first type of chevron-notched specimens to be tested, the knife-edge loaded rod and bar specimens have received more attention. In the next sections, the rod and bar specimens are reviewed. This review also includes the analytical round robin in which the rod and bar specimens were analyzed. In a later section, some recent results on the chevron-notched bend bars are also reviewed.

Barker [*5,6*] in the late 1970s, proposed the short-rod and bar specimens, Fig. 1, for determining plane-strain fracture toughness. These specimens are loaded by a knife-edge loading fixture [*5,7*] resulting in an applied line load, *P,* at location, *L,* as shown in Fig. 5*a*. Figure 5 shows the coordinate system used to define dimensions of the most commonly used rod and bar specimens. (Here the chevron notch intersects the specimen surface at $x = w$ or $\alpha_1 = 1$.)

Rod Specimens—Since 1977, the chevron-notched rod specimen, with $w/B = 1.45$, has been studied extensively. Figure 6 shows a comparison of the minimum normalized stress-intensity factor as a function of the year the result was published. The open symbols denote the method by which the values were

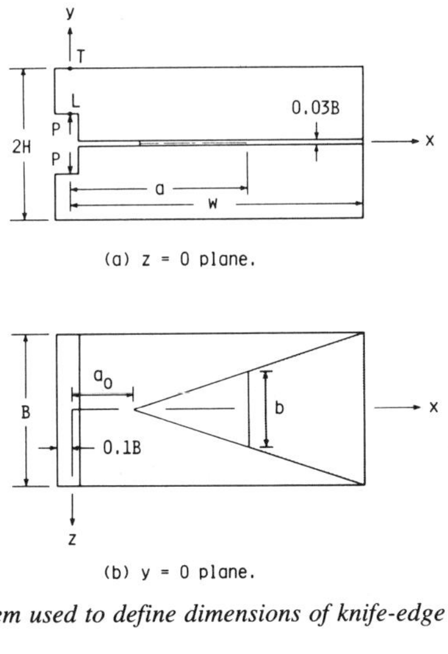

FIG. 5—*Coordinate system used to define dimensions of knife-edge loaded chevron-notched rod and bar specimens.*

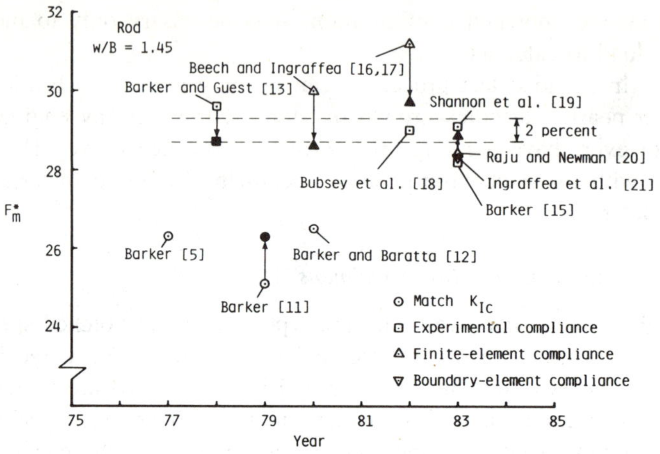

FIG. 6—*Comparison of minimum normalized stress-intensity factor for chevron-notched rod.*

obtained. Each method will be discussed. The solid symbols show the results of corrections that have been made by the author.

In 1977, Barker [*5*] used the K_{Ic}-value obtained from ASTM E 399 compact specimens made of 2014-T651 aluminum alloy to determine the minimum stress-intensity factor for the rod configuration by a "matching" procedure. The minimum stress-intensity factor was given by

$$K_m = \frac{P_{max} A}{\sqrt{B^3(1 - \nu^2)}} \tag{3}$$

where A is Barker's normalized stress-intensity factor that accounts for the configuration. By equating K_m to K_{Ic}, the value of A was 20.8. Equation 3 can be rewritten into the form

$$K_m = \frac{P_{max}}{B\sqrt{w}} F_m{}^* \tag{4}$$

where the value of $F_m{}^*$ is 26.3 ($\nu = 0.3$). (Equation 4 is the form commonly used for compact and knife-edge loaded specimens. The same form will be used herein.) Table 1 summarizes the minimum normalized stress-intensity factors obtained by various investigators; also listed are particular dimensions of the rod configuration used.

In 1979, Barker [*11*] replaced the term $(1 - \nu^2)$ in Eq 3 with unity without changing the value of A. Thus, the value of $F_m{}^*$ dropped by about 5%. The value of $F_m{}^*$ should have remained at 26.3 for $\nu = 0.3$.

Barker and Baratta [*12*] in 1980 extensively evaluated the fracture toughness of several steel, aluminum, and titanium alloys using the rod specimen and K_{Ic}-

TABLE 1—*Chronological development of minimum normalized stress-intensity factors* (F_m*) *for short chevron-notched rod.*

Investigator(s)	Year	Ref	w/B	a_0/w	a_1/w	F_m*
Barker[a]	1977	*5*	1.45	0.31[b]	0.96[b]	26.3
Barker and Guest[c]	1978	*13,14*	1.474	0.343[d]	0.992[d]	29.6
Barker	1979	*11*	1.45	0.31[b]	0.96[b]	25.1
Barker and Baratta[a]	1980	*12*	1.45	0.343[d]	0.992[d]	26.5
Beech and Ingraffea[e]	1980	*16*	1.5	0.35	1.0	31.4(30.0)[f]
Beech and Ingraffea[e]	1982	*17*	1.5	0.35	1.0	32.7(31.2)[f]
Bubsey et al[c]	1982	*18*	1.45	0.332	1.0	29.0
Barker[c]	1983	*15*	1.45	0.332	1.0	28.2
Shannon et al[c]	1983	*22*	1.45	0.332	1.0	29.1
Raju and Newman[e]	1983	*20*	1.45	0.332	1.0	28.4
Ingraffea et al[e,g]	1983	*21*	1.45	0.332	1.0	28.3

[a] F_m* determined from matching K_{Ic} from ASTM E 399 specimens.
[b] Dimensions estimated from photograph.
[c] F_m* determined from experimental compliance.
[d] Curved-sided chevron notch used in test, equivalent dimensions for straight-sided chevron notch used for table.
[e] F_m* determined from finite-element analysis and compliance.
[f] Values are from plane-strain (plane-stress) assumption.
[g] F_m* determined from boundary-element analysis and compliance.

values measured according to ASTM Standard Method of Test for Plane-Strain Fracture Toughness of Metallic Materials (E 399-78). They found that the critical stress-intensity factors, calculated from the rod specimen data using $F_m^* = 25.5$ [*12*], were consistently low, averaging about 6% below the K_{Ic}-values. They concluded that F_m^* for the test configuration used in their study should be increased by 4% to a value of 26.5.

Earlier, Barker and Guest [*13*] had conducted an experimental compliance calibration on the rod specimen and had obtained a value of F_m^* as 29.6. Their specimen, however, had a w/B ratio of 1.474 [*14*]. Subsequently, the value of F_m^* was corrected to a value corresponding to a w/B ratio of 1.45 by using a "constant moment" conversion described in Ref *15*. The corrected value of F_m^* (28.7) was about 3% lower than the compliance value from Ref *13*, as indicated in Fig. 6.

Beech and Ingraffea [*16,17*] were the first to rigorously numerically analyze a chevron-notched specimen. They used a three-dimensional finite-element method to determine stress-intensity factor distributions along the crack front and stress-intensity factors from compliance for the chevron-notched rod. The specimen they analyzed, however, differed from the proposed standard ($w/B = 1.45$; $\alpha_0 = 0.332$; and $\alpha_1 = 1$) specimen analyzed in the ASTM round robin in three ways: (1) the load line was at the front face of the specimen rather than at $0.05B$ into the specimen mouth, (2) the slot height ($0.03B$) was modelled (see Fig. 5*a*) as zero, and (3) the square- or V-shaped cutout at the load line was not modelled. (The effects of these differences in specimen configuration on stress-intensity factors are discussed in Ref *15* and will not be repeated here.) The stress-intensity factors reported in Refs *16* and *17* from their crack front evaluations were considerably lower (6 to 17%) than their values determined from a plane-strain compliance relation. They used their plane-strain compliance results to obtain a minimum stress-intensity factor. The value of F_m^* from Ref *17* was 4% higher than the value given in Ref *16*. The difference in these results was due to the manner by which the compliance derivative was evaluated. The values of F_m^* given in Table 1 were their plane-strain compliance values and, in parentheses, values obtained from a plane-stress compliance relation. The reason for using plane stress, herein, was that the displacements remote from the crack front are more nearly controlled by plane-stress conditions and, consequently, the plane-stress compliance relation would be more correct than using plane strain. (Also, all other results reported in Table 1, which were determined from compliance, were made with the plane-stress relation.) If the plane-strain compliance relation (with $\nu = 0.3$) had been used, the F_m^*-values would have been about 5% higher than the plane-stress values (square and triangular symbols) shown in Fig. 6.

Bubsey et al [*18*], Shannon et al [*19*], and Barker [*15*] used the experimental compliance (plane-stress) relation to evaluate stress-intensity factors for the short-rod specimen. Bubsey et al and Shannon et al used aluminum alloy specimens with w/B ratios of 1.5, 1.75, and 2 for a wide range in α_0. Their values in Table 1 and Fig. 6 were interpolated for $\alpha_0 = 0.332$ and extrapolated to $w/B = 1.45$

by using second degree polynomials in terms of α_0 and w/B, respectively. Because the proposed standard dimensions are quite close to those used in the experiments, the interpolation and extrapolation procedure is expected to induce only a small error (probably less than 2%). Barker [*15*], on the other hand, used fused quartz ($\nu = 0.17$) on specimens with $w/B = 1.45$. He reported a value of A as 23.38, therefore, $F_m{}^*$ would be about 28.2.

Raju and Newman [*20*], using a three-dimensional finite-element method, studied the effects of Poisson's ratio (ν) on stress-intensity factors for the rod specimen ($w/B = 1.45$). Their results indicated that a specimen with $\nu = 0.17$ (fused quartz) would have a stress-intensity factor about 2% lower than a specimen with $\nu = 0.3$ (aluminum alloy). Thus, if Barker [*15*] had used an aluminum alloy specimen, his experimental compliance value ($F_m{}^*$) would have been about 28.8.

Raju and Newman [*20*] and Ingraffea et al [*21*] determined the minimum stress-intensity factors for the rod specimen ($w/B = 1.45$) using compliance calculations from three-dimensional finite-element analyses. Each used the plane-stress compliance relation. Raju and Newman obtained a value of $F_m{}^*$ as 28.4 (as plotted in Fig. 6) and Ingraffea et al obtained a value of 28.3 (not plotted). The result from Raju and Newman, however, was estimated to be about 1.5% below the true solution based on a convergence study. Thus, the corrected value of $F_m{}^*$ would have been about 28.8.

Ingraffea et al [*21*] also used a boundary-element (boundary-integral) method to determine the minimum stress-intensity factor from compliance. They obtained a value of $F_m{}^*$ as 28.3 (as plotted in Fig. 6), the same as from their finite-element analysis. The results from Ingraffea et al [*21*] and Raju and Newman [*20*] were part of the analytical round robin, previously mentioned, and these results will be discussed and compared later.

A comparison of minimum stress-intensity factors for the rod specimen ($w/B = 1.45$) shows several interesting features. First, the method of using K_{Ic} to determine $F_m{}^*$ gives results that are about 8% below experimental and analytical compliance methods. Although the specimen used by Barker [*5,11*] and Barker and Baratta [*12*] was somewhat different than the proposed standard specimen, these differences are not expected to be significant (see Ref *15*, page 309). The specimens used in Refs *11* and *12* had chevron notches with curved sides instead of straight sides. Barker [*14*] argues that the calibration should be the same in a straight-sided and a curved-sided chevron-notched specimen, provided that the crack front length (b) and the rate of change in b is the same in both specimens at the minimum stress-intensity factor. He determined that the α_0 and α_1 for an "equivalent" straight-sided chevron-notched specimen should be 0.343 and 0.992, respectively. These values are quite close to those for the specimen analyzed in the ASTM round robin with straight-sided chevron notches. Therefore, at present, the 8% discrepancy in the values of $F_m{}^*$ cannot be explained from differences in specimen configuration.

One possible source of error in the K_{Ic} matching procedure may be due to the different loads used in each test procedure. In the K_{Ic} test, the 5% secant offset load, P_Q, is used to calculate K_{Ic}. The P_Q load is always less than or equal to P_{max}, the maximum test (failure) load. Whereas, in the chevron-notched specimen test, the maximum load is always used to calculate K_{Icv}. For example, if P_{max} was used to calculate K_{Ic} instead of P_Q, then K_{Ic} would tend to be higher than the current value. Thus, the value of $F_m{}^*$ would also tend to be higher than the current value (circular symbols in Fig. 6). This would make the value of $F_m{}^*$, determined from the K_{Ic} matching procedure, in closer agreement with the experimental and analytical compliance values shown in Fig. 6.

Second, the experimental [*13,15,18,19*] and the recent analytical [*20,21*] compliance determination of the minimum stress-intensity factor agree within about 3% of each other. Accounting for the fact that one of the analyses [*20*] was about 1.5% low, based on convergence studies, and that Ref *15* used fused quartz, which has a low value of Poisson's ratio so that a slightly lower value of $F_m{}^*$ would be expected (about 2%), the agreement generally is within about 1%. Thus, for the rod specimen with $w/B = 1.45$, $\alpha_0 = 0.332$, and $\alpha_1 = 1$ (straight-sided chevron) the value of $F_m{}^*$ is estimated to be 28.9 ± 0.3. The dashed lines in Fig. 6 show the expected error bounds on $F_m{}^*$.

Bar Specimens—Two types of chevron-notched bar specimens have been studied. In 1978, Barker [*6,15*] proposed a rectangular cross-sectioned bar specimen with an H/B ratio of 0.435 (see Fig. 1). This specimen was designed in such a way that the same minimum stress-intensity factor was obtained as for his rod specimen [*5*]. However, because the early compliance calibration for the rod specimen was about 8% low (see Fig. 6), it was not clear whether the bar and rod specimens now have the same value. Raju and Newman [*20*] analyzed both specimens and found that the compliance calibration for the rectangular bar specimen was about 3.8% lower than the rod specimen.

In 1980, Munz et al [*7*] proposed a square cross-sectioned bar specimen ($H/B = 0.5$). They conducted a very extensive experimental compliance calibration on bar specimens with $w/B = 1.5$ and 2 for α_0 ranging from 0.2 to 0.5 and $\alpha_1 = 1$. From these results, they obtained minimum values of stress-intensity factors for each configuration considered. Using the assumption that the change of compliance with crack length in a chevron-notched specimen was the same as that for a straight-through crack specimen, they obtained an equation that was identical to Eq 1 as

$$F^* = F\left(\frac{\alpha_1 - \alpha_0}{\alpha - \alpha_0}\right)^{1/2} = F\left(\frac{B}{b}\right)^{1/2} \tag{5}$$

for $\alpha_0 < \alpha \leq \alpha_1$. For specimens with an α_0 of about 0.2 and 0.35, the difference between experimental and analytical (Eq 5) minimum normalized stress-intensity factors was less than 1%. For an α_0-value of about 0.5, the difference was 3 to 3.5%. They concluded that Eq 5 should only be used to obtain minimum values because experimental and analytical values differed greatly at small crack-length-to-width (α) ratios near α_0.

Shannon et al [19] have developed minimum stress-intensity factor expressions for chevron-notched bar (square) and rod specimens with $\alpha_1 = 1$ and $\alpha_0 \leq 0.5$. These expressions were fitted to minimum stress-intensity factors determined from experimental compliance measurements. For the square-bar specimen, the w/B ratio was 1.5 or 2 and for the rod specimen, the w/B ratio was 1.5, 1.75, or 2.

The use of chevron-notched specimens with materials that have a rising crack-growth resistance curve may require stress-intensity factors as a function of crack length instead of using only the minimum value. Recently, Shannon et al [22] have developed polynomial expressions that give the stress-intensity factors and load-line displacements as a function of crack length for square-bar and rod specimens ($\alpha_1 = 1$). These expressions were obtained from experimental compliance measurements made for various w/B ratios. The w/B ratio for the square-bar specimen was, again, 1.5 or 2, and for the rod specimen was 1.5, 1.75, or 2. The expressions apply to α_0 between 0.2 and 0.4, and α varying from α_0 to 0.8. Some of these results will be compared with the results from the ASTM analytical round robin in the next section.

Analytical Round Robin on Chevron-Notched Rod and Bar Specimens

In 1981, plans were formulated for a cooperative test and analysis program on chevron-notched square-bar and round-rod specimens by an ASTM task group on Chevron-Notched Specimen Testing. Four configurations were selected: the square and round versions of a relatively short specimen ($w/B = 1.45$); and the square and round versions of a longer specimen ($w/B = 2$). These configurations were chosen so as to include as many features as possible of prior work [5–7]. The coordinate system used to define the specimens is shown in Fig. 5. The specimens were loaded by a knife-edge loading fixture that results in an applied load, P, at the load line, L in Fig. 5*a*. Specimens had either a square cutout [7] at the load line or a V-cutout [15] at the load line (not shown). The chevron notch, Fig. 5*b*, had straight sides and intersected the specimen sides at $x = w$ (or $\alpha_1 = 1$). The following table lists the dimensions of the specimens considered:

Specimen	w/B	a_0/w	H/B
Bar	1.45	0.332	0.5
Bar	2	0.2	0.5
Rod	1.45	0.332	0.5
Rod	2	0.2	0.5

The analysts were asked to calculate results for crack-length-to-width (a/w)

ratios of 0.4, 0.5, 0.55, 0.6, and 0.7. The information required from the analyses were:

1. *K*-distribution as a function of *z* and a/w (see Fig. 5*b*).
2. *K*-value from the plane-stress compliance relation as a function of a/w:

$$K = \left[\frac{EP}{b}\frac{dV_L}{da}\right]^{1/2} \tag{6}$$

3. Normalized displacements EVB/P at points L and T (see Fig. 5*a*) as a function of a/w.

The participants in the round robin were:

Investigators	Institution
A. R. Ingraffea, R. Perucchio, T. Y. Han, W. H. Gerstle, and Y. P. Huang	Cornell University
A. Mendelson and L. J. Ghosn	Case-Western Reserve University
I. S. Raju and J. C. Newman, Jr.	NASA Langley Research Center

The following table lists the investigators, the three-dimensional method(s) used in the analyses, and the particular configuration(s) analyzed:

			Rod		Bar	
Investigators	Method	w/B	1.45	2	1.45	2
Ingraffea et al [*21*]	finite-element		X			
	boundary-element		X			
Mendelson and Ghosn [*23*]	boundary-element					X
Raju and Newman [*20*]	finite-element		X	X	X	X

All analyses were conducted on models of specimens with the square cutout at the load line, as shown in Fig. 5*a*. The slot height (0.03*B*) shown in Fig. 5*a* was not modeled in any of the analyses (that is, the height was taken as zero).

Rod Specimen—Ingraffea et al [*21*] and Raju and Newman [*20*] determined the distribution of normalized stress-intensity factors along the crack front of a rod specimen ($w/B = 1.45$) with $\alpha = 0.55$ using boundary-element and finite-element methods, respectively. These results are compared in Fig. 7. The normalized stress-intensity factor (F^*) is plotted against $2z/b$. The center of the specimen is at $2z/b = 0$ and the crack intersects the chevron boundary at $2z/b = 1$, see insert. Ingraffea et al used only one element, a quarter-point

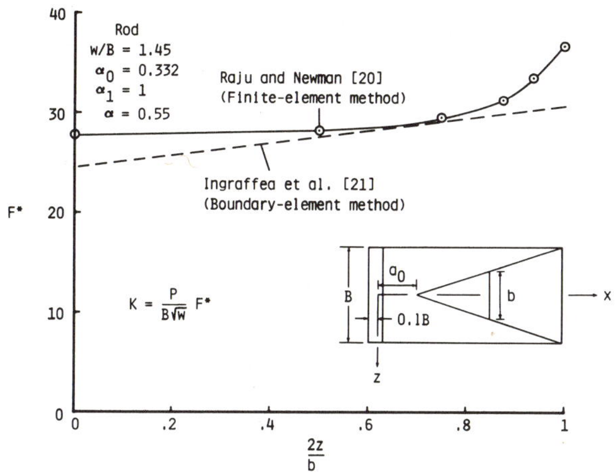

FIG. 7—*Comparison of distribution of normalized stress-intensity factors along crack front for short chevron-notched rod.*

singular element, to define one half of the crack front length ($b/2$); they showed a nearly linear distribution. On the other hand, Raju and Newman used five layers of singularity elements to define one half of the crack front, and they showed nearly constant stress-intensity factors for $2z/b < 0.5$. Their stress-intensity factors increased rapidly as $2z/b$ approached unity. The results from Raju and Newman were 0 to 16% higher than the results from Ingraffea et al. The difference is probably due to Ingraffea et al using only one element along the crack front.

A comparison of experimental and analytical load-point displacements for the short chevron-notched rod ($w/B = 1.45$) is shown in Fig. 8. The normalized

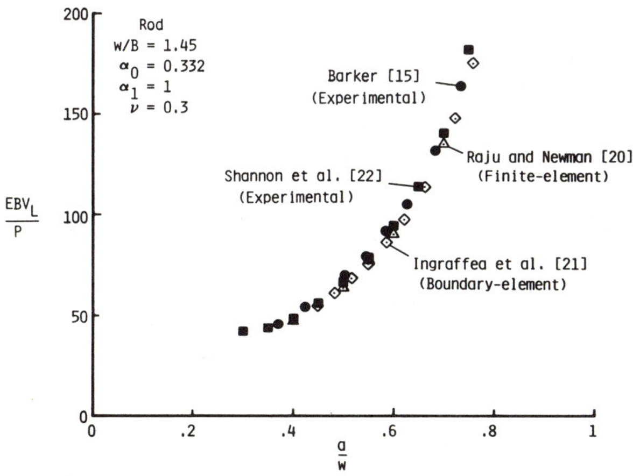

FIG. 8—*Comparison of experimental and analytical load-point displacements for short chevron-notched rod.*

displacement, EBV_L/P, is plotted against a/w. Load-point displacements (V_L) were either measured or calculated at $z = 0$ (see Fig. 5*b*) as a function of crack length. Because the experiments and analyses were conducted on materials with different Poisson ratios, the displacements have been adjusted, using results from Raju and Newman [*20*] on the Poisson effect, to displacements for a Poisson ratio of 0.3. Barker [*15*] measured load-point displacements on fused quartz ($\nu = 0.17$) using a laser-interferometric technique. His displacements have been reduced by 3% to compensate for the differences in Poisson ratios; his data are shown as circular symbols. In contrast, Shannon et al [*22*] measured displacements (V_T) at the top of aluminum alloy ($\nu = 0.3$) specimens (see Fig. 5*a*). They measured displacements for specimens with various values of α_0 ($0.2 \leq \alpha_0 \leq 0.4$) and with w/B equal to 1.5, 1.75, and 2. The results (square symbols) plotted in Fig. 8 were interpolated to $\alpha_0 = 0.332$ and extrapolated to $w/B = 1.45$, respectively, using second degree polynomials. These results agreed well with Barker's results.

Load-point displacements from Raju and Newman's finite-element analysis [*20*] and Ingraffea et al's [*21*] boundary-element analysis are also shown in Fig. 8. The displacements from Ingraffea et al have been reduced by 1% to compensate for a slight difference in Poisson's ratio. Both analytical results were from 4 to 6% below the experimental results. Based on beam theory [*24*], however, about 2% of this difference is caused by neglecting the notch ($0.03B$) made by a saw blade or chevron cutter (see Fig. 5*a*). These displacements were used by each investigator to determine the stress-intensity factors from the plane-stress compliance method. These results are described in the following section.

Experimental and analytical normalized stress-intensity factors (F^*), as functions of a/w, for the chevron-notched rod are compared in Fig. 9. (Note the use of a broken scale.) The experimental and analytical results were obtained from

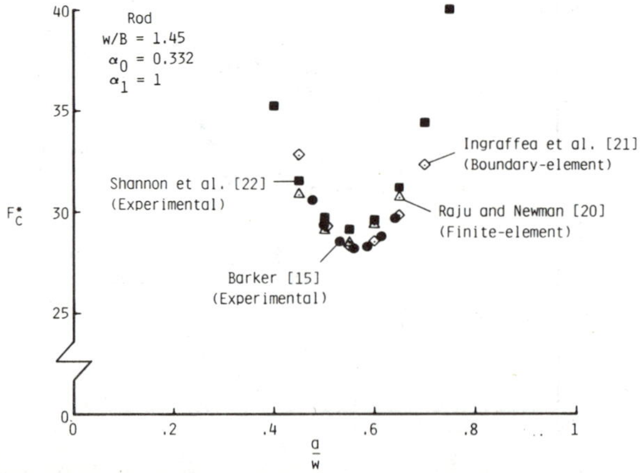

FIG. 9—*Comparison of experimental and analytical normalized stress-intensity factors for short chevron-notched rod.*

the plane-stress compliance relation (Eq 6) as

$$F_c^* = \frac{B\sqrt{w}}{P}\left[\frac{EP}{b}\frac{dV_L}{da}\right]^{1/2} = \left[\left(\frac{1-\alpha_0}{\alpha-\alpha_0}\right)\frac{dC^*}{d\alpha}\right]^{1/2} \tag{7}$$

where C^* is the normalized compliance, EBV_L/P. The load-point displacement (V_L) was either measured or calculated at $z = 0$ as a function of crack length. Barker [*15*] measured the load-point displacements on fused quartz ($\nu = 0.17$) using a laser-interferometric technique. The displacements were then fitted to an empirical equation in terms of crack length. This equation was differentiated to obtain the compliance derivative. Barker's results are shown as circular symbols. Shannon et al [*22*] measured displacements (V_T) at the top of aluminum alloy ($\nu = 0.3$) specimens. They assumed that dV_T/da was equal to dV_L/da to obtain stress-intensity factors. Again, these results were interpolated and extrapolated to $\alpha = 0.332$ and $w/B = 1.45$ using second degree polynomials. Shannon's results (square symbols) are a few percent higher than Barker's results. As previously mentioned, Raju and Newman [*20*] have shown by a three-dimensional stress analysis that there is a slight difference (about 2%) between stress-intensity factors for $\nu = 0.17$ and 0.3; these results agreed with the observed experimental differences.

The analytical results from Raju and Newman [*20*] and Ingraffea et al [*21*] are also shown in Fig. 9. Based on a convergence study [*20*], the analytical results are expected to lie about 1.5% below the "true" solution. The analytical results agreed well (within 3%) with the experimental results near the minimum value of F_c^*.

Figure 10 compares how analyses and test results (F_c^*) vary with a/w for the

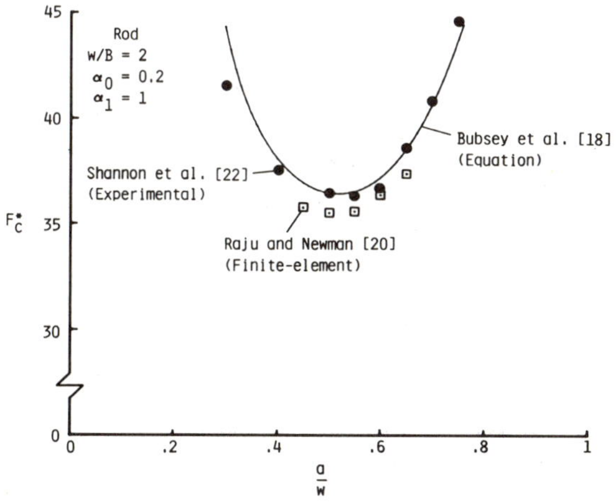

FIG. 10—*Comparison of experimental and analytical normalized stress-intensity factors for long chevron-notched rod.*

chevron-notched rod with $w/B = 2$. The solid curve represents an equation proposed by Bubsey et al [18] for the rod specimens. The equation they used was Eq 5 where F was the normalized stress-intensity factor for a straight-through crack in the same configuration [18].

Shannon et al's [22] results shown in Fig. 10 were obtained from Eq 7 using measured load-line displacements (V_T) on the rod specimen. Their results agreed well (within 1%) with the equation from Bubsey et al, except at small values of α. From previous work [7], it was recognized that Eq 5 overestimates values of $F_c{}^*$ for values of α approaching α_0. The finite-element results of Raju and Newman [20] were about 2.5% below the results from Bubsey et al and Shannon et al. Based on all of these results, the value of the minimum normalized stress-intensity factor ($F_m{}^*$) is estimated to be in the range 36.2 ± 0.4.

Bar Specimen—Mendelson and Ghosn [23], using the boundary-element method, and Raju and Newman [20], using the finite-element method, determined the distribution of boundary-correction factors along the crack front of a bar specimen with $w/B = 2$ and $\alpha = 0.55$. The results are compared in Fig. 11. Here F^* is plotted against $2z/b$. Mendelson and Ghosn, in contrast to Ingraffea et al [21], used five elements to define one half of the crack front length. Their elements were assumed to have either linear tractions or linear displacements. They determined F^*-values by using either crack-surface displacements or normal stresses near the crack front. For $2z/b < 0.9$, their results were 3 to 16% higher than the results from Raju and Newman, whereas the previous results from Ingraffea et al, using the same (boundary-element) method (Fig. 7), gave results on a rod specimen that were consistently lower than the results from Ref 20. The reason for the discrepancy between Refs 20 and 23 on stress-intensity factor distributions is not clear.

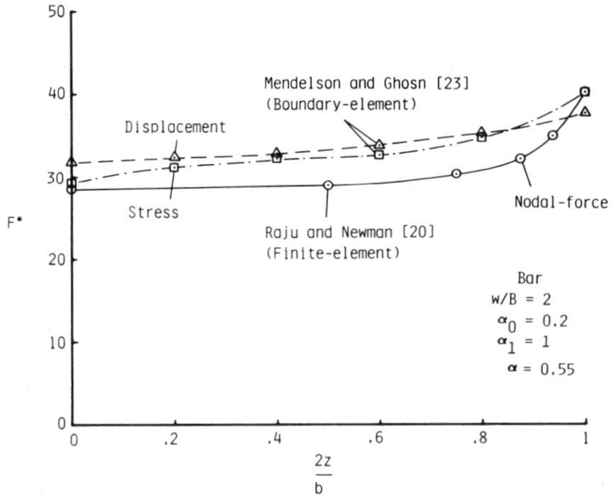

FIG. 11—*Comparison of distribution of normalized stress-intensity factors along crack front for long chevron-notched bar.*

Experimental and analytical load-point displacements at $z = 0$ for the chevron-notched bar with $w/B = 2$ are compared in Fig. 12. Normalized displacement is plotted against a/w. Shannon et al [22] measured displacements at the top of aluminum alloy specimens (circular symbols). The solid curve represents a polynomial equation from Ref 22 that was fitted to the experimental data. The finite-element results from Raju and Newman [20], $\nu = 0.3$, ranged from 3.5 to 6% lower than the experimental data. And the boundary-element results from Mendelson and Ghosn [23] were 8 to 11% lower than the experimental data. (Results from Ref 23, $\nu = 1/3$, were increased by 1% to compensate for the small difference in Poisson's ratio from $\nu = 0.3$.) Again, these displacements were used by each investigator to determine the stress-intensity factors from the plane-stress compliance method (Eq 7).

The normalized stress-intensity factors (F_c*), as functions of a/w, for the bar specimen with $w/B = 2$ are shown in Fig. 13. The experimental results and polynomial equation of Shannon et al [22] are shown as circular symbols and solid curve, respectively. The dashed curve shows an equation proposed by Munz et al [7] for bar specimens. For the chevron-notched specimen, Munz et al used Eq 5 where F was the normalized stress-intensity factor for a straight-through crack in the same configuration [7]. Again, Eq 5 overestimates F_c* for α approaching α_0. But for larger values of α, the equation underestimates F_c* based, at least, on the present experimental results [22].

The analytical results of Mendelson and Ghosn [23] and Raju and Newman [20] are also shown in Fig. 13. Near the minimum F_c*-value, the results from Mendelson and Ghosn were about 1.5% lower than the experimental results but overestimated F_c* on either side of the minimum. The results from Raju and Newman were about 2.5% lower than the experimental results. From all of the experimental and analytical results, the minimum F_m* is estimated to be 29.8 ± 0.3.

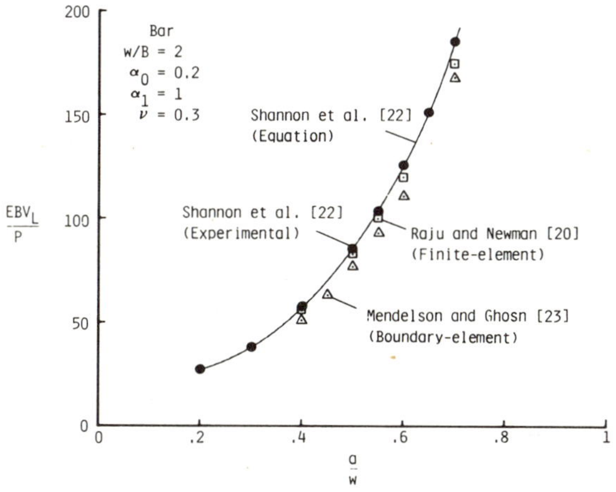

FIG. 12—*Comparison of experimental and analytical load-point displacements for long chevron-notched bar.*

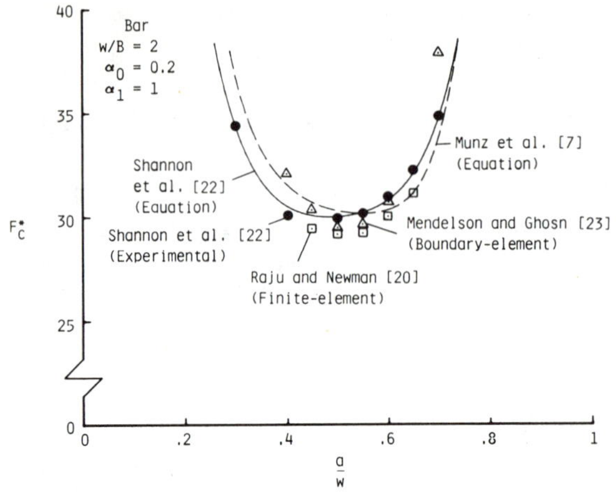

FIG. 13—*Comparison of experimental and analytical normalized stress-intensity factors for long chevron-notched bar.*

In Fig. 14, experimental and analytical normalized stress-intensity factors, as functions of a/w, are compared for the bar specimen with $w/B = 1.45$. The experimental results from Shannon et al [22] were, again, obtained by interpolation and extrapolation to $\alpha_0 = 0.332$ and $w/B = 1.45$ from results obtained from specimens with various α_0 and w/B ratios. The solid curve shows the equation proposed by Munz et al [7]. Near the minimum F_c^*, the equation agreed well with the experimental results (within 1%) but, again, overestimated results for a/w ratios less than about 0.55. The analytical results from Raju and Newman [20] were 0 to 1.5% lower than the experimental results. The minimum value

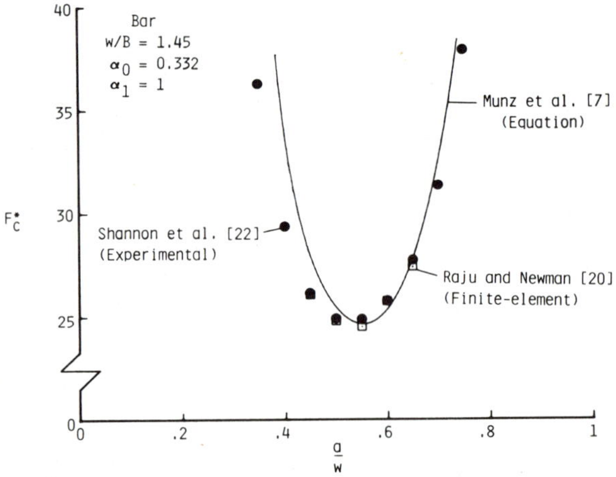

FIG. 14—*Comparison of experimental and analytical normalized stress-intensity factors for short chevron-notched bar.*

from Raju and Newman was 24.43, from Shannon et al was 24.85, and from Munz et al was 24.66. From these results, the minimum value of F_m* is estimated to be 24.8 ± 0.3.

Chevron-Notched Bend Bars

As previously mentioned, Nakayama [*1,2*], and Tattersall and Tappin [*3*] were the first to introduce and to determine fracture energies from chevron-notched bend bars. Pook [*4*] and Bluhm [*10*] were the first to provide approximate stress-intensity factors and compliance expressions, respectively, for these specimens. This section reviews the more recent experimental and analytical stress-intensity factor solutions that have been proposed for chevron-notched bend bars.

Munz et al [*25*] compared stress-intensity factors for various four-point bend specimens with $0.12 \leq \alpha_0 \leq 0.24$, $0.9 \leq \alpha_1 \leq 1$, and $w/B = 1$ or 1.25. Two analytical methods were studied. The first was by the use of Eq 7 wherein dC^*/da, the compliance derivative of the chevron-notched specimen, was assumed to be equivalent to $dC/d\alpha$, the compliance derivative of a straight-through crack. Under this assumption, Eq 7 reduces to Eq 5 or Pook's equation [*4*]. The second method was by using Bluhm's slice model [*10*]. Bluhm's slice model is probably more accurate than Pook's equation, but neither method has been substantiated by experimental compliance measurements or by more rigorous analytical (three-dimensional elasticity) methods. The slice model, however, was calibrated to experimental compliance measurements made on uncracked chevron-notch bend bars. A comparison of the two analytical results showed that the differences ranged from −5 to 10% for the particular configurations considered.

In 1981, Shih [*26*] proposed a "standard" chevron-notched bend-bar configuration for three-point loading with a major-span-to-width ratio (s/w) of 4. The w/B ratio was 1.82 with $\alpha = 0.3$ and $\alpha_1 = 0.6$. Shih [*26*] used the K_{Ic}-value from 7079-T6 aluminum alloy and the failure (maximum) load on the chevron-notched bend bars to estimate the minimum stress-intensity factor; this value is shown in Fig. 15 as the horizontal dashed line. The equation proposed by Pook [*4*] (upper solid curve) gave a minimum value very close to the value determined by Shih. Later, however, Shih [*27*] re-evaluated the minimum by testing 7079-T6 aluminum alloy compact specimens and chevron-notched specimens made from the same plate. The new K_{Ic}-value dropped by 19% from the old value and, consequently, the minimum value (F_m*) dropped to 10.17, as shown by the dash-dot line in Fig. 15.

Wu [*28*] used Eq 5 to determine the stress-intensity factors for three-point bend chevron-notched specimens. His equation gave essentially the same results (within 1%) as that shown for Pook in Fig. 15. Wu [*29*] also used Bluhm's slice model to determine specimen compliance and then used Eq 7 to determine F_c* as a function of α (or a/w). His equation was used herein to calculate F_c* in Fig. 15. Here the minimum value from Wu's equation was about 4% higher than the new minimum values proposed by Shih [*27*]. From these results, it is obvious

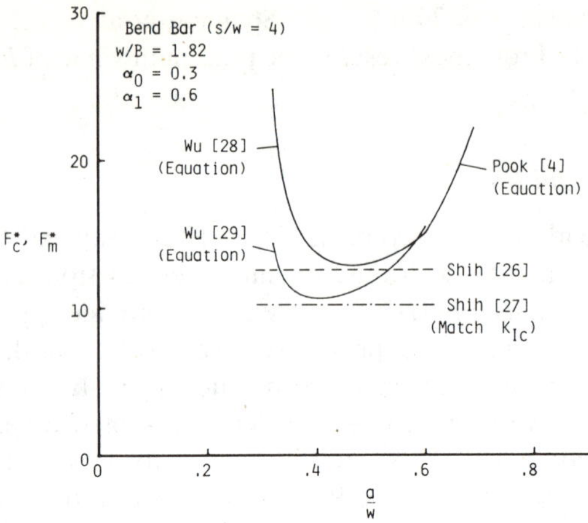

FIG. 15—*Comparison of normalized stress-intensity factors for chevron-notched three-point bend bar.*

that Pook's equation and Bluhm's slice model give drastically different values of stress-intensity factors, and that the determination of minimum values by matching K_{Ic} and K_m must be approached with caution.

Effects of Material Fracture Toughness Behavior

For a brittle material, a material which exhibits a "flat" crack-growth resistance curve as shown in Fig. 4, the use of a chevron-notched specimen to obtain K_{Icv} is well justified. But what if the material has a "rising" crack-growth resistance curve as shown in Fig. 16? Because most engineering materials, under nonplane-strain conditions, have rising crack-growth resistance curves or K_R-curves, the answer to this question is of utmost importance. The objective of

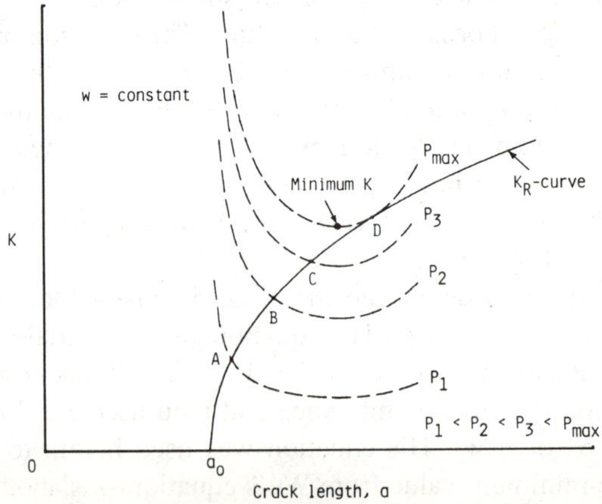

FIG. 16—*Fracture of "rising K_R-curve" material using a chevron-notched specimen.*

this paper, however, is not to answer this question, but to review some of the problems associated with using these specimens for such materials.

Figure 16 illustrates the application of the K_R-curve concept [*30*] to a material with a rising K_R-curve. The stress-intensity factor is plotted against crack length. The hypothetical K_R-curve (solid curve) begins at the initial crack length, a_0. The dashed curves show the crack-driving force curves for various values of applied load on a chevron-notched specimen (w = constant). As the load is increased, the crack grows stably into the material (point A, to B, to C, to D) until the load reaches P_{max}. At this load and crack length, crack growth becomes unstable (point D). As can be seen, the instability point (tangent point between crack-drive curve and K_R-curve) does not correspond to the minimum K-value (solid symbol). Consequently, the maximum load and minimum K-value cannot be used to compute the stress-intensity factor at failure, although the difference might be small. But if the specimen width is smaller than that used in Fig. 16, then the instability point would occur at a lower point on the K_R-curve. Conversely, the instability point for a larger width specimen would occur at a higher point on the K_R-curve. Thus, a specimen size (or width) effect exists and it has been the subject of several papers on chevron-notched specimens [*12,31–35*].

Discussion

Chevron-Notched Test Specimens

Many investigators have shown the advantages of using chevron-notched specimens for determination of plane-strain fracture toughness of brittle materials. The following table summarizes some of the advantages and disadvantages of these specimens:

Advantages	Disadvantages
Small specimens	Restricted to ''brittle'' materials
No fatigue precracking	Material thickness limitations
Simple test procedure	Notch machining difficulty
Maximum load test	
Screening test	
Notch guides crack path	
High constraint at crack front	

The chevron-notched specimens can be small because their width and height are of nearly the same size as their thickness (5 to 25 mm), so only a small amount of material is needed. Consequently, they are very useful as quality control specimens. They may be useful in alloy development programs where small amounts of material are produced. They can be also used to determine toughness profiles through the thicknesses of large plates. Because they require no fatigue precracking, they cost less than current fracture toughness specimens. For brittle materials, the test procedure is very simple; once the minimum stress-intensity factor has been obtained, it is only necessary to record the maximum failure load to calculate fracture toughness. Even for ductile materials, the specimens may be used in screening tests to rank materials.

The chevron notch tends to guide the crack path, and, therefore, these specimens can be used to test particular regions of a material such as heat-affected zones. The notch also constrains the crack front, which helps set up an approximate plane-strain condition around the crack front.

The major disadvantage in using chevron-notched specimens with the maximum load test procedure—for plane-strain fracture toughness testing—is that they are restricted to brittle materials, such as ceramics, rocks, high-strength metals, and other low toughness materials. Further studies are needed on more ductile materials to see if these specimens can be used for fracture-toughness evaluation. They are also limited in the thickness that can be tested. Thin materials, less than about 5 mm, cannot be easily tested.

Stress-Intensity Factors

Several methods have been used to determine stress-intensity factors and minimum stress-intensity factors for these specimens. In the first method, the minimum value was obtained by matching K_m to K_{Ic} from ASTM E 399 standard specimens. For the short-rod specimen, the minimum value obtained from K_{Ic}-matching [*5,11,12*] was about 8% below several experimental compliance calibrations and two recent three-dimensional elasticity solutions. In more recent applications of the K_{Ic}-matching procedure [*26,27*], the minimum values for a three-point bend specimen differed by about 20%. Thus, the K_{Ic}-matching procedure should be used with caution.

The second method is derived from the assumption that the change in compliance with crack length of the chevron-notch specimen is equal to the change in compliance of a straight-through crack specimen. The stress-intensity factors derived from this method match those from Pook's equation [*4*]. For the rod and bar specimens, researchers have shown that this method gives accurate values of minimum stress-intensity factors, but is unreliable on either side of the minimum. In contrast, this method gave very large differences on a three-point bend specimen. Again, this method must be used with caution.

The third, a more refined approximate method for chevron-notched specimens, is the slice model proposed by Bluhm [*10*]. This model has been used extensively on three- and four-point (chevron-notched) bend specimens. Munz et al [*7*] has used this model on chevron-notched bar specimens. The problem associated with this method is the ''shear-correction'' parameter (k) that must be determined from experimental compliance measurements. If the shear-correction parameter, k, is determined experimentally from uncracked chevron-notched specimens close to the desired configuration, then this method will probably give reliable results. But a systematic study to evaluate the accuracy of stress-intensity factors computed from the slice model has not been undertaken.

The fourth method is three-dimensional elasticity solutions, such as finite-element and boundary-integral equation methods. These methods can give accurate stress-intensity factors if care is taken especially in conducting convergence studies. These methods, however, tend to be expensive if a large number of solutions are desired.

The last method is experimental compliance calibration. This method can also give accurate stress-intensity factors if the tests are done carefully. But the method is limited to the particular specimen configurations studied. Coupled with Bluhm's slice model, this method may provide a reliable and inexpensive way of obtaining stress-intensity factors for a wide range of configuration parameters.

A summary of the consensus minimum normalized stress-intensity factor, F_m^*, for the four configurations considered in the analytical round robin and for the rectangular bar specimen [*6,15,20*] are shown in the following table.

Specimen	w/B	α_0	α_1	H/B	F_m^*
Bar	1.45	0.332	1	0.435	27.8 ± 0.3
Bar	1.45	0.332	1	0.5	24.8 ± 0.3
Bar	2	0.2	1	0.5	29.8 ± 0.3
Rod	1.45	0.332	1	0.5	28.9 ± 0.3
Rod	2	0.2	1	0.5	36.2 ± 0.4

The stress-intensity factor solutions for three- and four-point bend chevron-notched specimens have only been obtained from the K_{Ic}-matching procedure, Pook's equation, and Bluhm's slice model. Of these, the slice model is probably the most reliable. However, it is recommended that a detailed finite-element or boundary-element analysis, or careful experimental compliance calibrations, be performed on various chevron-notched bend bar configurations.

Conclusions

The historical development of chevron-notched fracture specimens and the stress-intensity solutions that have been proposed for these specimens was reviewed. The review covered the three- and four-point bend bars as well as the short-rod and bar specimens. The stress-intensity factor solutions and minimum stress-intensity value for these specimens had been obtained by using several different methods, either experimental or analytical. Results of a recent ASTM analytical round robin on the rod and bar specimens were summarized. Some problems associated with using these specimens for materials with rising crack-growth resistance curves were discussed. Based on this review, the following conclusions were drawn:

1. For the chevron-notched round-rod and bar specimens, the experimental compliance calibrations and the analytical (finite-element, boundary-element, and some approximate methods) calculations agreed within 3%. When the lower

bound convergence of the finite-element and boundary-element techniques were accounted for, the agreement was generally within about 1%.

2. Chevron-notched bend bars need further experimental and analytical stress-intensity factor calibrations. Although some recent stress-intensity factor solutions agreed within 5%, they were obtained from methods which have not been adequately substantiated.

3. Further studies are needed on using chevron-notched specimens with materials that exhibit a rising crack-growth resistance curve behavior.

References

[*1*] Nakayama, J., *Japan Journal of Applied Physics*, Vol. 3, 1964, pp. 422–423.

[*2*] Nakayama, J., *Journal of the American Ceramic Society*, Vol. 43, No. 11, Nov. 1965, pp. 583–587.

[*3*] Tattersall, H. G. and Tappin, G., *Journal of Materials Science*, Vol. 1, 1966, pp. 296–301.

[*4*] Pook, L. P., *International Journal of Fracture Mechanics*, Vol. 8, 1972, pp. 103–108.

[*5*] Barker, L. M., *Engineering Fracture Mechanics*, Vol. 9, 1977, pp. 361–369.

[*6*] Barker, L. M. in *Fracture Mechanics Applied to Brittle Materials, ASTM STP 678*, American Society for Testing and Materials, Philadelphia, 1979, pp. 73–82. (Proceedings Eleventh National Symposium Fracture Mechanics, Blacksburg, Va., June 1978.)

[*7*] Munz, D., Bubsey, R. T., and Srawley, J. E., *International Journal of Fracture*, Vol. 16, No. 4, 1980, pp. 359–374.

[*8*] Brown, W. F. and Srawley, J. E., *Plane Strain Crack Toughness Testing of High Strength Metallic Materials, ASTM STP 410*, American Society for Testing and Materials, Philadelphia, 1966.

[*9*] Freed, C. N. and Kraft, J. M., *Journal of Materials*, Vol. 1, No. 4, 1966, pp. 770–790.

[*10*] Bluhm, J. I., *Engineering Fracture Mechanics*, Vol. 7, 1975, pp. 593–604.

[*11*] Barker, L. M., *International Journal of Fracture*, Vol. 15, No. 6, 1979, pp. 515–536.

[*12*] Barker, L. M. and Baratta, F. I., *Journal of Testing and Evaluation*, Vol. 8, No. 3, 1980, pp. 97–102.

[*13*] Barker, L. M. and Guest, R. V., "Compliance Calibration of the Short-Rod Fracture Toughness Specimen," Terra Tek Report TR 78-20, April 1978.

[*14*] Barker, L. M., Discussion of "Compliance Calibration of the Short Rod Chevron-Notch Specimen for Fracture Toughness Testing of Brittle Materials," by Bubsey, R. T., Munz, D., Pierce, W. S., and Shannon, J. L., Jr., *International Journal of Fracture*, Vol. 19, 1982, pp. R3–R5.

[*15*] Barker, L. M., *Engineering Fracture Mechanics*, Vol. 17, No. 4, 1983, pp. 289–312.

[*16*] Beech, J. F. and Ingraffea, A. R., "Three-Dimensional Finite Element Calibration of the Short-Rod Specimen," Geotechnical Engineering Report 80-3, Cornell University, Ithaca, N.Y., 1980.

[*17*] Beech, J. F., and Ingraffea, A. R., *International Journal of Fracture*, Vol. 18, No. 3, 1982, pp. 217–229.

[*18*] Bubsey, R. T., Munz, D., Pierce, W. S., and Shannon, J. L., Jr., *International Journal of Fracture*, Vol. 18, No. 2, 1982, pp. 125–133.

[*19*] Shannon, J. L., Jr., Bubsey, R. T., Pierce, W. S., and Munz, D., *International Journal of Fracture*, Vol. 19, 1982, pp. R55–R58.

[*20*] Raju, I. S. and Newman, J. C., Jr., this publication, pp. 32–48.

[*21*] Ingraffea, A. R., Perucchio, R., Han, T. Y., Gerstle, W. H., and Huang, Y. P., this publication, pp. 49–68.

[*22*] Shannon, J. L., Jr., Bubsey, R. T., and Pierce, W. S., "Closed-Form Expressions for Crack-Mouth Displacements and Stress-Intensity Factors for Chevron-Notched Short Bar and Short Rod Specimens Based on Experimental Compliance Measurements," NASA Lewis Research Center (in preparation), 1984.

[*23*] Mendelson, A. and Ghosn, L. J., this publication, pp. 69–80.

[24] Timoshenko, S. and Goodier, J. N., *Theory of Elasticity,* second edition, McGraw-Hill, New York, 1951.
[25] Munz, D. G., Shannon, J. L., Jr., and Bubsey, R. T., *International Journal of Fracture,* Vol. 16, 1980, R137–R141.
[26] Shih, T. T., *Journal of Testing and Evaluation,* Vol. 9, No. 1, 1981, pp. 50–55.
[27] Shih, T. T., *Engineering Fracture Mechanics,* Vol. 14, No. 4, 1981, pp. 821–832.
[28] Wu Shang-Xian, *International Journal of Fracture,* Vol. 19, 1982, pp. R27–R30.
[29] Wu Shang-Xian, this publication, pp. 176–192.
[30] *Fracture Toughness Evaluation by R-Curve Method, ASTM STP 527,* D. E. McCabe, Ed., American Society for Testing and Materials, Philadelphia, 1973.
[31] Munz, D., Bubsey, R. T., and Shannon, J. L., Jr., *Journal of Testing and Evaluation,* Vol. 8, No. 3, 1980, pp. 103–107.
[32] Munz, D., *Engineering Fracture Mechanics,* Vol. 15, No. 1–2, 1981, pp. 231–236.
[33] Munz, D., Himsolt, G., and Eschweiler, J. in *Fracture Mechanics Methods for Ceramics, Rocks, and Concrete, ASTM STP 745,* S. W. Freiman and E. R. Fuller, Eds., American Society for Testing and Materials, Philadelphia 1981, pp. 69–84.
[34] Barker, L. M., this publication, pp. 117–133.
[35] Shannon, J. L., Jr., and Munz, D. G., this publication, pp. 270–280.

Ivatury S. Raju[1] *and J. C. Newman, Jr.*[2]

Three-Dimensional Finite-Element Analysis of Chevron-Notched Fracture Specimens

REFERENCE: Raju, I. S. and Newman, J. C., Jr., "**Three-Dimensional Finite-Element Analysis of Chevron-Notched Fracture Specimens,**" *Chevron-Notched Specimens: Testing and Stress Analysis, ASTM STP 855,* J. H. Underwood, S. W. Freiman, and F. I. Baratta, Eds., American Society for Testing and Materials, Philadelphia, 1984, pp. 32–48.

ABSTRACT: In this paper, stress-intensity factors and load-line displacements have been calculated for chevron-notched bar and rod-fracture specimens using a three-dimensional finite-element analysis. Both specimens were subject to simulated wedge loading (either uniform applied displacement or uniform applied load). The chevron-notch sides and crack front were assumed to be straight. Crack-length-to-specimen width ratios (a/w) ranged from 0.4 to 0.7. The width-to-thickness ratio (w/B) was 1.45 or 2. The bar specimen had a height-to-width ratio of 0.435 or 0.5. Finite-element models were composed of singularity elements around the crack front and 8-noded isoparametric elements elsewhere. The models had about 11 000 degrees of freedom. Stress-intensity factors were calculated by using a nodal-force method for distribution along the crack front and by using a compliance method for average values. The stress-intensity factors and load-line displacements are presented and compared with experimental solutions from the literature. The stress-intensity factors and load-line displacements were about 2.5 and 5% lower than the reported experimental values, respectively.

KEY WORDS: chevron notch, stress-intensity factor, cracks, finite-element analysis

Nomenclature

- a Crack length measured from load line
- a_o Initial crack length (to tip of chevron notch)
- b Length of crack front
- B Specimen thickness (diameter of rod specimen)
- E Young's modulus of elasticity
- E' Equals E for plane stress and $E/(1 - \nu^2)$ for plane strain
- F Boundary-correction factor determined from nodal-force method
- F_c Boundary-correction factor determined from compliance method

[1]Senior scientist, Vigyan Research Associates, Hampton, Va. 23666. Work performed under NASA contract NASI-17090.

[2]Senior scientist, NASA Langley Research Center, Hampton, Va. 23665.

F_m Minimum boundary-correction factor from compliance method
H Half of specimen height (radius of rod specimen)
K_I Stress-intensity factor (Mode I)
P Applied load
V_L Displacement at load point
V_T Displacement at top of specimen along load line
w Specimen width
x, y, z Cartesian coordinates
ν Poisson's ratio

The chevron-notched specimens [*1,2*], shown in Fig. 1, are small fracture toughness specimens being considered for use in standard tests by the American Society for Testing and Materials (ASTM) Committee E-24. Because they are small (5 to 25 mm thick) and because they require no fatigue precracking, they are well suited for quality control and materials toughness evaluation specimens. Currently, these specimens can only be used for high-strength alloys, ceramics, and other such low-toughness brittle materials. Further advances in elastic-plastic fracture mechanics are needed to use these specimens for ductile materials.

The unique features of a chevron-notched specimen, over conventional fracture-toughness specimens are: (1) the extremely high-stress concentration at the tip of the chevron notch, and (2) the development of a minimum stress-intensity factor as the crack grows. The high-stress concentration at the tip of the chevron notch causes a crack to initiate at a low applied load, eliminating the need to precrack a specimen, a costly and time consuming procedure. The minimum stress-intensity factor allows the fracture toughness to be evaluated from this failure (maximum) load without the need to make a load-displacement record, such as currently used in the ASTM Test Methods for Plane-Strain Fracture Toughness of Metallic Materials (E 399-83).

Experimental compliance calibrations of the chevron-notched bar (short-bar) and rod (short-rod) specimens have been done by Barker and Guest [*3*], Munz et al [*4*], Bubsey et al [*5*], Shannon et al [*6*], and Barker [*7*] for the determination of stress-intensity factors. In addition to the experimental calibrations, several analytical attempts have been made. Munz et al [*4*] used a quasi-analytical

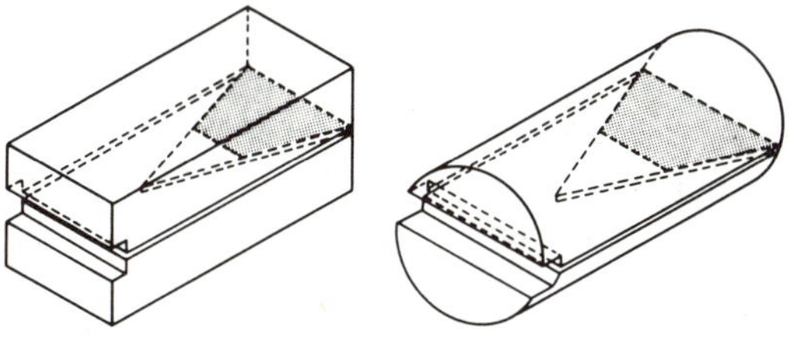

(a) Bar. (b) Rod.

FIG. 1—*Chevron-notched bar and rod specimens.*

procedure (slice model) developed by Bluhm [*8*] to analyze the chevron-notched bar specimen. Again, they determined stress-intensity factors from the compliance method. But, the experimental and analytical compliance methods give only an "average" stress-intensity factor along the crack front for each crack configuration considered. More rigorous three-dimensional analyses are required to determine stress-intensity factor variation along the crack front. Beech and Ingraffea [*9*] used a three-dimensional finite element method to determine stress-intensity factor distributions along the crack front and stress-intensity factors from analytical compliance for the chevron-notched rod ($w/B = 1.5$). Their crack front evaluations of stress-intensity factors, however, were in considerable disagreement (6 to 17%) with their values determined from compliance. But their analytical compliance values were in good agreement with experimental compliance results.

In this paper, stress-intensity factors and load-line displacements have been calculated by a three-dimensional finite-element analysis [*10*] for chevron-notched bar (square and rectangular) and rod-fracture specimens. The specimens were subjected to simulated wedge loading (either uniform applied displacement or uniform applied load). The chevron-notched sides and crack front were assumed to be straight. Crack-length-to-specimen width ratios (a/w) ranged from 0.4 to 0.7. The width-to-thickness ratio (w/B) was 1.45 or 2. The bar specimens had a height-to-width ratio (H/B) of 0.435 or 0.5. Stress-intensity factors were calculated by using a nodal-force method [*10*] for distributions along the crack front and by using a compliance method for average values. The minimum stress-intensity factors for five particular configurations were evaluated. Stress-intensity factors and load-line displacements are presented and compared with experimental solutions from the literature.

Analysis

Stress-intensity factors and load-line displacement for the chevron-notched bar and rod specimens, shown in Fig. 1, were obtained by using a three-dimensional finite-element analysis [*10*]. In this analysis, Poisson's ratio was assumed to be 0.3. The coordinate system used to define the chevron-notched specimens is shown in Fig. 2. The specimens are loaded by a knife-edge loading fixture [*4*] that results in an applied load, P, at point L, as shown in Fig. 2*a*. Specimens may have either a square notch [*4*] at the load line or a V-notch [*7*] at the load line (not shown). Only the square notch detail was considered herein. The slot height ($0.03B$) is for a saw blade to cut the chevron-shaped notch. In the present model, the slot height was assumed to be zero. The chevron was modeled and was assumed to have straight sides. Initial crack length, a_o, is the distance from the load line to the chevron tip (see Fig. 2*b*). The crack length, a, and specimen width, w, are measured from the load line. The crack front (b) was assumed to be straight. Crack-length-to-specimen width ratios (a/w) ranged from 0.4 to 0.7.

The following table gives the specimen dimensions of configurations analyzed herein:

Specimen	w/B	a_o/w	H/B
Bar	1.45	0.332	0.435
Bar	1.45	0.332	0.5
Bar	2	0.2	0.5
Rod	1.45	0.332	0.5
Rod	2	0.2	0.5

The configurations with $H/B = 0.5$ have been selected for possible standardization by ASTM Committee E-24.

Finite-Element Idealization

Two types of elements (isoparametric and singular [*10*] were used in combination to model the specimens. Figure 3*a* shows a typical finite-element model for the chevron-notched bar. The model idealized one quarter of the specimen and employed about 11 000 degrees of freedom (2 960 elements). The isoparametric eight-noded hexahedron elements were used everywhere except at the crack front, where eight singularity elements shaped like pentahedrons were used. The singularity elements produced a square-foot singularity in stress and strain at the crack front. A typical finite-element pattern on the crack plane is

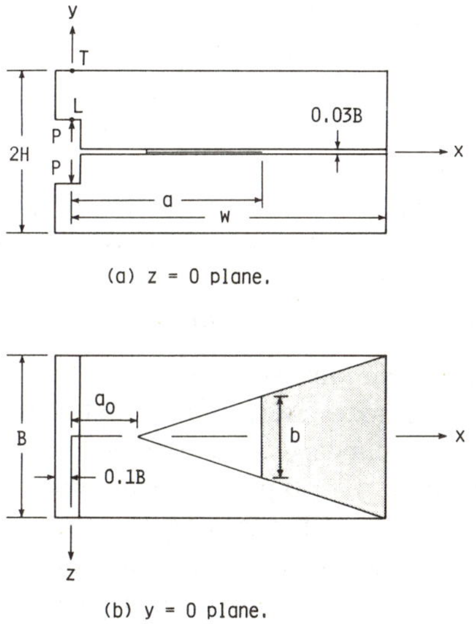

FIG. 2—*Coordinate system used to define dimensions of chevron-notched specimens.*

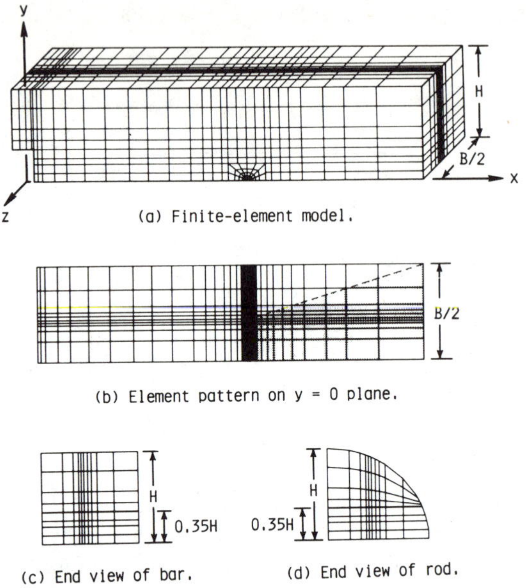

FIG. 3—*Finite-element idealization of chevron-notched bar and rod specimens.*

shown in Fig. 3*b*. This view shows the crack plane for an a/w ratio of 0.55. One half of the specimen thickness (B) was modeled with 10 layers. Figures 3*c* and *d* show an end view of the bar and rod specimen, respectively. The notch height was at $0.35H$.

To evaluate the finite-element mesh pattern used around the crack front in the three-dimensional models, two- and three-dimensional finite-element analyses of through-the-thickness edge cracks subjected to wedge loading were also analyzed. The two-dimensional analysis used a mesh pattern identical to the front view ($z = 0$ plane) shown in Fig. 3*a*. The three-dimensional analysis used the same model as that used for the chevron-notched specimens except that the singularity elements extended all the way across the specimen thickness.

Boundary Conditions and Applied Loading

Symmetry boundary conditions were applied on the $z = 0$ plane (see Fig. 3). On the $y = 0$ plane, all nodes were free except those that lie in the shaded region. Here, symmetry boundary conditions were applied. (The intent of the fixed-node condition on the $y = 0$ plane was to prescribe zero V-displacements for the shaded area. Because of the rectangular mesh idealization in the $y = 0$ plane, however, the $V = 0$ condition was only approximately achieved at locations along the edge of the shaded area. This is approximate because the chevron edge (edge of the shaded area) crossed elements that had one or more free nodes.) The specimens were subjected to wedge loading at point L in Fig.

$2a$. The loading was either a uniform applied load or a uniform applied displacement across the thickness.

Stress-Intensity Factors

Two methods were used to obtain stress-intensity factors. In the first method, the stress-intensity factor distributions along the crack front from the finite-element models were obtained by using a nodal-force method, details of which are given in Refs *10* and *11*. In this method, the nodal forces normal to the crack plane and ahead of the crack front are used to evaluate the stress-intensity factors.

The Mode I stress-intensity factor, K_I, at any point along the crack front was taken to be

$$K_I = \frac{P}{B\sqrt{w}} F\left(\frac{a}{w}, \frac{z}{b}\right) \tag{1}$$

where F was determined from the nodal-force method.

In the second method, an "average" stress-intensity factor along the crack front was obtained from specimen compliance as

$$K_I = \left(\frac{E'}{b}\frac{dU}{da}\right)^{1/2} \tag{2}$$

for the applied load case where $E' = E$ for plane stress or $E' = E/(1 - \nu^2)$ for plane strain. The total strain energy of the specimen, U, was calculated by

$$U = 4 \sum_{i=1}^{n} P_i V_i/2 \tag{3}$$

where P_i and V_i are the load and displacement, respectively, for the n nodes along the load line in the finite-element models. The stress-intensity factor from compliance was written as

$$K_I = \frac{P}{B\sqrt{w}} F_c\left(\frac{a}{w}\right) \tag{4}$$

and, therefore, equating Eqs 2 and 4 gives

$$F_c = \frac{B\sqrt{w}}{P}\left(\frac{E'}{b}\frac{dU}{da}\right)^{1/2} \tag{5}$$

The dU/da in Eq 5 was determined from the values of U evaluated at different crack lengths, a. Consider three crack lengths ($a_i < a_j < a_k$) and their corresponding total strain energies, U_i, U_j, and U_k. The strain energy was fitted to a second degree polynomial in terms of crack length as

$$U = \beta_1 + \beta_2 a + \beta_3 a^2 \tag{6}$$

The dU/da at crack length a_j was determined by

$$\left.\frac{dU}{da}\right|_{a_j} = \beta_2 + 2\beta_3 a_j \tag{7}$$

This slope was used in Eq 5 to evaluate the stress-intensity boundary-correction factor at crack length a_j.

Results and Discussion

In this section, two- and three-dimensional analyses are used to evaluate the accuracy of the finite-element model presented earlier (see Fig. 3). Next, a convergence study is presented for the chevron-notched configuration. Then, the stress-intensity factor variations along the crack front and the stress-intensity factors determined from the analytical compliance method (Eq 5) are presented for various chevron-notched configurations. Finally, the stress-intensity factors and load-line displacements from the present analyses are compared with experimental solutions from the literature. No comparisons are made with Beech and Ingraffea [*9*] finite-element analysis because different w/B ratios were considered.

Two- and Three-Dimensional Through-the-Thickness Crack Configurations

The finite-element idealization shown in Fig. 3 was evaluated by analyzing two- and three-dimensional through-the-thickness crack configurations. These evaluations consisted of studying convergence of stress-intensity factors and load-line displacements with mesh refinement in the $z = 0$ plane and in the thickness direction. A two-dimensional edge-crack configuration, like a double-cantilever beam specimen, was used to arrive at an adequate mesh refinement in the $z = 0$ plane and a three-dimensional through-the-thickness crack configuration was used to determine the mech refinement in the thickness direction.

Two-Dimensional Configuration—The finite-element mesh pattern on the $z = 0$ plane in Fig. 3*a* was used to model a wedge-loaded edge-cracked plate under plane-strain conditions. The results from this analysis are compared with the results from a boundary-collocation analysis [*12*] in Fig. 4. The boundary-collocation analysis was conducted on an edge-cracked plate with the same dimensions as those used in the finite-element analysis except that the square-notch detail at the load point was not modeled. The model used in the collocation

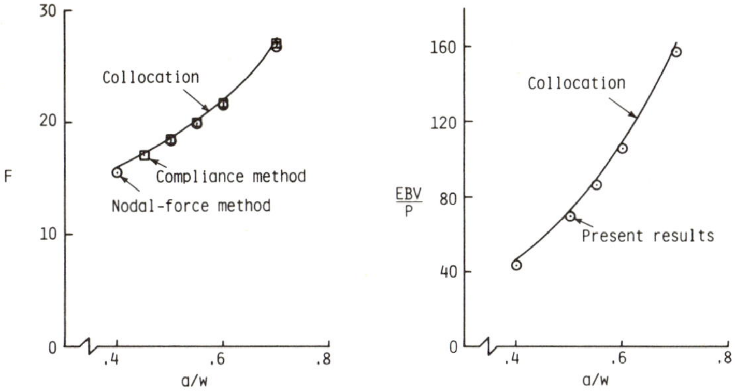

(a) Boundary-correction factors. (b) Crack-surface displacements.

FIG. 4—*Comparison of boundary-correction factors and crack-surface displacements from boundary collocation* [11] *and two-dimensional finite-element analyses of an edge-cracked plate.*

analysis was subjected to a line load acting over a small segment of the crack surface at $x = 0$. The solid curves in Figs. 4*a* and *b* show the boundary-correction factor and the normalized load-point displacement, respectively, from the collocation analysis as a function of a/w. The symbols in Fig. 4*a* show stress-intensity factors calculated from the finite-element analysis using the nodal-force and compliance methods. The correction factors evaluated from the nodal-force method were about 2% lower than those calculated from the collocation analysis, whereas those obtained from the compliance method were about 1.5% lower. The normalized load-point displacements obtained from the finite-element analysis (symbols in Fig. 4*b*) were about 4% lower than those calculated from the collocation analysis. Because the results from finite-element and boundary-collocation analyses agreed well, the mesh pattern along the $z = 0$ plane in Fig. 3*a* was considered sufficient for use in the three-dimensional models.

Three-Dimensional Configurations—To evaluate the three-dimensional models, a through-the-thickness crack in a square-bar configuration was analyzed with 2-, 4-, and 8-equal layers through one half of the thickness. Stress-intensity boundary-correction factors, determined from the nodal-force method, are shown in Fig. 5. The results in the interior of the specimen ($2z/B < 0.75$) agreed within a few percent for all three models. The correction factors decreased from the middle of the specimen ($2z/B = 0$) to its lowest value at the intersection of the crack with the free surface. The value at the free surface, however, varied with the number of layers (or layer thickness). Hartranft and Sih [*13*] have shown that the crack-front singularity differs from the square-root singularity in a very thin "boundary layer" near the free surface and that the stress-intensity factors drop off rapidly and equal zero at the surface. Thus, the finite-element method employed here cannot adequately evaluate the stress-intensity factors in this

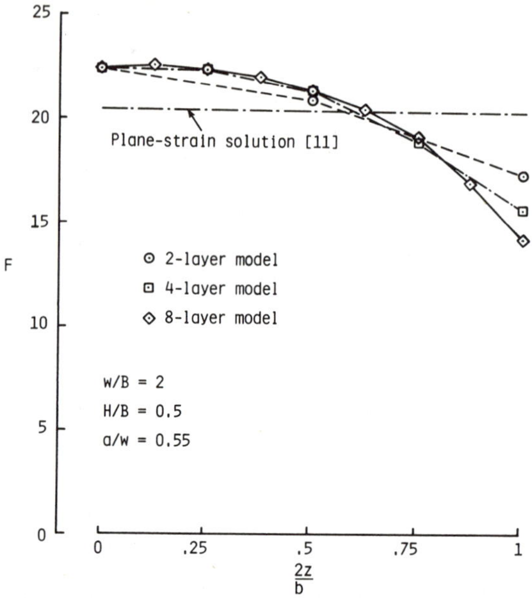

FIG. 5—*Distribution of boundary-correction factors along crack front for through crack in an edge-cracked plate using various three-dimensional finite-element models.*

"boundary layer." But the "average" stress-intensity correction factors across the thickness for all three models were in good agreement (2%) with the plane-strain solution [*12*].

Chevron-Notched Configurations

Convergence—The convergence study in Fig. 5 showed that a four-layer model is adequate and yields accurate stress-intensity factors along most of the crack front for through-the-thickness crack configurations. However, for more complex configurations, such as a chevron-notched specimen, the number of layers needed along the crack front may be greater than four. Therefore, two models were considered for a chevron-notched bar configuration ($w/B = 2$, $a/w = 0.55$). The first model had 10 layers across half the specimen thickness with 5 unequal thickness layers along the crack front. This model is shown in Fig. 3. The second model had 18 layers, with 8 unequal thickness layers along the crack front. A comparison between the stress-intensity factor distributions along the crack front for the 10- and 18-layer models is shown in Fig. 6. The center of the specimen is at $2z/b = 0$. The stress-intensity factors for the two models are nearly constant for $2z/b < 0.5$ but increase rapidly as the $2z/b$ approaches unity (edge of chevron). Results from the two models agreed well for $2z/b < 0.9$. At the chevron-notched location, however, the results were sensitive to layer thickness. Again, as observed in the preceding section on the "boundary layer" effect, the finite-element analysis cannot adequately evaluate the stress-intensity factors at locations where the crack front intersects another boundary. But these results do show that the 10-layer model is sufficient to model the chevron-notched configurations.

Loading Conditions—Because the chevron-notched specimens are loaded with either a knife-edged fixture [*4*] or a pressurized flat jack [*14*], two types of loading conditions were applied to some of the bar and rod configurations ($w/B = 2$, $a/w = 0.5$ and 0.55). The loadings were either a uniform applied load or a uniform applied displacement along the load line. The displacement variations along the load line for the applied load cases were very small (less than 0.6% from the average). For the same total applied load, the displacement variations along the load line for the applied load case were within 0.6% of the displacement from the applied displacement case. Likewise, for the same total applied load, the stress-intensity factors for the two types of applied loading were in excellent agreement (0.1%). Thus, the type of applied loading has no significant effect on the results. Consequently, all crack configurations considered herein were subjected to a uniform applied loading.

Bar and Rod Configurations—The stress-intensity correction factor distributions along the crack front for the square-bar and rod configurations are given in Tables 1 and 2 for various a/w ratios. Some typical results for the bar configuration ($w/B = 2$; $H/B = 0.5$) are shown in Fig. 7 for various a/w ratios. Results for $a/w = 0.55$ are not shown for clarity. The distributions as a function of $2z/b$ are similar for all a/w ratios with the lowest values occurring at the

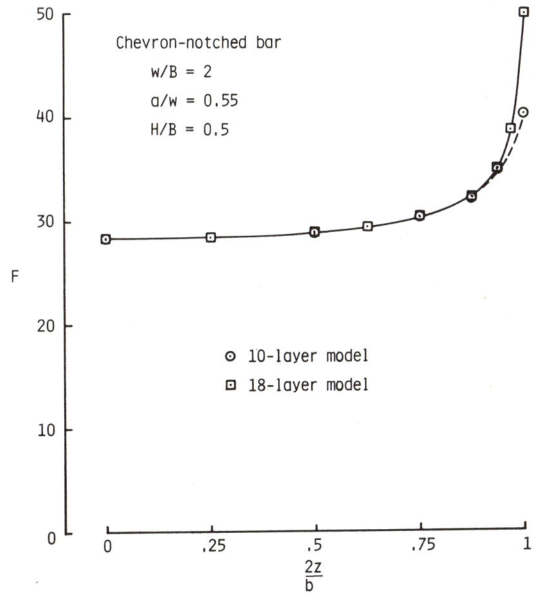

FIG. 6—*Distribution of boundary-correction factors along crack front in chevron-notched bar for two finite-element models.*

TABLE 1—*Boundary-correction factor,* F, *distributions for chevron-notched bar (square) specimens.*

(*a*) $KBw^{1/2}/P$ for w/B = 1.45

	a/w				
2*z/b*	0.4	0.5	0.55	0.6	0.7
0.0	27.95	23.83	23.50	24.45	30.37
0.5	28.82	24.19	24.08	24.96	30.76
0.75	30.59	25.69	25.46	26.23	31.84
0.875	32.45	27.49	27.19	27.90	33.46
0.9375	33.56	29.49	29.33	30.17	36.09
1.0	36.66	32.30	32.30	33.38	40.17

(*b*) $KBw^{1/2}/P$ for w/B = 2

	a/w				
2*z/b*	0.4	0.5	0.55	0.6	0.7
0.0	28.28	27.98	28.43	29.33	33.86
0.5	29.14	28.60	28.93	29.71	33.96
0.75	31.11	30.16	30.27	30.80	34.54
0.875	33.48	32.21	32.13	32.46	35.78
0.9375	36.09	35.00	34.89	35.17	38.43
1.0	41.42	40.26	40.17	40.48	44.12

TABLE 2—*Boundary-correction factor,* F, *distributions for chevron-notched rod specimens.*

(*a*) $KBw^{1/2}/P$ for w/B = 1.45

	a/w				
2*z/b*	0.4	0.5	0.55	0.6	0.7
0.0	33.52	27.97	27.73	28.84	34.19
0.5	34.53	28.64	28.24	28.87	34.13
0.75	36.60	30.22	29.55	29.89	34.47
0.875	38.77	32.17	31.30	31.44	35.54
0.9375	40.07	34.37	33.57	33.70	37.80
1.0	43.70	37.51	36.76	37.01	41.56

(*b*) $KBw^{1/2}/P$ for w/B = 2

	a/w				
2*z/b*	0.4	0.5	0.55	0.6	0.7
0.0	34.68	34.62	35.30	36.44	41.28
0.5	35.59	35.12	35.55	36.42	40.57
0.75	37.74	36.55	36.55	36.93	39.85
0.875	40.40	38.63	38.26	38.22	40.11
0.9375	43.40	41.65	41.13	40.86	42.11
1.0	49.61	47.58	46.91	46.46	47.40

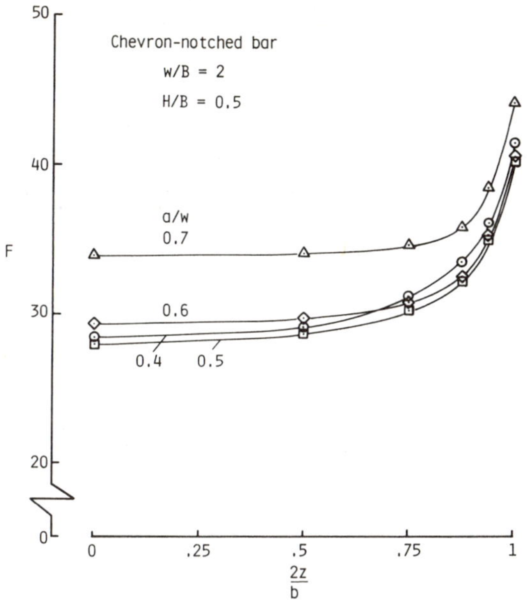

FIG. 7—*Distribution of boundary-correction factors along crack front in chevron-notched bars with various crack-length-to-width ratios.*

center of the specimen ($2z/b = 0$) and the highest values at the intersection of the crack with the chevron notch ($2z/b = 1$). These values were about 40% higher than the values at the center of the specimens.

Because of the rising stress-intensity factor distribution from the center of specimen to the edge of the chevron notch, the crack should grow more at the edges of the chevron notch than at the center of the specimen, thus causing a reverse-thumbnailing effect. Experimental results from Ref *7* confirm this observation.

A comparison between the stress-intensity factor distributions obtained from the three-dimensional finite-element method and from the compliance method is shown in Fig. 8. These results are for the square-bar configuration ($w/B = 2$) with $a/w = 0.5$. This configuration gave the lowest stress-intensity factors for all the a/w ratios considered. The solid symbols show the distribution as a function of $2z/b$. The dashed-dot and dash-double-dot lines show values determined for the compliance method (Eq 5) assuming either plane-stress or plane-strain conditions, respectively. The plane-strain value was about 5% higher than the plane-stress value. An experimentally determined compliance value [*6*] assuming plane-stress conditions is shown as the dashed line. The experimental value is about midway between the numerical values for plane stress and plane strain. But based on the previous two-dimensional results, the numerical values from compliance are estimated to be about 1.5% lower than the "true" values. Thus, the experimental value and the "corrected" numerical plane-stress value

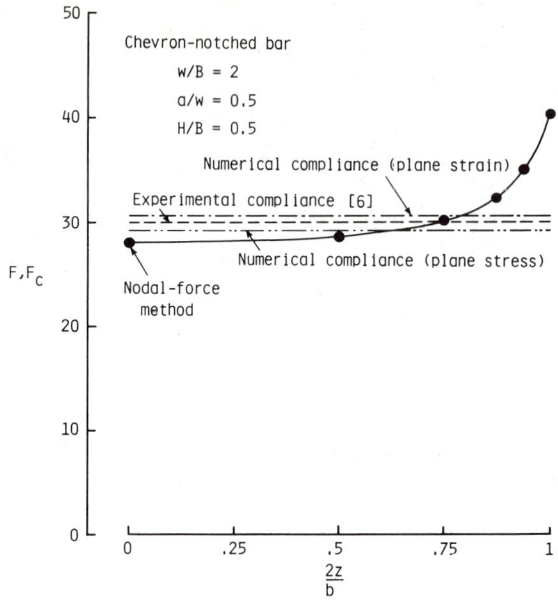

FIG. 8—*Comparison of boundary-correction factors from nodal-force and compliance methods for chevron-notched bar.*

($1.015F_c$) would be in very good agreement (about 1%). However, the use of the compliance method is, in itself, an approximation. The state-of-stress throughout the specimen is not either purely plane stress or purely plane strain. But the induced error is probably less than 2%.

Stress-intensity correction factors (F_c) determined from compliance (plane stress) for the five configurations considered are shown in Fig. 9 for various a/w ratios. For each configuration, these results were fitted to a third degree polynomial equation in terms of a/w to find the minimum value of the correction factor, F_m. The minimum values are shown as solid symbols. The following table compares these minimum values and those obtained experimentally in Ref *6*.

Specimen	w/B	$(a/w)_m$	F_m	$1.015\ F_m$	Y_m[a]	Percent Difference
Bar[b]	1.45	0.55	27.36	27.77	. . .	. . .
Bar[c]	1.45	0.54	24.43	24.80	24.85	−0.2
Bar[c]	2	0.52	29.13	29.57	29.91	−1.1
Rod	1.45	0.55	28.43	28.86	29.11	−0.9
Rod	2	0.52	35.40	35.93	36.36	−1.2

[a]Reference 6 uses Y_m to denote correction factor.
[b]$H/B = 0.435$.
[c]$H/B = 0.5$.

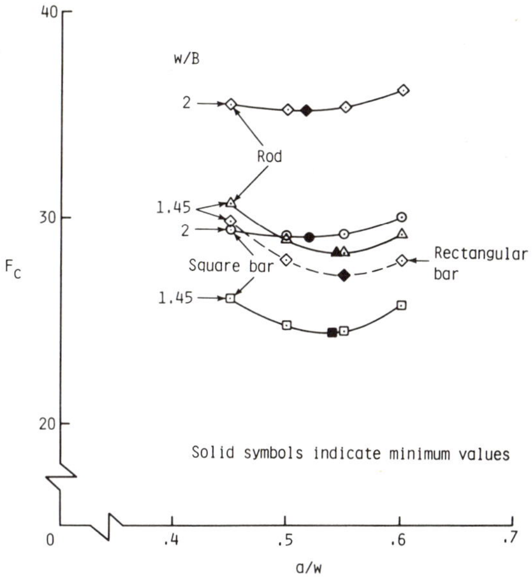

FIG. 9—*Comparison of boundary-correction factors from numerical compliance method for chevron-notched bar and rod specimens.*

The $(a/w)_m$-value is the crack-length-to-width ratio where the minimum F-value, F_m, occurred in the compliance analysis. The F_m-values determined from the finite-element analysis are estimated to be about 1.5% lower than the "true" solution because the potential energy method gives a lower bound solution and because of comparisons made between finite-element and boundary-collocation analyses (see Fig. 4*a*). Thus, the "corrected" numerical results for both the square-bar and rod specimens are about 1% lower than the experimental values [*6*].

Barker [*2*] selected the rectangular bar specimen (H/B = 0.435) to have the same compliance derivative as the rod specimen (w/B = 1.45) and, consequently, the same boundary-correction factor; that is, F_m was equal to 26.3 for both specimens. The present finite-element results gave a value of F_m as 27.77. This value was close to the finite-element results obtained on the rod specimen with w/B = 1.45 but was about 4% higher than Barker's value. Based on the current analysis, the recommended minimum value is 27.8 for the rectangular bar specimen with H/B = 0.435.

Table 3 gives the normalized displacements, EBV/P, at the midplane ($z = 0$) of the specimen for the load point (L) and for the top of specimen (T) as a function of a/w (see Fig. 2*a*). Some typical numerical results at the top of specimen are compared with experimental results in Fig. 10 for the rod specimen with w/B = 2. The circular and square symbols show experimental [*5*] and numerical results, respectively. These results are consistent with the comparisons made on two-dimensional analyses in Fig. 4*b*, in that the finite-element results

TABLE 3—*Normalized displacements as a function of* a/w *for chevron-notched square-bar and rod specimens*

(*a*) EBV_L/P at midplane ($x = z = 0$; $y = 0.35H$)

Type	w/B	a/w 0.4	0.5	0.55	0.6	0.7
Bar[a]	1.45	35.5	47.6	56.2	67.3	103.0
Bar[a]	2	55.5	82.6	99.8	119.0	174.5
Rod	1.45	46.9	63.7	75.5	90.2	135.1
Rod	2	76.3	116.1	141.4	171.3	249.8

(*b*) EBV_T/P at midplane ($x = z = 0$; $y = H$)

Type	w/B	a/w 0.4	0.5	0.55	0.6	0.7
Bar[a]	1.45	33.9	46.0	54.6	65.5	101.3
Bar[a]	2	54.0	81.1	98.3	118.5	173.0
Rod	1.45	45.1	61.9	73.6	88.3	133.2
Rod	2	74.7	114.4	139.7	169.6	248.1

[a]Square bar ($H/B = 0.5$).

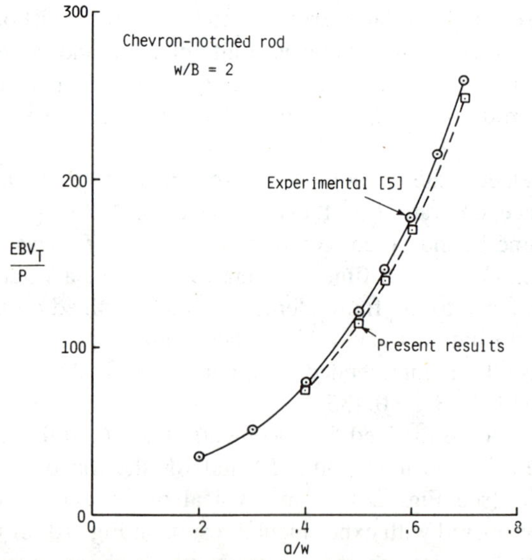

FIG. 10—*Comparison of experimental and calculated load-line displacements for chevron-notched rod.*

were about 4 to 6% lower than the experimental results. Based on beam theory [*15*], however, about 2% of this difference is caused by neglecting the slot height (0.03*B*) made by a saw blade or chevron cutter (see Fig. 2*a*).

Effect of Poisson's Ratio—Most experimental compliance results reported in the literature and the analyses reported herein were made with a Poisson's ratio of 0.3. However, Barker [7] used fused quartz which has a Poisson's ratio of 0.17. Therefore, to evaluate the effect of Poisson's ratio on stress-intensity factors a very limited study was made using the rod configuration with $w/B = 1.45$ and $a/w = 0.55$. Four different Poisson's ratios, 0.0, 0.17, 0.3, and 0.49, were used in the three-dimensional analyses. The following table shows the normalized stress-intensity factor at midplane ($z = 0$), the average normalized stress-intensity factor, and the load-line displacements for various Poisson's ratios.

Poisson's Ratio, ν	$\left.\frac{KBw^{1/2}}{P}\right\|_{z=0}$	$\left.\frac{KBw^{1/2}}{P}\right\|_{avg}$	$\frac{EBV_L}{P}$	$\frac{EBV_T}{P}$
0.0	26.33	28.03	79.2	77.5
0.17	26.92	28.49	77.9	76.1
0.3	27.73	29.20	75.5	73.6
0.49	27.99	29.12	64.4	63.1

The normalized stress-intensity factors at the midplane are higher for higher Poisson's ratios, and they change as much as 6% as the Poisson's ratio changes from 0 to 0.49. The average normalized stress-intensity factors show similar trends but with a smaller change, about 4%. These results indicate that a specimen with $\nu = 0.17$ (fused quartz) would have a stress-intensity factor about 2.5% lower than a specimen with $\nu = 0.3$.

In contrast to the stress-intensity factors, the load-line displacements are lower for higher Poisson's ratios. Also, as Poisson's ratio changes from 0 to 0.3, the change in the load-line displacements is about 5%. But as Poisson's ratio changes from 0.3 to 0.49, the load-line displacements change by as much as 15%.

Conclusions

Three-dimensional elastic finite-element analyses were used to obtain stress-intensity factors and crack-opening displacements for chevron-notched fracture specimens. Two types of specimens, a chevron-notched bar and rod, were subjected to simulated wedge loading (either uniform load or uniform displacement). The bar specimens had a height-to-width ratio of 0.435 or 0.5. In the analyses, the crack fronts and chevron-notched sides were assumed to be straight and the slot height for the chevron cutter was taken as zero. The crack-length-to-specimen-width ratio (a/w) ranged from 0.4 to 0.7. The width-to-thickness ratios (w/B) were 1.45 or 2. Stress-intensity factor variations along the crack front for these configurations were obtained by a nodal-force method. Also, "average" stress-intensity factors were obtained by a compliance method. Based on these analyses, the following conclusions were made:

1. The type of loading, either uniform load or uniform displacement, has no significant effect on stress-intensity factors and displacements.

2. The calculated load-line displacements at the top of the specimens are about 5% lower than reported experimental values.

3. The stress-intensity factor is lowest at the midplane of the specimen and highest at the intersection of the crack with the chevron notch. For most of the crack front, however, the stress-intensity factor is nearly constant. The rise occurs in the close vicinity of the chevron notch.

4. The "average" stress-intensity factor obtained from the three-dimensional finite-element compliance method (plane-stress) is about 2.5% lower than reported experimental values for both the square bar and rod specimens.

5. The a/w ratio at which the minimum stress-intensity factor occurred was between 0.5 and 0.55 for all chevron-notch configurations analyzed.

References

[*1*] Barker, L. M., *Engineering Fracture Mechanics,* Vol. 9, 1977, pp. 361–369.

[*2*] Barker, L. M., *Fracture Mechanics Applied to Brittle Materials, ASTM STP 678,* American Society for Testing and Materials, Philadelphia, 1979, pp. 73–82.

[*3*] Barker, L. M. and Guest, R. V., "Compliance Calibration of the Short Rod Fracture Toughness Specimen," Terra Tek Report TR 78–20, 1978.

[*4*] Munz, D., Bubsey, R. T., and Srawley, J. E., *International Journal of Fracture,* Vol. 16, No. 4, 1980, pp. 359–374.

[*5*] Bubsey, R. T., Munz, D., Pierce, W. S., and Shannon, J. L., Jr., *International Journal of Fracture,* Vol. 18, No. 1, 1982, pp. 125–132.

[*6*] Shannon, J. L., Jr., Bubsey, R. T., Pierce, W. S., and Munz, D., *International Journal of Fracture,* Vol. 19, No. 3, 1982, pp. R55–R58.

[*7*] Barker, L. M., *Engineering Fracture Mechanics,* Vol. 17, No. 4, 1983, pp. 289–312.

[*8*] Bluhm, J. I., *Engineering Fracture Mechanics,* Vol. 7, 1975, pp. 593–604.

[*9*] Beech, J. F. and Ingraffea, A. R., *International Journal of Fracture,* Vol. 18, No. 3, 1982, pp. 217–229.

[*10*] Raju, I. S. and Newman, J. C., Jr., "Three-Dimensional Finite-Element Analysis of Finite-Thickness Fracture Specimens," NASA TN D-8414, National Aeronautics and Space Administration, Washington, D.C., 1977.

[*11*] Raju, I. S. and Newman, J. C., Jr., *Engineering Fracture Mechanics,* Vol. 11, No. 4, 1979, pp. 817–829.

[*12*] Newman, J. C., Jr., *Fracture Analysis, ASTM STP 560,* American Society for Testing and Materials, Philadelphia, 1974, pp. 105–121.

[*13*] Hartranft, R. J. and Sih, G. C., *International Journal of Engineering Science,* Vol. 8, No. 8, 1970, pp. 711–729.

[*14*] Barker, L. M., *Fracture Mechanics of Ceramics,* Vol. 3, R. C. Bradt, D. P. H. Hasselman, and F. F. Lange, Eds., Plenum, New York, 1978, p. 483.

[*15*] Timoshenko, S. and Goodier, J. N., *Theory of Elasticity,* second edition, McGraw-Hill, New York, 1951.

Anthony R. Ingraffea,[1] *Renato Perucchio,*[2] *and Tao-Yang Han,*[2]
Walter H. Gerstle,[2] *and Yih-Ping Huang*[2]

Three-Dimensional Finite and Boundary Element Calibration of the Short-Rod Specimen

REFERENCE: Ingraffea, A. R., Perucchio, R., Han, T.-Y., Gerstle, W. H., and Huang, Y.-P., **"Three-Dimensional Finite and Boundary Element Calibration of the Short-Rod Specimen,"** *Chevron-Notched Specimens: Testing and Stress Analysis, ASTM STP 855,* J. H. Underwood, S. W. Freiman, and G. I. Baratta, Eds., American Society for Testing and Materials, Philadelphia, 1984, pp. 49–68.

ABSTRACT: The results of numerical calibration of a proposed standard short-rod specimen are reported. Three-dimensional finite and boundary element approaches were used to compute compliance and average stress-intensity factor as a function of crack length. Local variation of stress intensity along the crack front was computed for nine crack lengths. Calibration constants for fracture toughness evaluation were found to be: $\bar{Y}_{min} = 28.3$ and $A_{min} = 23.4$. A convergence study indicated that these values are lower bounds accurate to within 1%. Stress intensity was found to be minimum at the specimen centerline and maximum at the chevron edge for each crack length analyzed. At the critical crack length, the edge value was about 20% higher than the centerline value for an assumed straight crack front. Comparison of the boundary and finite element approaches indicated that, for the same stress intensity factor accuracy, the boundary element solution time was about 15% less than that of the finite element method.

KEY WORDS: short rod, calibration, compliance, stress-intensity factor, finite element, boundary element

The work reported in this paper is in response to a call by ASTM E24.01.05 Task Group on Fracture Specimen Design to perform analytical calibration of short-rod and short-bar specimens. The first author had previously performed [*1*] three-dimensional finite element calibration of a short-rod specimen. Considerable research had been also done at Cornell University into the usefulness of the short rod for fracture toughness measurements of rock [*2,3*] and portland cement concrete [*4,5*].

It was decided, therefore, to pursue numerical calibration of one of the newly proposed short-rod geometries, shown in Fig. 1, which is very similar to that

[1]Associate professor and manager, Experimental Research, Cornell University, Ithaca, N.Y. 14853.
[2]Graduate students, Cornell University, Ithaca, N.Y. 14853.

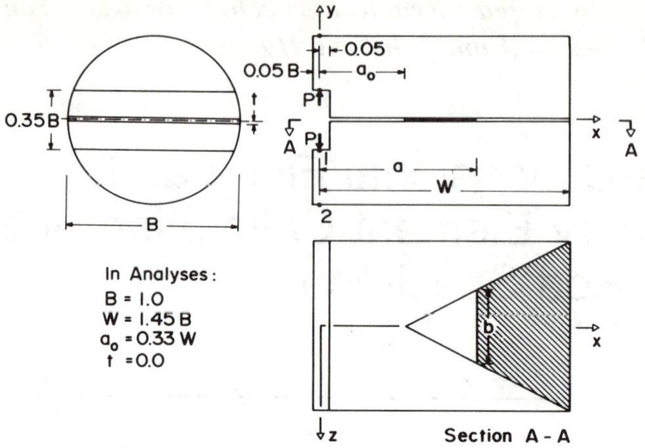

FIG. 1—*Short-rod geometry used in the present study.*

used in all previous work at Cornell. Reported herein are results of both three-dimensional finite and boundary element investigations of this specimen. The primary objectives of these investigations were:

1. To compute compliances as points 1 and 2 (Fig. 1) as a function of a/w.
2. To compute an average stress-intensity factor as a function of a/w according to

$$\bar{Y} = \frac{K_I B\sqrt{W}}{P} = \sqrt{W}\left[\frac{1}{2b}\frac{d(C_1EB)}{d(a/B)}\right]^{1/2} \tag{1}$$

where

$\bar{Y}$ = average, normalized stress-intensity factor,
K_I = average stress-intensity factor along a crack front,
C_1 = compliance at point 1 (Fig. 1),
E = Young's modulus, and
P = applied load.

3. To compute calibration constants, $\bar{Y}_{min}$ and A_{min}, for the specimen according to

$$K_{Ic} = \frac{P_{max}\bar{Y}_{min}}{B\sqrt{W}} \tag{2}$$

where

K_{Ic} = Mode I, plane strain fracture toughness,
P_{max} = maximum P in a test for K_{Ic},

and according to

$$K_{Ic} = \frac{P_{max} A_{min}}{B^{3/2}} \tag{3}$$

where

A_{min} = minimum value of average, normalized stress-intensity factor as defined by Barker [*6,7*].

4. To compute local stress-intensity factor distribution as a function of z and a/W.

All of these objectives were met and results reported here. Their validity is assessed by way of convergence studies and by comparisons with previous numerical and experimental results.

A secondary objective of the investigation was a comparison of the boundary element method (BEM) [*8*] and the finite element method (FEM) for calibration of fracture specimens requiring three-dimensional modeling. This effort produced a valid comparison of computer central processing unit (CPU) time and memory statistics for the two methods.

Convergence Studies

Finite Element Analysis

The present investigation began with the results of Beech and Ingraffea [*1*]. They performed three-dimensional finite elment analyses of a specimen only slightly different from that shown in Fig. 1 in that W/B was 1.50 rather than the present 1.45. Beech and Ingraffea used three meshes (shown in Fig. 2 of Ref *1*) employing the twenty-noded isoparametric brick element and increasing number of nodes to produce the compliance curves shown in Fig. 2. Equation 1 was used to compute average stress-intensity factors for each mesh. The calibration constant, A_{min}, used in Eq 3 was also computed. Results are summarized in Table 1. Fifth order polynomials were least-square fit to the results shown in Fig. 2 to produce those in Table 1.

It can be seen in Table 1 that, as expected, $\bar{Y}_{min}$ and A_{min} values were converging from below and more rapidly than typical compliance values and that the critical crack length had essentially stabilized. It was therefore decided that only one finite element mesh would be used for the present study, that it would also use the twenty-noded isoparametric brick element, but that it would have approximately twice as many nodes as the third mesh of Beech and Ingraffea [*1*]. Figure 3 shows the exterior surfaces of the generic finite mesh (shown here for $a/W = 0.59$) used in the present study. The mesh, hereafter called FEMI, had 867 nodes and differed from the ideal geometry shown in Fig. 1 in two ways: the slot thickness, t, was zero, and material to the left of the load line was not

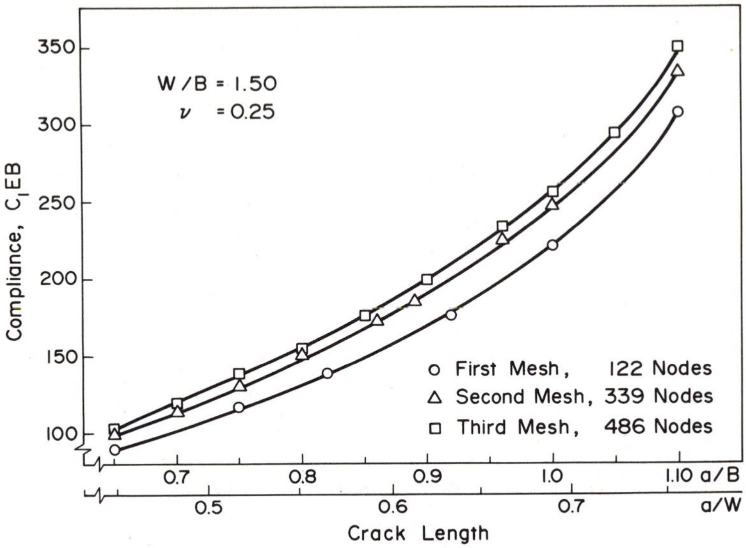

FIG. 2—*Compliance results obtained by Beech and Ingraffea* [1] *using three-dimensional finite element analyses.*

modeled. While the latter difference certainly has no effect on results, Barker [7] has shown that there is a slight dependence of calibration on slot thickness.

Nine crack lengths were modeled with meshes of the type shown in Fig. 3. In each case the crack front was assumed straight and normal to the *x*-axis. Compliance results are shown in Fig. 4 and Table 2.

Boundary Element Analysis

The compliance results from FEM1 were used as a guide in the convergence study for the boundary element approach. Five generic boundary element meshes, shown in Fig. 5, were used in this study. All used six- and eight-noded isoparametric elements. Meshes BEM1 and BEM2 had the same end-facet and curved surface discretization, but BEM2 had an additional row of elements between the load-line and the crack front. It can be seen in Figs. 5*a* and *b* that the geometry of the structure, in particular the quarter-circular surface, was being represented by two, piecewise quadratic patches with one patch subtending most of the quarter-circular arc.

TABLE 1—*Summary of results from Beech and Ingraffea* [1], W/B = *1.50.*

Mesh	$\bar{Y}_{min}$ [a]	A_{min} [a]	$(a/W)_c$
1	28.0	22.8	0.56
2	30.4	24.9	0.61
3	31.0	25.2	0.63

[a] Plane stress assumption.

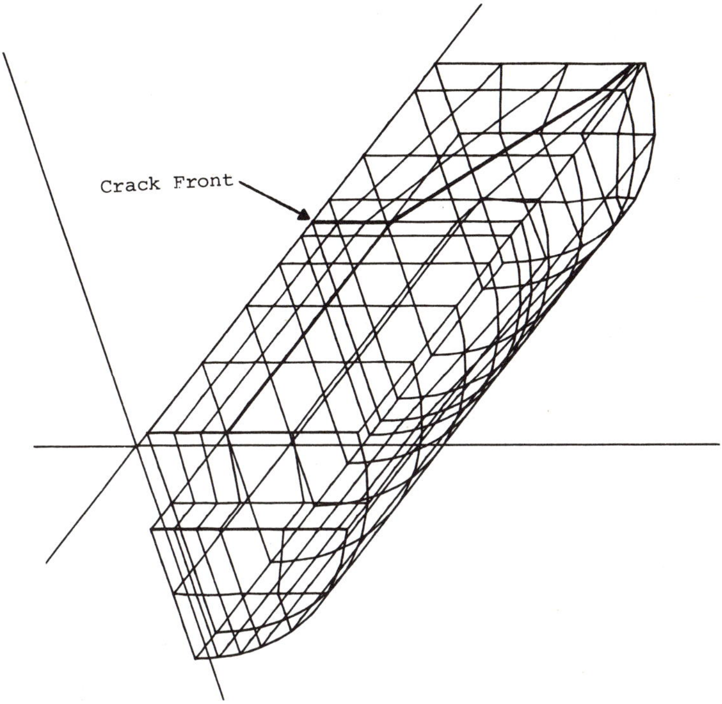

FIG. 3—*Mesh FEM1. Only surface traces of elements shown for clarity.*

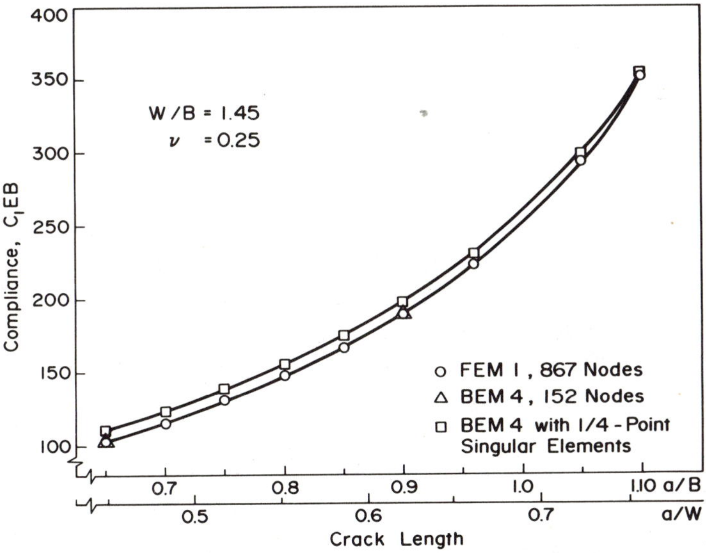

FIG. 4—*Compliance results obtained in the present study.*

TABLE 2—*Compliances,* C_1EB,[a] *computed from meshes FEM1 and BEM4.*[b]

a/W	FEM1	BEM4
0.448	104.26	110.56
0.482	116.85	123.76
0.517	131.10	138.44
0.552	147.76	155.24
0.586	167.24	174.60
0.621	190.38	197.36
0.662	222.79	230.24
0.724	293.74	299.12
0.759	352.54	354.34

[a] $\nu = 0.25$.
[b] With ¼-point, singular elements.

In meshes BEM3 and BEM4, end-facet and circular surface discretization is improved. Figures 5*b* and *c* show that BEM4 differs from BEM3 with the addition of an extra row of elements between load line and crack front and the breakdown into triangles of some elements around the crack front. BEM5 further improves curved surface representation and increases the number of element layers ahead of and behind the crack front. As of this writing, BEM5 has been used only for a crack length $a/W = 0.55$.

Figure 6 presents the results of the convergence study. These can be discussed with respect to the effects of geometry and bending mode representation. It can

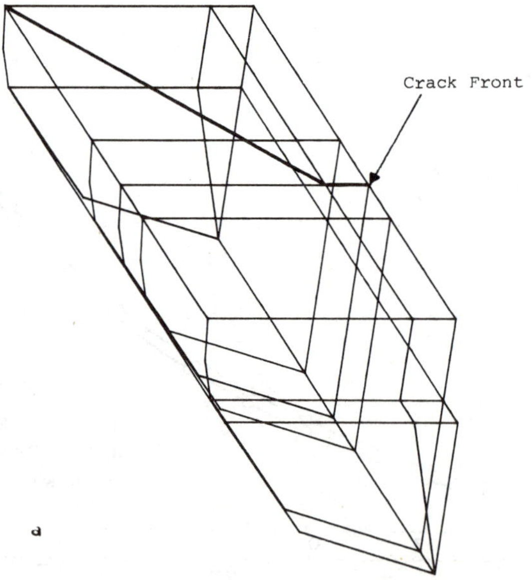

FIG. 5—*Boundary element meshes used in the present study* (a) *BEM1,* (b) *BEM2,* (c) *BEM3,* (d) *BEM4, and* (e) *BEM5.*

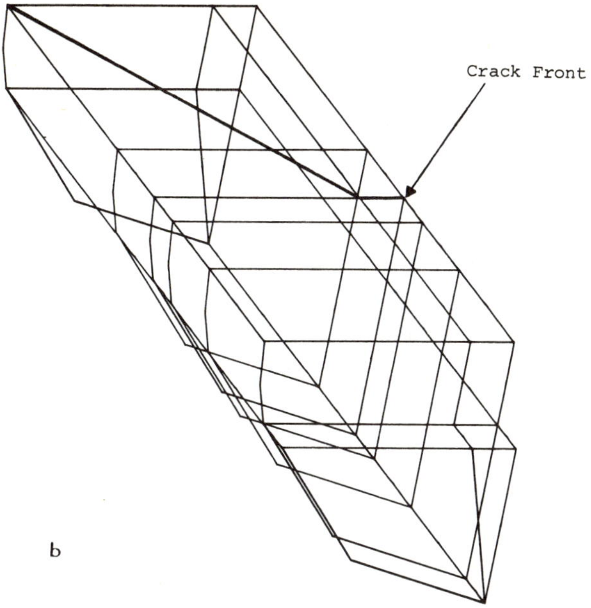

b

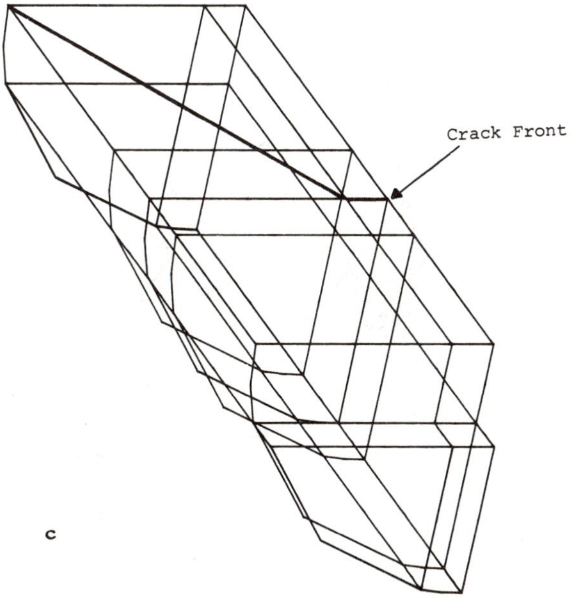

c

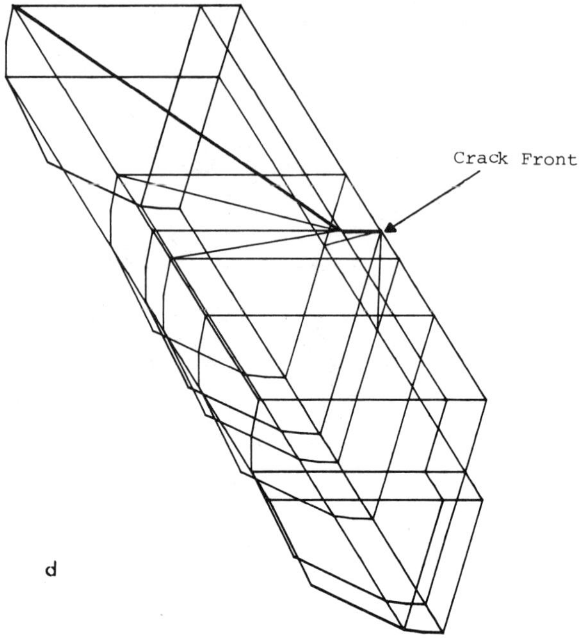

d

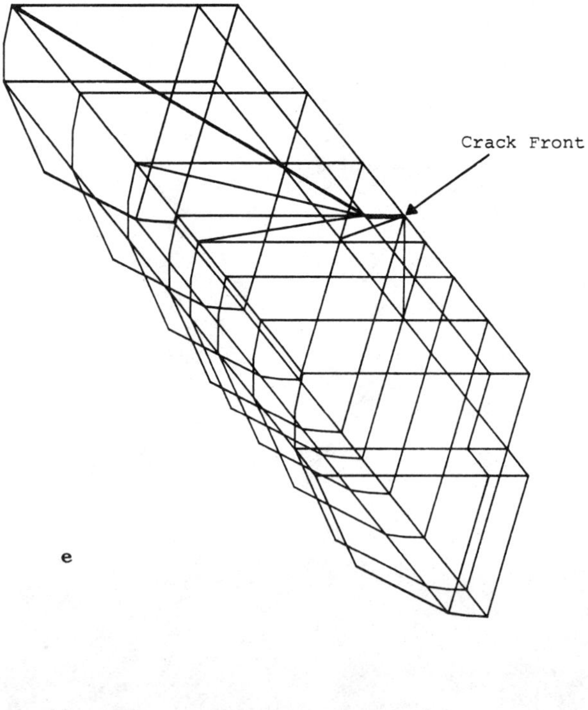

e

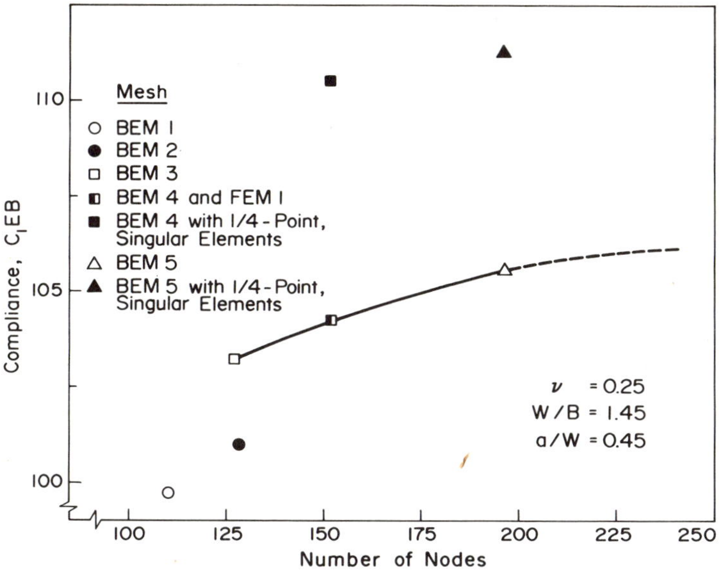

FIG. 6—*Compliance results from the boundary element meshes shown in Fig. 5.*

be seen that inaccuracy generated by coarse geometry representation, the difference between BEM1 and BEM3 or BEM2 and BEM4 results, was larger than that generated by insufficient bending mode modeling, the difference between BEM1 and BEM2 or BEM3 and BEM4. There was less than a 1% increase in compliance in going from BEM3 to BEM4, and, surprisingly, the BEM4 results agreed with the FEM1 results to the third decimal place for both $a/W = 0.45$ and $a/W = 0.62$. A convergence trend is established with the result from BEM5 which showed that a 30% increase in nodes caused only a 1% increase in compliance. Figure 6 shows that a conservative extrapolation of the BEM3-BEM4-BEM5 result trend would produce a compliance only about 2% higher than that obtained with the BEM4 mesh. It was therefore decided, given existing time constraints, that BEM4 would be the baseline mesh for subsequent boundary element analyses.

The first of these was an examination of the effect on compliance of the singular traction field and $\sqrt{r}$ displacement variation near the crack front. These characteristics were not included in the finite element analyses of the present study. Figure 6 shows that, when the triangular elements around, and the rectangular elements along the crack front in BEM4 and BEM5 were appropriately modified to become ¼-point and traction singular [*9*], the computed compliance increased, as expected. This increase was about 5% at $a/W = 0.45$; due to St. Venant effect, the increase was only 1% at $a/W = 0.76$. This study indicated, however, that the crack front singularity should be modeled.

It was decided, therefore, to perform analyses with BEM4, with ¼-point, singular elements, for nine crack lengths. Figure 4 and Table 2 summarize compliance results from these analyses.

Compliance Results

Fifth order polynomials were least-square fit to the results shown in Fig. 4. The resulting expressions were

From FEM1

$$C_1EB = -3049.0 + 20124.7(a/B) - 52052.3(a/B)^2 + 67368.7(a/B)^3 - 43334.7(a/B)^4 + 11192.9(a/B)^5 \quad (4)$$

From BEM4 with ¼-point, traction singular elements

$$C_1EB = -2721.1 + 17377.4(a/B) - 43561.7(a/B)^2 + 55023.0(a/B)^3 - 34716.2(a/B)^4 + 8855.8(a/B)^5 \quad (5)$$

Both of these expressions are for $0.65 \leq a/B \leq 1.10$.

The displacement computed at point 2, Fig. 1, was always less than that of point 1; however, the difference varied from a maximum of 2.2% of the point 1 value at $a/W = 0.45$ to a minimum of 0.5% at $a/W = 0.76$.

Since all analyses were performed under load control, a variation in crack-mouth-opening-displacement (CMOD) with z along the load-line was computed. The extreme case was again for $a/W = 0.45$ for which CMOD at $z = 0$ was about 4% less than that at $z = B/2$.

Stress-Intensity Factor Results

Average Stress-Intensity Factor, $\bar{Y}(a/W)$

Using Eqs 1, 4, and 5 average, normalized stress intensity factors, $\bar{Y}(a/W)$, were computed under a plane-stress assumption. These are shown in Fig. 7. It is not clear that the plane stress assumption is more correct than plane strain for this structure. Rather, plane stress was assumed to be consistent with other workers in reporting results of the round-robin analytical calibration of the specimen. The consequences of plane strain assumption can be easily determined by substituting E' for E in Eq 1, where

$$E' = E(1 - \nu^2) \quad (6)$$

The effect is a uniform shift in the $\bar{Y}$-curve by few percent.

It can be seen that, in terms of $\bar{Y}_{min}$ and $(a/W)_c$ values, the results of the FEM and BEM approaches are nearly identical. Calibration constant, A_{min}, for use in Eq 3 was also computed. Table 3 presents a summary of results for the average stress-intensity approach.

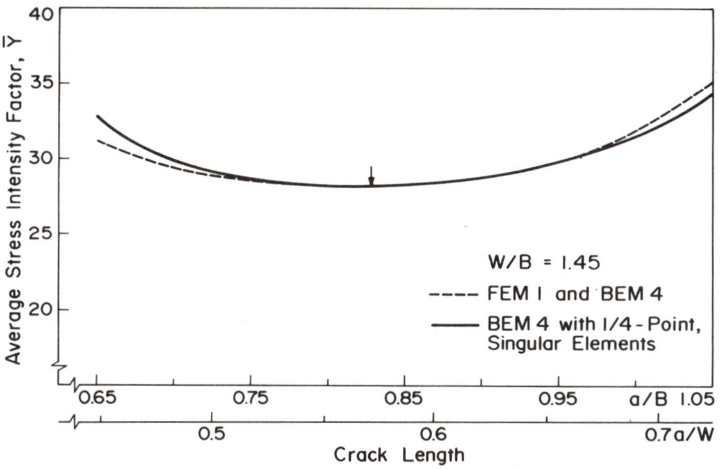

FIG. 7—*Average stress-intensity factor variation with* a/W *computed from compliances shown in Fig. 4.*

An investigation was performed into the sensitivity of the results shown in Table 3 to the order of polynomial used to fit the computed compliances. The results of this investigation are shown in Table 4 where it is seen that results are not sensitive to curve fitting model.

As noted previously, a variation in CMOD under the line load was computed. This observation calls into question the validity of Eq 1, based on two-dimensional notions, for the inherently three-dimensional problem under consideration. Rather than C_1EB which is the normalized work done by a load concentrated at point 1, stress-intensity factor computations should be based on work done by the line load actually used in the analyses. Such computations were performed and the results indicated, in the most extreme case, less than a 1% increase in work done over that obtained with the lumped load simplification. The change in $\bar{Y}$ which would result from inclusion of this modification would be negligible.

Local Stress-Intensity Factor Variation, Y(z/b)

Stress-intensity factors were computed pointwise along each of the nine crack fronts considered using the method of Ingraffea and Manu [*10,14,15,16*] and BEM4 with ¼-point, traction singular elements. Plane strain was assumed along

TABLE 3—*Summary of critical stress-intensity factor results,* W/B = *1.45.*

Mesh	$\bar{Y}_{min}{}^{b}$	$A_{min}{}^{b}$	$(a/W)_c$
Standard BEM4/FEM1	28.32	23.44	0.572
BEM4[a]	28.28	23.40	0.566

[a]With ¼-point, traction singular elements.
[b]Plane stress assumption.

TABLE 4—*Effect of polynomial order on results from mesh BEM4.*[a]

Order	$\bar{Y}_{min}$[b]	$(a/W)_c$
4th	28.12	0.579
5th	28.28	0.566
6th	28.26	0.566

[a]With ¼-point, traction singular elements.
[b]Plane stress assumption.

the entire crack front. It is known that the relative size of crack front elements somewhat influence accuracy of stress intensity factor computation with the method employed. Consequently, the length in the x-direction, L, of the boundary element immediately behind the crack front, the element used for stress-intensity factor computation, was generated to be always $0.16a$.

Figure 8 shows a comparison of average stress intensity with local values for selected crack lengths. It can be seen that, for short and long crack lengths, the average value computed from compliance is higher than the average of the locally computed values. It is probably fortuitous that, for the critical crack length, $a/$

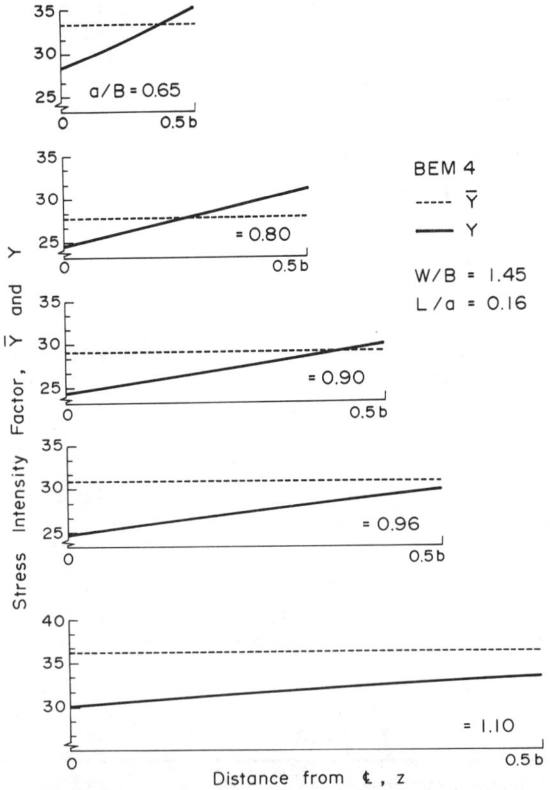

FIG. 8—*Variation of stress-intensity factor with z for selected crack lengths.*

$B = 0.80$, there is an excellent agreement between the two averages. The inconsistency in agreement is most likely due to inaccuracy in the locally computed values in that a single element comprised the entire crack front. Even so, the difference in average values never exceeded 12% of $\bar{Y}$ for all crack lengths considered.

Figure 8 also reveals that the highest local value of stress intensity is at the outer edge of the crack front, while the lowest value is at the specimen centerline for all crack lengths considered. The variation was nearly linear, although the element used allows up to quadratic order change. For the shortest crack length modeled, the edge value was about 25% higher than the centerline value, while this difference had decreased to about 10% for the longest crack length.

Discussion

Discussion of the results presented previously will center on three topics: (1) their accuracy when compared to other numerical and experimental results, (2) the implications of stress-intensity factor variation along the crack front, (3) and a comparison of the FEM and BEM approaches.

Accuracy of Results

To the authors' best knowledge, no analytical solutions had been obtained for the geometry of Fig. 1 before those produced for the present round-robin investigation. It is obvious, however, that $\bar{Y}$ should be less at all crack lengths for the present geometry than that analyzed by Beech and Ingraffea [*1*]. A first approximation to the amount of decrease in $\bar{Y}_{min}$ can be obtained by noting that, among other parameters, $\bar{Y}_{min}$ should be directly proportional to the moment arm, a_c [7]. With this observation and using the results shown in Tables 1 and 3, it would be expected that for the present geometry $\bar{Y}_{min}$ would be 27.1. This estimate is only about 4% lower than the computed value. This comparison establishes the internal consistency of the results obtained at Cornell University for two geometries using different numerical methods.

The present results can be compared to the experimental results of Bubsey et al [*11*]. They measured compliance variation with crack length on a range of short-rod geometries. Table 5 presents a comparison of present results with those interpolated and slightly extrapolated from the Bubsey et al [*11*] study.

Present results can be also compared to experimental findings of Barker [7]. This comparison is shown in Table 5. In his study, Barker showed the influence on specimen calibration of a number of geometrical variations from a "standard" configuration. Further discussion of the results shown in Table 5 centers on Barker's findings and, to be consistent with his notation, the use of his adjustment factors on the value of A_{min}.

First the A_{min}-value of Bubsey et al [*11*] will be compared to that found here. With one exception, all of the geometrical characteristics of the specimen analyzed here can be interpolated into and slightly extrapolated from those of the

TABLE 5—*Comparison of present results with experimental results of Bubsey et al* [11] *and Barker* [7].

	C_1EB @ $(a/W) = 0.56$	$\bar{Y}_{min}^c$	A_{min}^c	$(a/W)_c$
Present result	155.3	28.3	23.4	0.566
Bubsey et al[a]	154.5	29.2	24.3	0.56
Barker	167.5[b]	28.3	23.4	0.56 [b]

[a]Obtained by linear interpolation and extrapolation.
[b]Obtained by linear interpolation.
[c]Plane stress assumption.

specimens tested by Bubsey et al [*11*]. That exception is the chevron slot thickness, t. While they used a value of 0.021B for this parameter, it was assumed zero in the present analyses. Barker [7] has shown that decreasing t decreases A_{min}. Although his results are not conclusive, their extrapolation to zero thickness would cause a reduction in A_{min} to about 24.0. This is about 3% higher than the A_{min} found in the present study.

Barker's [7] "standard" specimen differs from that analyzed here in four ways: it has a chevron with curved sides, a V-notch, a larger effective chevron angle, and a slot thickness of 0.031B. He argues convincingly that the first and second differences have negligible effects on calibration. Barker's experiments show that a decrease in chevron angle increases A_{min}, while, as has been previously mentioned, a decrease in slot thickness decreases A_{min}. For the particular geometry differences which exist between his "standard" and the present geometry, these influences appear to cancel each other. Consequently, the difference in A_{min}-values remains as shown in Table 5, Barker's value being the same as that found by the present analyses. It should be noted that the relatively large differences in compliance shown in Table 5 is likely due to the V-notch and finite slot thickness used in Barker's specimens.

A final estimate on the accuracy of the present result can be made by noting that the 6% increase in compliance at $a/W = 0.55$ obtained in going from FEM1 to BEM4 with singularity modeled resulted in virtually no increase in A_{min}. It is estimated, therefore, that a further increase of about 2% in compliance to convergence as indicated in Fig. 6 would cause an overall increase in A_{min} of less than 1% with the same curve fitting procedure.

Stress-Intensity Variation Along Crack Front

All of the analyses reported by Beech and Ingraffea [*1*] and in the present study assumed a straight crack front normal to the longitudinal axis. However, both studies produced results which indicated that stress intensity varied, sometimes considerably, along the assumed crack front. This would imply non-uniform propagation, and some inaccuracy in the use of the numerical results. Figure 9 contains a sequence of photographs showing crack front evolution in a short-rod specimen of polystyrene. It is obvious that propagation along the centerline is relatively retarded for short crack lengths, in agreement with the implications of Fig. 8. However, it is also seen in Fig. 9 that the crack front gradually straightens and, ultimately, thumbnails slightly. In contrast, although Fig. 8 shows a gradual trend towards uniformity of stress intensity with increasing crack length, thumbnailing is not predicted.

This inconsistency can be explained by noting that the numerical results are based on linear elastic assumptions, while plastic zone development is occurring in the test specimen. Apparently, although some stress triaxiality is being generated by the chevron notch near the end of the crack front, an effectively higher degree of plane strain is gradually developing along the interior of the crack front. In effect, two phenomena are working against each other near the end of the crack front: higher stress-intensity accelerating growth and larger plastic zone retarding growth. Of course, the retardation effects in this competition are material dependent. One would expect that front shape would straighten less quickly for more brittle materials and that thumbnailing might not occur. In support of this contention is Fig. 6 of Barker [*7*] which shows crack front evolution in a specimen of fused quartz, a nearly ideal brittle material. This figure shows a tendency towards straightening of the front but no thumbnailing.

Interestingly, in nonchevron specimens like the standard SE(B) (after fatigue crack growth), highest stress intensity occurs on the midplane [*10*], lowest near the end of the crack front. Consequently, thumbnailing is exacerbated, and a straight through crack should never occur in such a specimen composed of a material which exhibits crack front plasticity.

Comparison of BEM and FEM Approaches

A secondary objective of the present effort was a comparison of three-dimensional BEM and FEM approaches for calibration of fracture toughness testing geometries. Such a comparison became particularly valid when it was discovered that the FEM1 and BEM4 meshes produced virtually identical displacement results. Table 6 summarizes manpower and CPU time requirements for each approach with these meshes. A high-level interactive graphics preprocessor [*12,13*] was used to input both models. Both models used isoparametric quadratic order (geometry and displacement for the FEM; geometry, displacement and traction for the BEM) elements. All computations were done on a Digital Equipment Corporation VAX 11/780 super-minicomputer using a floating-point accelerator and under the VMS operating system. Direct Gauss elimination was used to solve the equations; a skyline-banded approach was used for the FEM equations, while operations were performed on the entire coefficient matrix for the BEM equations.

It can be seen that the BEM approach required considerably less storage and total CPU time to produce the same displacement solution as the FEM approach. Since only displacements were required to obtain the compliances and local values of stress-intensity factors in the present study, stress recovery time was not included in the total solution time for the FEM approach.

Neither FEM1 nor BEM4 represent fully converged solutions in displacements; however, Fig. 6 indicates that a boundary element mesh with about 50% more

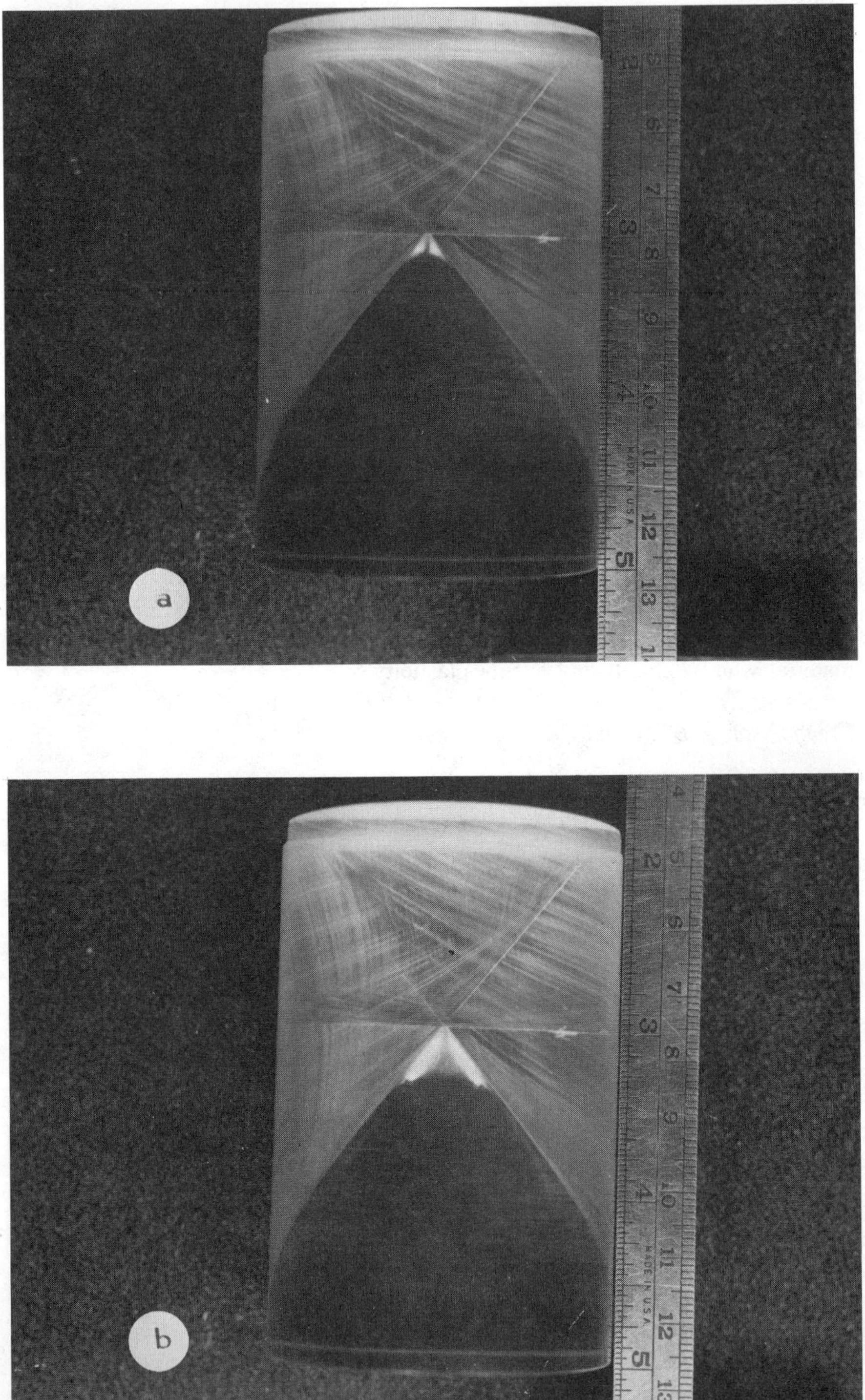
a
b

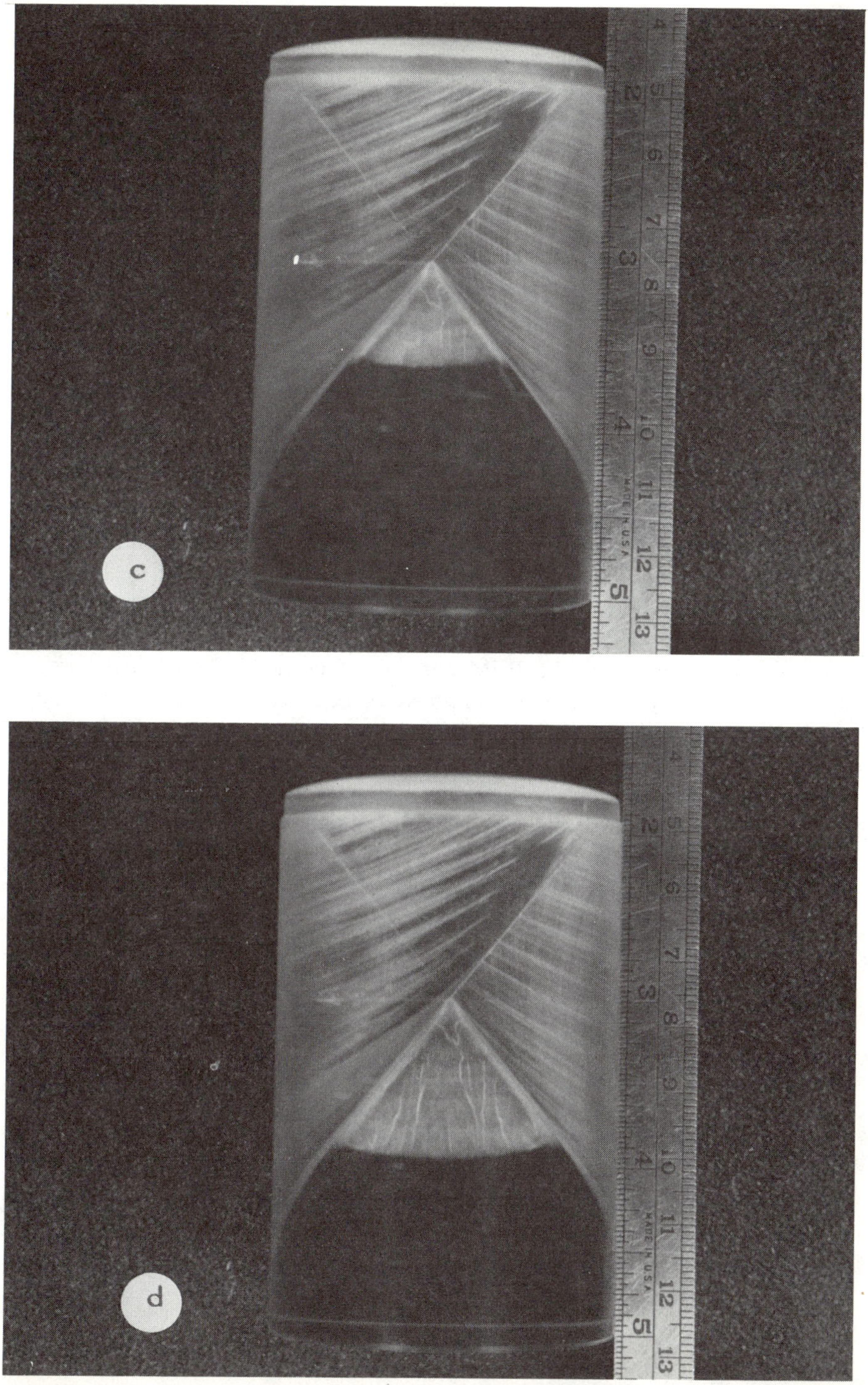

FIG. 9—*Crack front evolution,* (a–d), *in polystyrene short-rod specimen.*

TABLE 6—*Comparison of typical BEM4 and FEM1 solution.*

Mesh	FEM1	BEM4
Preprocessing time (man-minutes)	~40	~40
Number of elements	145	54
Number of nodes	867	152
Number of equations	2 430	456
Equation formulation time (CPU-s)[a]	2 060	1 315
Number of nonzero matrix elements	524 147	207 936
Equation solution time (CPU-s)[a]	1 380	1 615
Total solution time[b] (CPU-s)	3 440	2 930

[a]These values varied slightly (±3 to 4%) from run to run.
[b]Does not include stress recovery for FEM1.

nodes than BEM4 would probably be fully converged. Based on the equivalence between BEM4 and FEM1, this means a finite element mesh of the same element type as used in FEM1 would need about 80% more nodes to fully converge. This is because, for a given decrease in element size, the number of nodes will increase one order faster for a volume discretization (FEM) than for a surface discretization (BEM). Consequently, the present efficiency advantage enjoyed by the BEM approach would probably increase for fully converged solutions.

A further advantage of the BEM approach, one not yet exploited here, is related to crack *propagation* modeling. It would be most interesting in the present problem, for example, to allow the crack front to evolve numerically according to a growth rule (for example, constant K_I at all points along the front). This would require sequential remeshing of only a portion of a surface for the BEM approach, a process easily approached using interactive graphics. The authors feel strongly that the advent of such a capability would rid the calibration of inherently three-dimensional geometries such as the short rod of the unnecessary and inaccurate vestiges of two-dimensional thinking, such as the compliance derivative technique.

Conclusions

Three-dimensional finite and boundary element analyses were performed on the short-rod geometry shown in Fig. 1. Convergence studies for both approaches were performed. The variation of compliance with crack length was computed and given in Eqs 3 and 4. The variation of average stress intensity factor with crack length was obtained using computed compliance changes through Eq 1. Using both approaches, the calibration constant for use in Eq 2 was computed to be

$$\bar{Y}_{min} = \frac{K_{Ic}\, B\sqrt{W}}{P_{max}} = 28.3$$

under a plane stress assumption and at a critical crack length

$$(a/W)_c = 0.57$$

The corresponding value of calibration constant for use in Eq 3 is

$$A_{\min} = \frac{K_{\mathrm{Ic}} B\sqrt{B}}{P_{\max}} = 23.4$$

These values are lower bounds estimated to be accurate to within 1%. Both values were also found to be in good agreement with available experimental results.

Pointwise variation of stress-intensity factor along the crack front was also investigated. Traction singular, ¼-point boundary elements were used to compute stress intensity factors functionally with respect to crack front location for each crack length investigated. At each crack length it was found that maximum stress-intensity factor occurs at the point where the crack front intersects the edge of the chevron notch, the minimum value at the specimen centerline. At the critical crack length the maximum value was found to be about 25% higher than the centerline value, and very good agreement was found between the average stress-intensity factor computed from compliance change and the average of the local values computed from the ¼-point singular elements.

Comparison of BEM and FEM approaches revealed that, for nearly identical displacement solutions, significant savings of computer time and storage accrued to the BEM method.

Acknowledgments

The authors would like to thank Professor Donald Greenberg for continued support of their work in the Program for Computer Graphics at Cornell University, Professor John F. Abel for his guidance in developing the finite-element preprocessor which helped to make this effort efficient and fun, and Yih-Ping Huang who developed the finite-element analysis program.

Development of all software used in this study was partially supported by NSF Grant No. CME79-16818.

References

[*1*] Beech, J. and Ingraffea, A. R., *International Journal of Fracture*, Vol. 18, No. 3, 1982, pp. 217–229.

[*2*] Ingraffea, A. R., Gunsallus, K. L., Beech, J. F., and Nelson, P., *Proceedings of the 23rd U.S. Symposium on Rock Mechanics*, Berkeley, Calif., Aug. 1982, pp. 463–470.

[*3*] Ingraffea, A. R., Gunsallus, K. L., Beech, J. F., and Nelson, P. P., this publication, pp. 152–166.

[*4*] Catalano, D. and Ingraffea, A. R., "Concrete Fracture: A Linear Elastic Fracture Mechanics Approach," Department of Structural Engineering Report 82-1, School of Civil and Environmental Engineering, Cornell University, Ithaca, N.Y., 1982.

[5] Ingraffea, A. R., "An Experimental Study of Propagation of Cracks Near Interfaces in Rock," Department of Structural Engineering Report 82-4, School of Civil and Environmental Engineering, Cornell University, Ithaca, N.Y., Feb. 1982.
[6] Barker, L. M., *Engineering Fracture Mechanics,* Vol. 9, No. 2, 1977, pp. 361–369.
[7] Barker, L. M., *Engineering Fracture Mechanics,* Vol. 17, No. 4, 1983, pp. 289–312.
[8] Perucchio, R., "An Integrated Boundary Element Analysis System with Interactive Graphics for Three-Dimensional Linear Elastic Fracture Mechanics," Ph.D. thesis, Cornell University, Ithaca, N.Y., May 1983.
[9] Cruse, T. A. and Wilson, R. B., "Boundary-Integral Equation Method for Elastic Fracture Mechanics," AFOSR-TR-78-0355, 1977.
[10] Ingraffea, A. R. and Manu, C., *International Journal for Numerical Methods in Engineering,* Vol. 15, No. 10, 1980, pp. 1427–1445.
[11] Bubsey, R. T., Munz, D., Pierce, W. S., and Shannon, J. L., Jr., *International Journal of Fracture,* Vol. 18, No. 2, Feb. 1982, pp. 125–133.
[12] Perucchio, R., Ingraffea, A. R., and Abel, J. F., *International Journal for Numerical Methods in Engineering,* Vol. 18, No. 6, 1982, pp. 909–926.
[13] Perucchio, R. and Ingraffea, A. R., *Journal of Computer and Structures,* Vol. 16, No. 1–4, 1983, pp. 153–166.
[14] Henshell, R. D. and Shaw, K. G., *International Journal for Numerical Methods in Engineering,* Vol. 9, 1975, pp. 495–509.
[15] Barsoum, R. S., *International Journal for Numerical Methods in Engineering,* Vol. 10, 1976, pp. 25–28.
[16] Blandford, G., Ingraffea, A. R., and Liggett, J. A., *International Journal for Numerical Methods in Engineering,* Vol. 17, 1981, pp. 387–401.

Alexander Mendelson[1] *and Louis J. Ghosn*[1]

Three-Dimensional Analysis of Short-Bar Chevron-Notched Specimens by the Boundary Integral Method

REFERENCE: Mendelson, A. and Ghosn, L. J., "**Three-Dimensional Analysis of Short-Bar Chevron-Notched Specimens by the Boundary Integral Method,**" *Chevron-Notched Specimens: Testing and Stress Analysis, ASTM STP 855,* J. H. Underwood, S. W. Freimen, and F. I. Baratta, Eds., American Society for Testing and Materials, Philadelphia, 1984, pp. 69–80.

ABSTRACT: An analysis was performed, of the three-dimensional elastic problem of the chevron-notched short-bar specimen, using the boundary integral equation method. This method makes use of boundary surface elements only in obtaining the solution. Results were obtained for various notch geometries, and included displacement and stress fields. Load line displacements, and stress-intensity factors determined both from strain calculations and compliance calculations were compared with experimental values obtained at NASA Lewis Research Center. Good agreement was obtained.

KEY WORDS: fracture mechanics, chevron notch, boundary integral method, stress-intensity factor, three-dimensional elasticity, numerical methods

Nomenclature

a	Crack length
a_o	Distance between the load line and the apex of the v-notch
b	Crack front length ($b = B(a - a_o)/(W - a_o)$)
B	Specimen width
C	Compliance ($C = V_{LL}/P$)
C'	Normalized compliance ($C' = EV_{LL}B/P$)
E	Modulus of elasticity
G	Shear modulus
H	Half the specimen height
K_I	Mode I stress-intensity factor
K_{Ic}	Critical Mode I stress-intensity factor
n_i	Component of the outward unit normal in i-direction
P	Total load, or point in continuum
r	Distance

[1]Professor and graduate assistant, respectively, Department of Civil Engineering, Case Western Reserve University, Cleveland, Ohio 44106.

S Bounding surface
t_i Traction components
u_i Displacement components
V u_2, displacement in y-direction
V_{LL} Load line displacement
W Specimen length
x' Distance between the end of the specimen and the load line
Y^* Normalized stress-intensity factor ($Y^* = K_I B\sqrt{W}/P$)
α_i a_i/W
δ_{ij} Kronecker delta
θ $u_{j,j}$
σ_{ij} Stress tensor
ν Poisson's ratio

The usefulness of the chevron-notched specimens (Fig. 1) is by now well known and requires no further elaboration. For generalized application the analysis of the specimen over a range of specimen proportions and chevron notch configurations is, of course, necessary. In the present paper the results of such an analysis applied to the chevron-notched bar specimen using the boundary integral equation (BIE) method to solve this three dimensional elastic problem is presented.

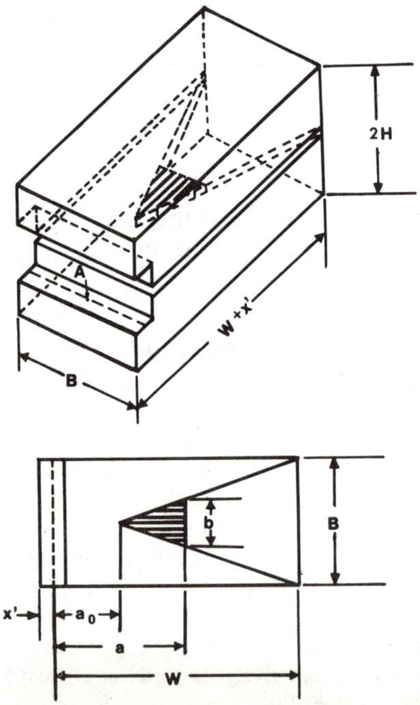

FIG. 1—*Chevron-notched short-bar specimen.*

In recent years the BIE method has been applied and used extensively for fracture mechanics problems. Its major advantage lies in the fact that nodal points are needed only on the boundary of the region under consideration instead of throughout the interior as required by finite difference or finite element methods. This results in a much smaller number of unknowns, the dimensions of the problem being effectively reduced by one.

The Boundary Integral Equation Method

The Navier equations of equilibrium for the three-dimensional problem in elasticity are

$$\nabla^2 u_i + \frac{1}{1 - 2\nu}\theta_{,i} = 0$$
$$\theta = u_{j,j} \qquad i,j = 1,2,3 \tag{1}$$

where u_i are the displacements, ν is poisson's ratio, and, the usual tensor notation is used.

A solution of Eq 1 can be obtained by making use of the singular solution of the Navier equations due to a point load, as given for example in Love's classical book on elasticity [*1*], and also making use of Betti's reciprocal theorem. An integral equation known also as Somigliana's identity is then obtained, [2]

$$\lambda u_i(P) = \int_S (U_{ij}t_j - T_{ij}u_j)dS \tag{2}$$

where the tensors U_{ij} and T_{ij} are given by

$$U_{ij} = \frac{1}{16\pi(1 - \nu)G}\frac{1}{r}[(3 - 4\nu)\delta_{ij} + r_{,i}r_{,j}] \tag{3}$$

$$T_{ij} = -\frac{1}{8\pi(1 - \nu)}\frac{1}{r^2}\left\{[(1 - 2\nu)\delta_{ij} + 3r_{,i}r_{,j}]\frac{\partial r}{\partial n} + (1 - 2\nu)(r_{,j}n_i - r_{,i}n_j)\right\} \tag{4}$$

The coefficient λ is equal to 1, if P is an interior point and is equal to ½ if P is a point on the surface, t_j and u_j are the surface tractions and displacements, respectively, and the integration in Eq 2 is performed over the surface S of the body. For a mixed boundary value problem, the t_j will be known over part of the bounding surface and the u_j over the rest, and the integral Eq 2 is solved for those parts of t_j and u_j which are unknown on the surface.

The solution is obtained by dividing the boundary surface into a series of surface elements, usually triangular or rectangular. The unknown functions may be assumed either constant or to vary linearly [*3*], parabolically or in a higher polynomial fashion over each element, and the integrals in Eq 2 are replaced by sums over all the elements. If the surface is divided into m elements, resulting in n nodal points we get $3n$ simultaneous equations to solve. It is to be noted that because the unknowns of the problem appear only on the surface, the number of unknowns to be determined will generally be an order of magnitude less than for corresponding finite element or finite difference formulations.

Once the tractions and displacements are known over the complete surface, the displacements can be determined at any interior point from Eq 2 with $\lambda = 1$. The stresses can be calculated at any interior point from, Ref *2*

$$\sigma_{ij}(P) = \int_S (V_{ijk}t_k - T_{ijk}u_k)dS \qquad (5)$$

where the third order tensors are given by

$$V_{ijk} = -G\left(U_{ik,j} + U_{jk,i} + \delta_{ij}\frac{2\nu}{1-2\nu}U_{lk,l}\right) \qquad (6)$$

$$T_{ijk} = -G\left(T_{ik,j} + T_{jk,i} + \delta_{ij}\frac{2\nu}{1-2\nu}T_{lk,l}\right) \qquad (7)$$

Thus, the displacements and stresses at any interior point are obtained by quadratures.

Numerical Procedure

As previously indicated, the first step was to divide the surface of the body under consideration into rectangular and triangular elements. In addition to the chevron-notched specimen shown in Fig. 1, the single edge-cracked tension specimen (SECT), shown in Fig. 2 was also analyzed. This was done as a check on the equations and the computer program, since results for the SECT specimen were available by other methods. A typical surface mesh for the chevron-notched specimen is shown in Fig. 3. Several mesh sizes were used to determine the effect of mesh size in the vicinity of the crack front on the numerical results.

The unknown surface displacements and tractions were assumed to vary linearly over the surface elements. The details of the numerical integration procedures used are described in the appendix.

Results for Single-Edge-Cracked Tension Specimen

The SECT specimen has been analyzed by the finite element method in Ref *4* and by the method of lines in Ref *5*. Figure 4 shows the variation through the thickness of the dimensionless stress-intensity factor $K_I/(\sigma\sqrt{a\pi})$, obtained by these two methods as well as the BIE method. It is seen that very good agreement is obtained by the three methods.

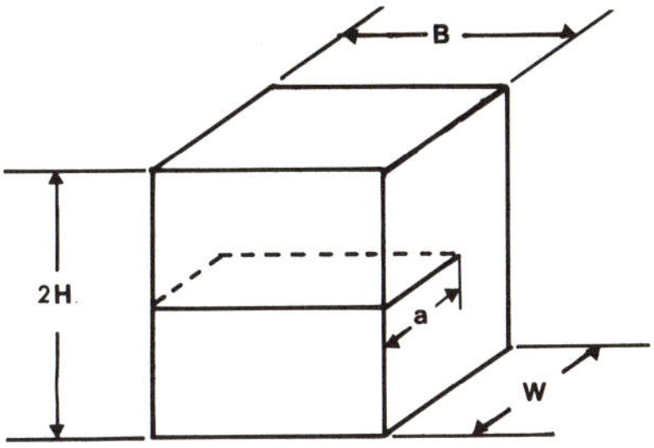

FIG. 2—*Single-edge-notched tension specimen.*

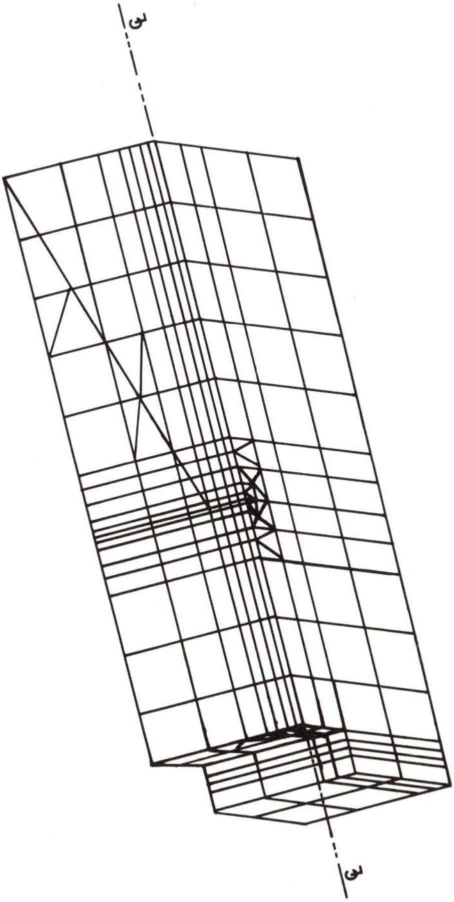

FIG. 3—*Typical surface mesh for chevron-notched specimens.*

Results for Chevron-Notched Specimen

Having checked out the equations and computer program on the SECT specimen, the chevron-notched bar specimen was analyzed in some detail. The basic dimensions of the specimen were

$$W/B = 2, \quad a_o/B = 0.4, \quad 2H/B = 1, \quad x'/B = 0.1, \quad \alpha_o = a_o/W = 0.2$$

The square grip groove shown in Fig. 1 is also modelled, having the dimensions recommended by the ASTM E24.01.05 task group. The total height of the groove is 0.35 B, and its depth is 0.15 B. Since the chevron-notched specimens is usually loaded by means of a knife edge fixture, a uniform traction along the specimen width was assumed at a distance x' from the end of the specimen (Fig. 1). Calculations were performed for different values of $\alpha = a/W$.

Figure 5 shows the dimensionless displacement EBV_{LL}/P at point A at the center of the loading line (Fig. 1) as a function of $\alpha = a/W$. Also, shown are the experimentally measured values obtained at NASA Lewis Research Center [*6*]. The calculated displacement are consistently lower than the measured values. Part of this discrepancy is probably due to the finite width slot cut into the specimens in forming the chevron notch. This is estimated to increase the compliance by about 5% from simple beam theory where the displacement changes with the cube of the height. Although it is possible to model the finite slot width in obtaining the displacement and stress distributions, the calculation of stress-intensity factors requires a sharp crack.

Using the results shown in Fig. 5, the stress-intensity factor K_I was computed from the compliance by the relation [*7*]

$$K_I = \left[\frac{EP}{2b}\frac{\partial V_{LL}}{\partial a}\right]^{1/2} = \frac{P}{B\sqrt{W}}\left[\frac{1}{2}\frac{1-\alpha_o}{\alpha-\alpha_o}\frac{\partial C'}{\partial \alpha}\right]^{1/2} = \frac{P}{B\sqrt{W}}Y^* \qquad (8)$$

where the dimensionless compliance C' is given by $C' = EV_{LL}B/P$.

The slope of the compliance curve was obtained by first fitting the average normalized compliance curve, since the load line displacements were not constant through the thickness, by a least square fit given by

$$\ln C' = \ln\frac{EV_{LL}B}{P} = 1.674 + 12.518\alpha - 16.313\alpha^2 + 9.965\alpha^3 \qquad (9)$$

The results of the normalized stress intensity factor, Y^* are plotted in Fig. 6, together with experimental values from Ref *6*. The minimum occurs between 0.5 and 0.55. A difference of 6% is observed between the experimental stress-intensity factor and the analytical value at $a/W = 0.5$.

This calculation gives the average value of the stress-intensity factor K_I, based on the compliance of the specimen. The variation of K_I along the crack front

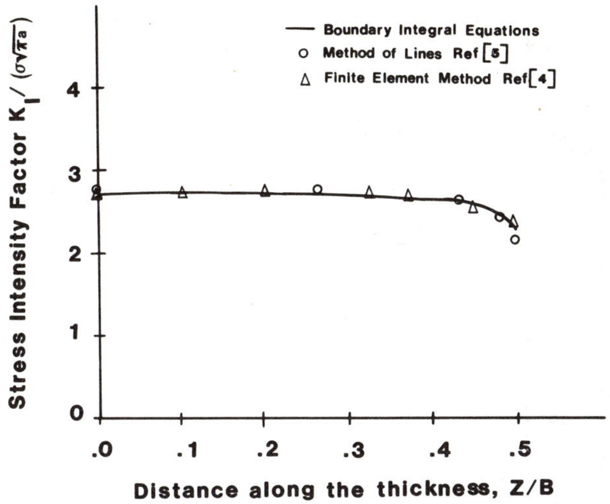

FIG. 4—*Variation of the nondimensional stress-intensity factor through the thickness,* z/B, *for the single-edge-notched tension specimen.*

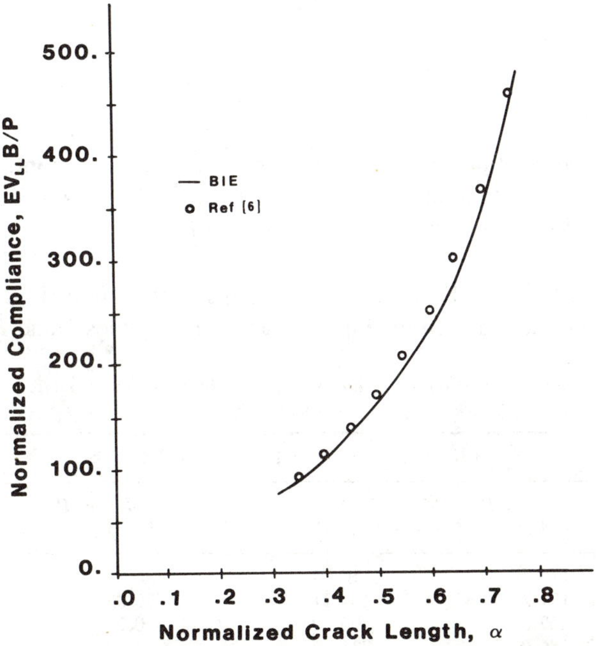

FIG. 5—*Variation of the nondimensional compliance,* C′ = EV B/P, *at the center of the load line, as function of the crack length* α.

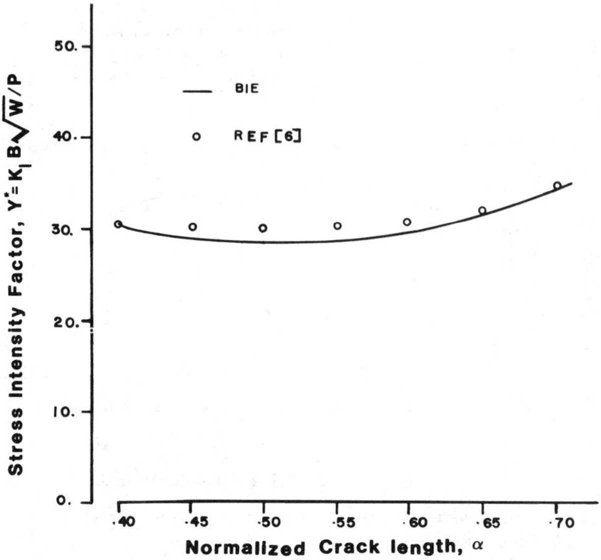

FIG. 6—*Nondimensional stress-intensity factor from compliance,* $Y = K_I B\sqrt{W}/P$, *as function of the crack length* α.

can be obtained from its basic definition using the computed displacements, that is

$$K_I = \lim_{r \to 0} \frac{V\sqrt{2\pi}\,E}{4(1 - \nu^2)\sqrt{r}} \tag{10}$$

This definition implies plane strain condition along the crack front. The results of this point-wise calculation are shown in Fig. 7 for various values of α. As seen in Fig. 7, K_I is appreciably higher at the surface than at the center of the specimen.

A comparison between Y^* from the compliance calculation (Fig. 6) and the average Y^* along the crack front (Fig. 7), (as well as Tables 1 and 2), shows a

TABLE 1—*Dimensionless compliance and stress-intensity factor as a function of* a/W *for chevron-notched specimen.*

$W/B = 2,\ 2H/B = 1,\ a_o/B = 0.4$				
a/W	$C' = EV_{LL}B/P$, Ref *6*	C'_A, BIE	$Y^* = K_I B\sqrt{W}/P$, Ref *6*	Y^*, BIE
0.40	113.5	110.04	30.54	30.50
0.50	171.3	164.31	30.03	28.33
0.55	207.8	200.30	30.24	28.74
0.60	250.0	236.96	30.82	29.58
0.70	368.4	346.38	34.79	34.61

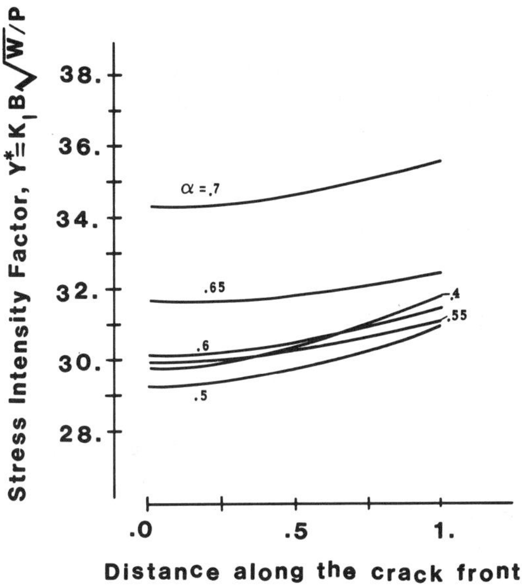

FIG. 7—*Variation of the nondimensional stress-intensity factor through the thickness,* 2z/b, *for various values of the crack length* α.

maximum difference of 5% at α equal 0.5. Therefore, the critical stress-intensity factor for the short-bar specimen with $W/B = 2$ and $a_0/B = 0.4$ is taken as the average of the minimum Y^* of the two methods, that is

$$Y^*_{min} = K_{Ic}B\sqrt{W}/P = 29.1 \tag{11}$$

The results of Figs. 5, 6, and 7 have been summarized in Tables 1 and 2. Calculations for other geometries including the round bar specimen are in progress and will be reported at a later date.

TABLE 2—*Variation of the dimensionless stress-intensity factor,* $Y^* = K_I B\sqrt{W}/P$, *along the crack front.*

	$W/B = 2$, $2H/B = 1$, $a_o/B = 0.4$					
	2Z/b					
a/W	0.000	0.500	0.750	0.875	1.000	Y^*, avg
0.40	29.79	30.35	30.94	31.32	31.79	30.53
0.50	29.28	29.76	30.22	30.53	30.93	29.90
0.55	29.88	30.29	30.62	30.84	31.01	30.36
0.60	30.10	30.54	30.85	31.15	31.57	30.63
0.70	34.29	34.78	34.94	35.23	35.79	34.81

Summary

An analysis was performed of the chevron-notched short-bar specimen using the boundary integral equation method. The method makes use of boundary surface elements only in obtaining the solution. Load line displacements and stress-intensity factors were obtained for specimens with ratios of $B/W = 2H/W = 1/2$ and $a_o/W = 0.2$ for a/W ratios ranging from 0.4 to 0.7. The value of the critical stress-intensity factor obtained from the analytical solutions agreed with experimental results showing a minimum Y^* equal to 29.1 at $a/W = 0.5$.

Acknowledgments

This work was sponsored by the National Aeronautics and Space Administration Langley Research Center under Grant No. NAG-1-304. The authors also wish to thank John L. Shannon, Jr., and his colleagues at NASA Lewis Research Center for their furnishing of the experimental data and the many helpful discussions and their interest in the progress of the investigation.

APPENDIX

Numerical Solution of the Integral Equations

General analytical solutions to the integral equations are not available, and it is therefore necessary to solve the equations numerically. The integral equations have the form

$$C_{ij}(P)u_j(P) + \int_S T_{ij}(P,Q)\, u_j(Q)\, dS(Q) = \int_S U_{ij}(P,Q)\, t_j(Q)\, dS(Q) \tag{12}$$

where u_j and t_j are the displacement and traction vectors, respectively, P is the source point indicating the location at which the force acts, and Q is the field point denoting the actual boundary point.

The integrals are Cauchy principal value integrals where $C_{ij}(P)$ is a field of constants depending on the smoothness of boundary at P. $C_{ij}(P)$ is equal $\frac{1}{2}\,\delta_{ij}$ if the surface is smooth at P. For the case where P is at an edge or a corner [*3*]

$$C_{ij}(P) = -\int_S T_{ij}(P,Q)\, dS(Q) \tag{13}$$

The numerical solution for the integral equations are found by discretizing the boundary into elements. In the computer program used in the present work, the surface is represented by triangular and rectangular elements. The traction and displacement acting on each element are linear functions of the traction and displacement at the corners of the element.

For triangular elements

$$t_i(\xi) = C^k(\xi)\, t_i^k \tag{14}$$

$$u_i(\xi) = C^k(\xi)\, u_i^k \tag{15}$$

where

$$C^k(\xi) = \tfrac{1}{3} + (F_{k2}\xi_{1m} - F_{k1}\xi_{2m})/2A - (F_{k2}\xi_1 - F_{k1}\xi_2)/2A \tag{16}$$

ξ_1, ξ_2 are local in-plane coordinates of the field point Q, ξ_{1m}, ξ_{2m} are local in-plane coordinates of the centroid of the mth element. F_{k1}, F_{k2} are the projections of the distance between two adjacent nodes in local coordinates and $k = 1, 2, 3$, and A is the area of the triangle.

For rectangular elements

$$t_i(\xi) = N^k(\xi)\, t_i^k \tag{17}$$

$$u_i(\xi) = N^k(\xi)\, u_i^k \tag{18}$$

where

$$N^1 = (1 - \xi_1)(1 - \xi_2)/4 \qquad N^2 = (1 + \xi_1)(1 - \xi_2)/4$$
$$N^3 = (1 + \xi_1)(1 + \xi_2)/4 \qquad N^4 = (1 - \xi_1)(1 + \xi_2)/4 \tag{19}$$

Where k now has a range of 1 to 4.

If the surface is represented by m triangular elements and n rectangular elements the equations become

$$\begin{aligned} C_{ij}(P)\, u_j(P) &+ \sum_{b=1}^{m}\sum_{k=1}^{3} u_j(Q^{bk}) \int_{\Delta S} T_{ij}(P,Q)\, C^k(\xi)\, J(\xi)\, d\xi \\ &+ \sum_{b=1}^{n}\sum_{k=1}^{4} u_j(Q^{bk}) \int_{\Delta S} T_{ij}(P,Q)\, N^k(\xi)\, J(\xi)\, d\xi \\ &= \sum_{b=1}^{m}\sum_{k=1}^{3} t_j(Q^{bk}) \int_{\Delta S} U_{ij}(P,Q)\, C^k(\xi)\, J(\xi)\, d\xi \\ &+ \sum_{b=1}^{n}\sum_{k=1}^{4} t_j(Q^{bk}) \int_{\Delta S} U_{ij}(P,Q)\, N^k(\xi)\, J(\xi)\, d\xi \end{aligned} \tag{20}$$

where $J(\xi)$ is the well-known Jacobi function. The terms $u_j(Q^{bk})$ or $t_j(Q^{bk})$, respectively, are the corner values of displacements and tractions of the kth node within the bth element.

For $Q^{bk} \neq P$, a 4×4 Gaussian quadrature formula is used to evaluate the integration numerically.

For $Q^{bk} = P$, in a triangular element the integrals are evaluated in closed form by a change to cylindrical coordinate (r,θ) [*3*]

$$\lim_{\epsilon \to 0} \int_{\theta_1}^{\theta_2} \int_{\epsilon}^{r} T_{ij}(P,Q)\, C^k(\xi)\, J(\xi)\, d\xi \tag{21}$$

$$\lim_{\epsilon \to 0} \int_{\theta_1}^{\theta_2} \int_{\epsilon}^{r} U_{ij}(P,Q)\, C^k(\xi)\, J(\xi)\, d\xi \tag{22}$$

However, for rectangular elements a special singular Gauss quadrature is used, derived in Ref *8* for an integral with a $1/r$ singularity.

When the integrals are calculated for P at a node, then $C_{ij}(P)$ is obtained by summing the $\int_s T_{ij}\, ds$ terms.

The integral equations then result in a system of $3x(m + n)$ linear algebraic equations to be solved for the unknown boundary tractions or displacements.

The use of both triangular and rectangular elements is necessitated due to the use of a fine mesh near the crack front and a coarse mesh further away. The triangular elements are thus used as transition elements.

References

[1] Love, A. E. H., *A Treatise on the Mathematical Theory of Elasticity*, Dover, New York, 1944.

[2] Mendelson, A., "Boundary-Integral Methods in Elasticity and Plasticity," NASA TND-7418, National Aeronautics and Space Administration, Washington, D.C. Nov. 1973.

[3] Cruse, T. A., *Computers and Structures*, Vol. 4, 1974, pp. 741–754.

[4] Raju, L. S. and Newman, J. C., Jr., "Three-Dimensional Finite-Element Analysis of Finite-Thickness Fracture Specimens," NASA TND-8414, National Aeronautics and Space Administration, Washington, D.C., May 1977.

[5] Mendelson, A. and Alam, J., *International Journal of Fracture*, Vol. 22, 1983, pp. 105–116.

[6] Shannon, J. L., Jr., NASA Lewis Research Center, private communication.

[7] Munz, D., Bubsey, R. T., and Srawley, J. E., *International Journal of Fracture*, Vol. 18, No. 1, 1982, pp. 125–132.

[8] Cristescu, M. and Loubignac, G., *Advances in Boundary Element Methods,* C. A. Brebbia, Ed., Pentech Press, London, 1978, pp. 375–389.

R. J. Sanford[1] *and R. Chona*[1]

Photoelastic Calibration of the Short-Bar Chevron-Notched Specimen

REFERENCE: Sanford, R. J. and Chona, R., "**Photoelastic Calibration of the Short-Bar Chevron-Notched Specimen,**" *Chevron-Notched Specimens: Testing and Stress Analysis, ASTM STP 855,* J. H. Underwood, S. W. Freiman, and F. I. Baratta, Eds., American Society for Testing and Materials, Philadelphia, 1984, pp. 81–97.

ABSTRACT: The geometric shape function calibration of the short-bar chevron-notched specimen has been determined using a hybrid experimental-numerical technique. The results are in good agreement with those obtained by other methods. The hybrid method, which is rapid and low in cost compared to purely numerical methods, also provides additional information about the stress state in the neighborhood of the crack tip. This additional information has been used to determine the size and shape of the singularity dominated zone.

KEY WORDS: fracture mechanics, chevron-notched specimen, photoelasticity, crack-tip stress fields, local collocation

The analysis of fringe patterns around crack tips in photoelastic models has been used for over one-quarter century [*1*] to determine the stress-intensity factor in both static and dynamic studies [*2*]. If engineering accuracy is sufficient, relatively simple techniques can be applied to estimate the stress-intensity factor. However, the fringe pattern contains a wealth of other information which can be also extracted if the full field nature of the pattern is considered. Recently developed algorithms [*3–5*] for analyzing photoelastic fracture patterns allow the fracture parameters of interest to be evaluated both accurately and reliably. This paper presents a new hybrid experimental-numerical technique that uses "local collocation" in a region surrounding the crack tip to determine the shape function needed to account for the effects of specimen geometry in stress-intensity factor relations. The method previously has been applied to the calibration of the three-point-bend specimen geometry [*6*] and has been shown to provide accurate results rapidly, at low cost, with minimal computer requirements.

The results reported here were obtained by applying this technique to a two-dimensional model representing the midplane of a chevron-notch, short-bar fracture specimen. A procedure similar to that of Munz, Bubsey, and Srawley [*7*]

[1]Professor and instructor respectively, Department of Mechanical Engineering, University of Maryland, College Park, Md. 20742.

was then used to obtain a calibration for the three-dimensional geometry. In addition to obtaining the desired geometric shape function, the photoelastic results have been also used to characterize the nonsingular stress contributions to the crack-tip stress field and to determine the size and shape of the singularity-dominated zone [*8*].

Experimental Procedure

A two-dimensional photoelastic model of the midthickness plane of a short-bar, chevron-notched specimen was prepared from a sheet of 6.25-mm-thick polycarbonate material. The in-plane dimensions of the photoelastic model were 147 by 254 mm. The geometry of the specimen is shown in Fig. 1 and has an aspect ratio of W/H = 3.33, which corresponds to the geometry suggested by Barker [*9*]. Figure 2 shows an isochromatic fringe pattern representative of those recorded in this geometry. The specimen was loaded through two disks cut from the same sheet of photoelastic material as the specimen. Since the fringe order at the center of the disk is proportional to the applied load [*10*], this value can be measured in lieu of the load. This method of loading, called autocalibration, has the advantage of eliminating any need for calibration of the model material, measurement of the model thickness, or the applied load, thereby removing these variables as potential sources of error in the analysis. The crack was modeled in the specimen by a carefully cut 1-mm wide slot which was produced using a fine jeweler's saw. Fringe patterns were recorded at several different load levels for each slot length as the slot was extended from an a/W of 0.33 to an a/W of 0.90 in intervals of 0.05 a/W. Figure 3 shows a series of photographs illustrating the changes in the isochromatic fringe pattern as the slot was extended from one end of the specimen to the other. Data points were taken from within

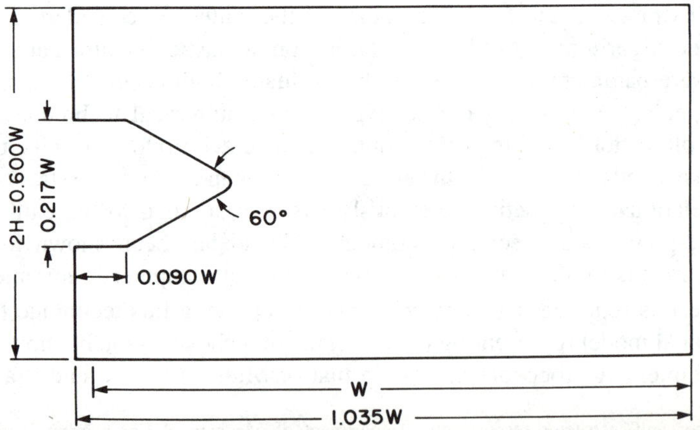

FIG. 1—*The geometry of the photoelastic model of the midplane of the short-bar chevron-notched specimen.*

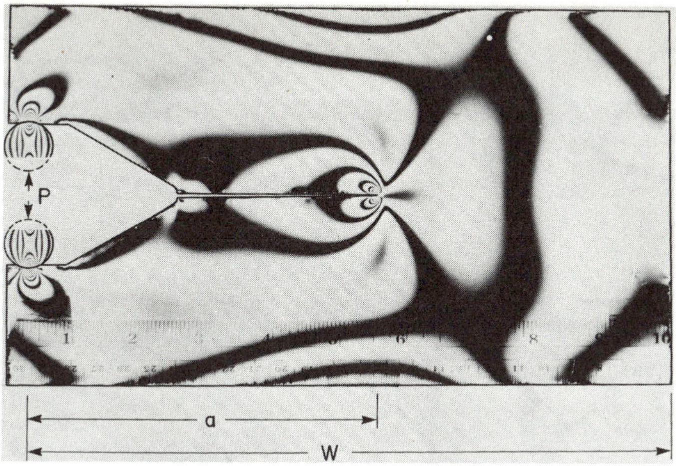

FIG. 2—*An isochromatic fringe pattern representative of those recorded in the model geometry of Fig. 1; load applied through autocalibration disks.*

the circular region indicated on each fringe pattern and analyzed using the procedure described in the next section.

Local Collocation Technique

The analysis procedure which was used to determine the stress field parameters has been termed the local collocation method because of its similarity to the boundary collocation method used in numerical analysis. Both methods start from the general solution to the crack problem and determine the unknown coefficients by matching information from selected regions. Unlike boundary collocation, however, the local collocation technique utilizes experimental measurements in the neighborhood of the crack tip where the influence of the singular term is large. In the particular example discussed in this paper, the use of near crack tip information produces a solution of sufficient accuracy with, at most, 10 to 12 unknown coefficients rather than the 150 or more that are needed when only remote boundary information is available. As a by-product of reducing the number of unknowns required for an accurate determination of the stress-intensity factor, the computational requirements are dramatically reduced. A typical computation of the type detailed later in the paper requires less than 35 central processing unit seconds (cpus) on a Univac 1108 as compared to over 3000 cpus for boundary collocation or boundary integral methods.

The stress field representation used is derived from an Airy stress function of the form

$$F = Re\bar{\bar{Z}} + yIm\bar{Z} + yIm\bar{Y} \tag{1}$$

and is a generalized form [*11*] of the familiar Westergaard stress function [*12*].

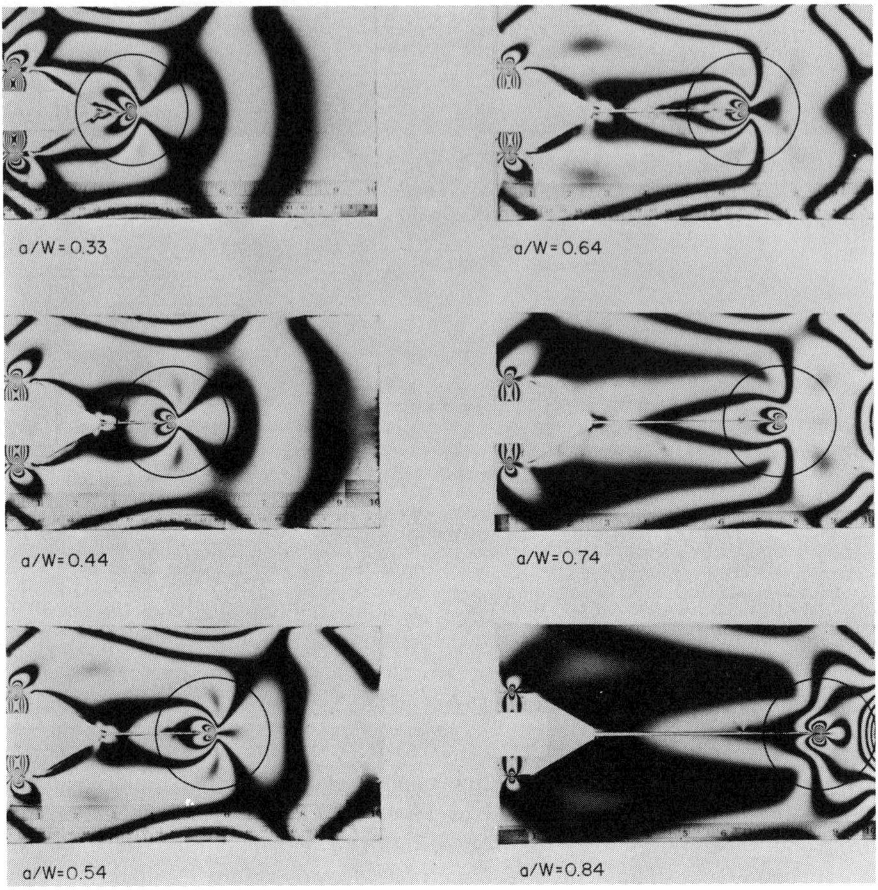

FIG. 3—*A sequence of photographs showing the changes in the isochromatic fringe pattern as the slot was extended across the specimen.*

Equation 1 can be used to obtain expressions for stresses, strains, or displacements, as needed, depending on the type of experimental measurements available. Photoelastic techniques, which have been used here, provide contours of constant maximum in-plane shear stress, τ_{max}, which can be expressed in terms of the cartesian stress components as

$$\tau^2_{max} = \frac{1}{4}(\sigma_y - \sigma_x)^2 + \tau_{xy}{}^2 \qquad (2)$$

From Eq 1, the cartesian stress components can be written as

$$\begin{aligned} \sigma_x &= ReZ - yImZ' - yImY' + 2ReY \\ \sigma_y &= ReZ + yImZ' + yImY' \qquad (3) \\ \tau_{xy} &= -yReZ' - yReY' - ImY \end{aligned}$$

Finally, the stress-optic law is used to relate τ_{max} to the photoelastic fringe order, N, so that

$$Nf_\sigma/2t = \tau_{max} \tag{4}$$

where f_σ is the stress-optic constant for the model material and t is the model thickness. It has previously been demonstrated [8] that suitable choices for the functions $Z(z)$ and $Y(z)$ which satisfy the boundary conditions for the opening-mode crack problem are

$$Z(z) = \frac{A_o}{\sqrt{W}} (z/W)^{-1/2} \sum_{n=o}^{n=N} \frac{A_n}{A_o} W^n (z/W)^n$$

$$= \frac{K}{\sqrt{2\pi W}} (z/W)^{-1/2} \sum_{n=o}^{n=N} A'_n (z/W)^n \tag{5}$$

$$Y(z) = \frac{A_o}{\sqrt{W}} (z/W)^{-1/2} \sum_{m=o}^{m=M} \frac{B_m}{A_o} W^{m+1/2} (z/W)^{m+1/2}$$

$$= \frac{K}{\sqrt{2\pi W}} (z/W)^{-1/2} \sum_{m=o}^{m=M} B'_m (z/W)^{m+1/2}$$

Combining Eq 2 to 5 yields a nonlinear equation of the form

$$(Nf_\sigma/2t)^2 = D^2 + T^2 \tag{6}$$

where

$$D = \frac{K}{\sqrt{2\pi W}} (r/W)^{-1/2} \left\{ \sum_{n=o}^{n=N} (n - 1/2\, A'_n (r/W)^n \sin\theta \sin (n - 3/2)\, \theta \right.$$

$$\left. + \sum_{m=o}^{m=M} B'_m (r/W)^{m+1/2} [m \sin\theta \sin (m - 1)\, \theta - \cos (m\theta)] \right\} \tag{7}$$

$$T = \frac{K}{\sqrt{2\pi W}} (r/W)^{-1/2} \left\{ \sum_{n=o}^{n=N} (n - 1/2\, A'_n (r/W)^n \sin\theta \cos (n - 3/2)\, \theta \right.$$

$$\left. + \sum_{m=o}^{m=M} B'_m (r/W)^{m+1/2} [m \sin\theta \cos (m - 1)\, \theta + \sin (m\theta)] \right\}$$

Equations 6 and 7 relate the fringe order, polar coordinates (crack-tip coordinates), and the unknown coefficients. Note that $A'_o \equiv 1$ and that the stress-intensity factor, $K = A_o \sqrt{2\pi}$.

The next step in the procedure is to take a region around the crack tip from the model geometry that is being analyzed, extract a large number of individual data points, and determine the coordinates and fringe order at each point. These data points are then used as inputs to an over-determined system of nonlinear equations of the form of Eq 6 and solved in a least-squares sense for the unknown coefficients, A_n and B_m. As a final check, the best-fit set of coefficients is used to reconstruct the fringe pattern over the region of data acquisition, to ensure that the computed solution set does, in fact, predict the same stress distribution as that observed experimentally. Figure 4 shows a typical isochromatic fringe pattern from the chevron-notched geometry ($a/W = 0.54$), together with the data points used for the analysis and the reconstructed pattern corresponding to a seven coefficient least-squares best-fit solution.

For each photoelastic pattern studied data points were taken over a circular region of radius, $r/W = 0.155$ and analyzed using 2 through 12 parameter models. The use of successively higher parameter models provides a means of assessing the convergence of the series solution to a true result. The data set generally consisted of 100 to 120 data points distributed over the entire data acquisition region, which was selected so as to include fringes that showed clear higher-order term effects.

The multiple-point least squares method [*3,4*] offers several advantages not available with other procedures for analyzing the stress field surrounding the tip of a crack under various types of loading. Among these are: increased freedom in selection of data points, the use of redundant data to minimize effects of experimental error, and applicability to a wide variety of mathematical models. An extension of this method [*5*], which combines statistical analysis and sampling procedures with the least squares method provides increased accuracy in parameter determination and a measure of parameter variability. This procedure consists of taking a large number of randomly selected subsets (typically 100) of 30 data points each from the full set of data points and determining the best-fit coefficient set for each subset. These results are then tabulated, the mean and standard deviation computed and histograms for each coefficient constructed. The results from the sampled least-squares method for the fringe pattern shown in Fig. 4 are given in Fig. 5, which shows the variability of each of the first six series coefficients in a seven coefficient solution. The means and standard deviations for this same data set are tabulated in Table 1. The results shown in the remainder of the paper correspond to the mean values from sampled least-squares analyses.

Geometric Shape Functions

For straight-through cracks in two-dimensional geometries subjected to point loads, P, the stress-intensity factor can be written in the form

$$K = \frac{P}{B\sqrt{W}} \cdot Y\left(\frac{a}{W}\right) \tag{8}$$

where B is the speciman thickness and W is a characteristic in-plane dimension, typically the distance from the load line to the end of the specimen as measured along the crack path. All of the effects of specimen geometry are imbedded in the dimensionless shape function Y.

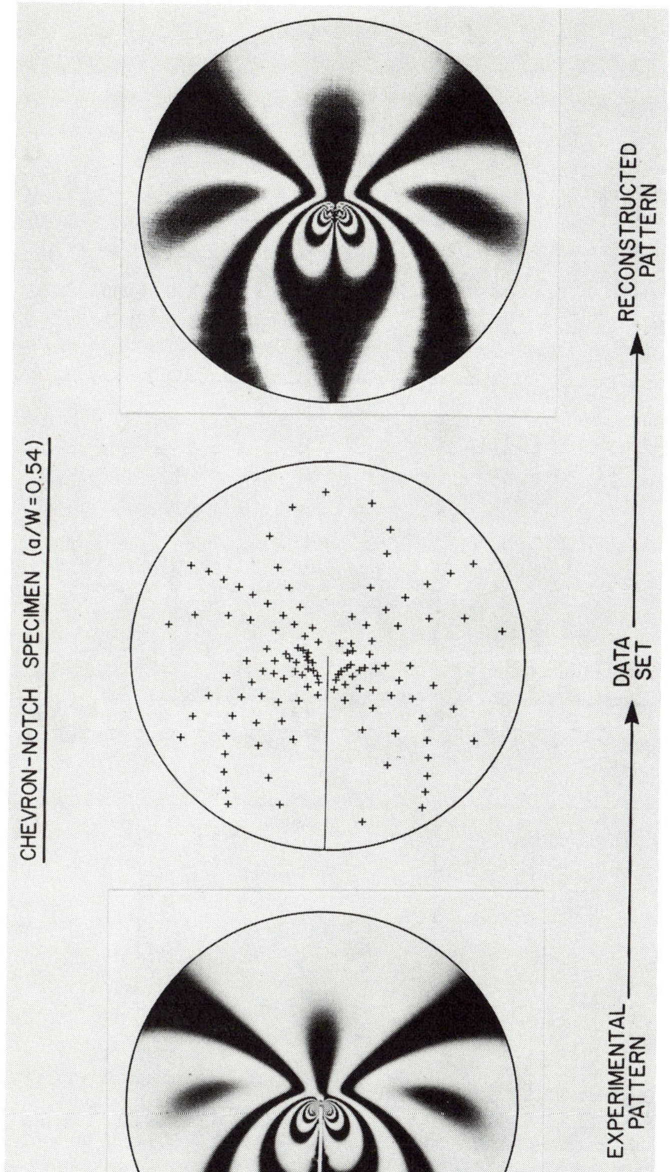

FIG. 4—*An experimental fringe pattern, the data points used for analysis, and the reconstructed fringe pattern corresponding to a seven-coefficient least-squares best-fit solution.*

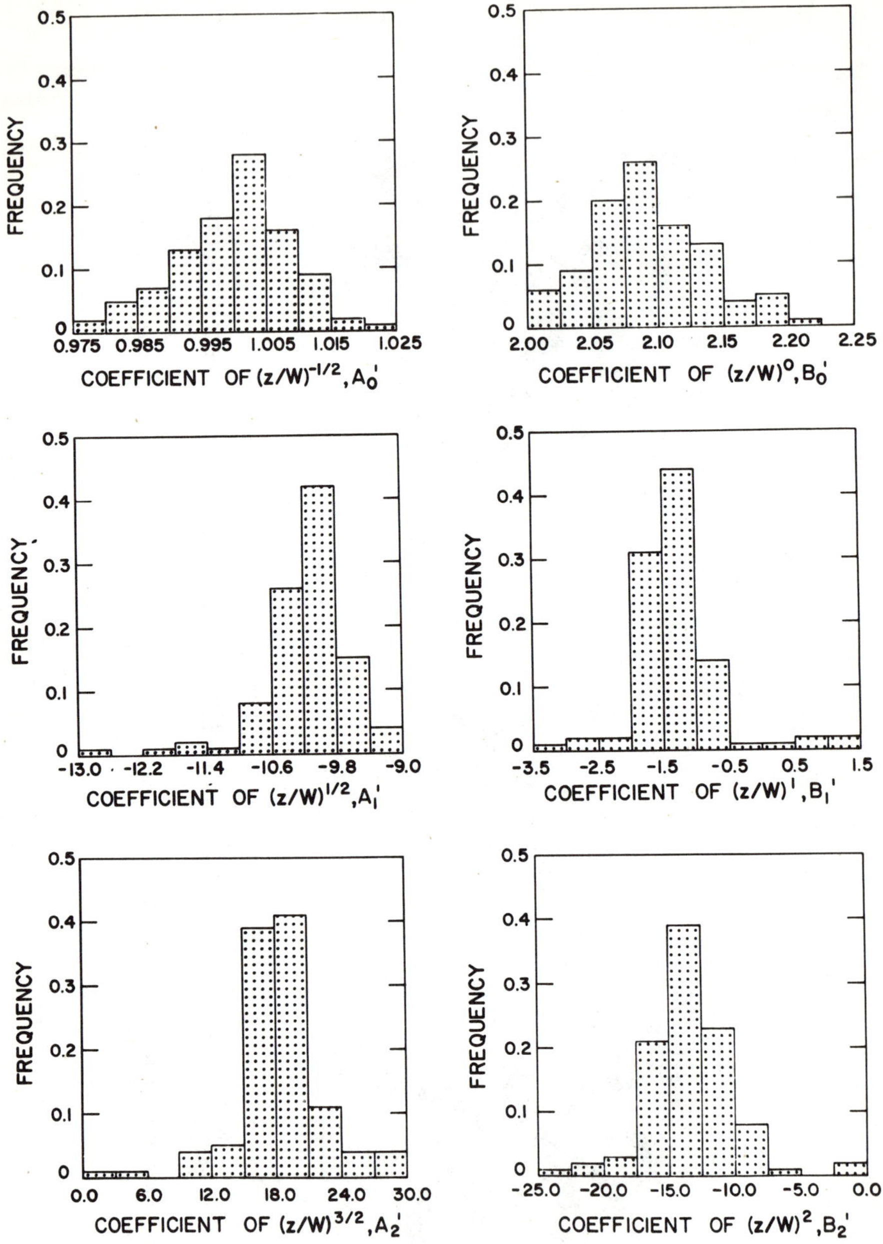

FIG. 5—*The results from a sampled least-squares analysis for the first six coefficients of the general solution for the illustrative example shown in Fig. 4.*

TABLE 1—*Parameter values associated with a sampled least squares analysis of the isochromatic fringe pattern shown in Fig. 4.*

Parameter	Symbol	Mean	Standard Deviation
Coefficient of $(z/W)^{-1/2}$	A'_o	1.00	0.013
Coefficient of $(z/W)^0$	B'_o	2.09	0.049
Coefficient of $(z/W)^{1/2}$	A'_1	−10.16	0.551
Coefficient of $(z/W)^1$	B'_1	−1.26	0.811
Coefficient of $(z/W)^{3/2}$	A'_2	18.48	4.481
Coefficient of $(z/W)^2$	B'_2	−13.35	3.772
Coefficient of $(z/W)^{5/2}$	A'_3	17.14	9.019
Fringe order error	$\overline{\vert\Delta n\vert}$	1.61%	0.43%

The stress-intensity factor at each crack length was determined from the photoelastic fringe patterns using the local collocation technique described earlier, and normalized through Eq 8 to obtain the shape function, Y. These results are shown in Fig. 6 as a function of a/W for the two-dimensional calibration and tabulated in Table 2. The variability in K corresponding to a $\pm 2\sigma$ confidence level ranges from ±1.8 to ±3.4%. Also shown in Fig. 6 and Table 2 is the approximate solution suggested by Munz et al [7], which was obtained by an exponential fitting of two limiting solutions (the very short and very deep crack). The Munz solution is an approximation; however, it does serve as a baseline reference and has been shown for the purposes of comparison only.

The results for Y shown in Fig. 6 and Table 2 apply only for the case of a straight through crack, that is, a two-dimensional specimen geometry. The corresponding shape function, Y^*, for the chevron-notched geometry can be esti-

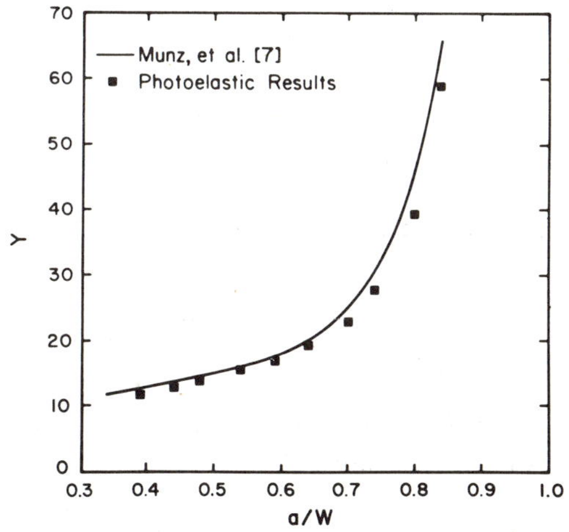

FIG. 6—*The dimensionless geometric shape function,* Y(a/W), *from local collocation and the approximate solution of Ref 7 for* W/H = *3.33 and the through-crack geometry.*

TABLE 2—*The through-crack geometric shape function,* Y(a/W), *from photoelastic calibration and the approximate method of Ref 7 for* W/H = *3.33.*

a/W	Photoelastic Results	Approximate Solution [7]	Ratio[a]
0.39	11.70	12.52	0.935
0.44	12.86	13.58	0.947
0.48	13.83	14.44	0.958
0.54	15.24	15.83	0.963
0.59	16.79	17.30	0.970
0.64	19.21	19.62	0.979
0.70	22.89	24.70	0.927
0.74	27.64	30.39	0.909
0.80	39.11	45.39	0.862
0.84	58.81	64.13	0.917

[a]Ratio = photoelastic results/approximate solution.

mated using the method proposed by Munz et al [*7*]. For any specimen geometry the energy, ΔU, available to extend the crack by an amount Δa is

$$\Delta U = \frac{P^2}{2}\frac{dC}{da}\Delta a \tag{9}$$

where C is the compliance of the specimen. Similarly, the energy, ΔW, necessary to create a new fracture surface of area, ΔA, is

$$\Delta W = \frac{K_c^2}{E'}\Delta A \tag{10}$$

where K_c is the critical stress-intensity factor and $E' = E$ for plane stress or $E' = E/(1 - \upsilon^2)$ for plane strain. During stable crack extension these energies are equal; hence, the compliance can be expressed as

$$\frac{dC}{da} = \frac{2Y^2}{E'WB^2}\frac{\Delta A}{\Delta a} \tag{11}$$

For a straight through crack, $\Delta A/\Delta a = B$, the overall specimen thickness, thus

$$\left(\frac{dC}{da}\right)_s = Y^2\frac{2}{E'WB} \tag{12}$$

On the other hand, for the chevron notched geometry, $\Delta A/\Delta a = b$, the width of the crack front as illustrated in Fig. 7

$$\left(\frac{dC}{da}\right)_{CN} = (Y^*)^2\frac{2b}{E'WB^2} \tag{13}$$

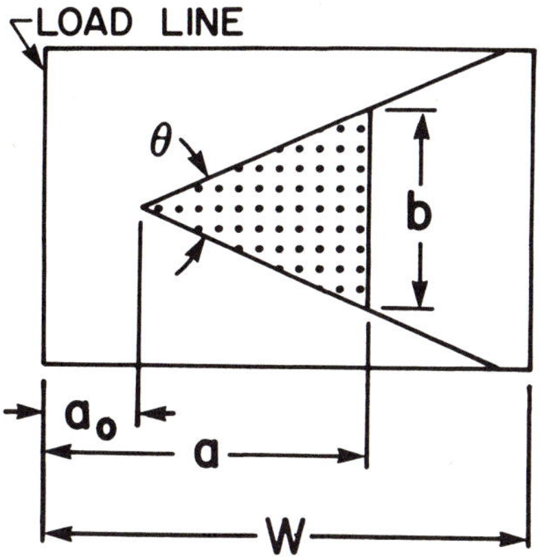

FIG. 7—*Chevron parameters used to relate the change in compliance for through-crack and chevron-notched configurations.*

The width of the crack front can be expressed in terms of crack length, a, and the chevron angle, θ, as

$$b = 2(a - a_o) \tan \frac{\theta}{2} \tag{14}$$

where a_o is the length to the apex of the chevron. If dC/da for both specimen geometries is the same [7], then the desired shape function for the chevron-notched geometry can be expressed in terms of the shape function for the straight crack geometry and the chevron parameters as

$$Y^* = Y \sqrt{\frac{B}{W} \left[\frac{1}{2\left(\frac{a}{W} - \frac{a_o}{W}\right) \tan \frac{\theta}{2}} \right]} \tag{15}$$

The result of applying Eq 15 to the photoelastic calibration is shown in Fig. 8 and tabulated in Table 3. Also shown in Fig. 8 and Table 3 is the approximate shape function of Munz, modified by the same factor. The minimum value of Y^* from the photoelastic calibration is 2% lower than the approximate solution, which is consistent with compliance calibration results [7] and detailed three dimensional finite element results [13] for this geometry.

Additional Results and Observations

The photoelastic calibration method has the additional advantage that it provides more than just an accurate K-calibration for the specimen geometry being studied. The analysis of the isochromatic fringe pattern provides, in addition to

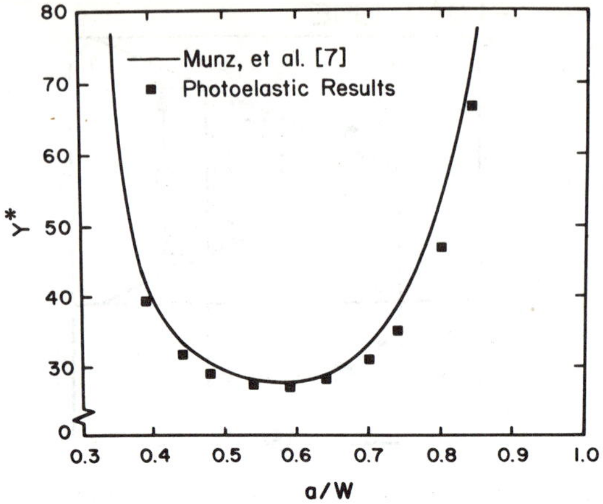

FIG. 8—*The dimensionless geometric shape function,* Y*(a/W), *from photoelastic calibration and the approximate solution of Ref* 7 *for* W/H = *3.33 and the chevron-notched geometry.*

the leading (singular) term, values for all the additional terms that are retained in the series expansion of Eq 5. Figure 9 shows the results obtained for the variation with a/W of the first four nonsingular terms. As one would expect, these coefficients also vary in a systematic fashion as the crack is extended across the specimen, with the contribution of the nonsingular terms to the crack-tip stress field becoming larger and larger as the crack approaches the specimen boundary.

The use of a reasonable number of these coefficients (generally 6 to 8) allows the stress field around the crack tip to be represented very accurately in a region that is of the size of that used for data acquisition in this study. By comparing the contribution of the nonsingular terms in this quasi-complete solution to the contribution of the singular term, it is possible to define the zone in which the crack tip singularity is indeed the dominant parameter.

TABLE 3—*The chevron-notch geometric shape function,* Y*(a/W), *from photoelastic calibration and the approximate solution of Ref* 7 *for* W/H = *3.33.*

a/W	Photoelastic Results	Approximate Solution [7]	Ratio[a]
0.39	39.35	42.10	0.935
0.44	31.74	33.50	0.947
0.48	29.17	30.45	0.958
0.54	27.11	28.16	0.963
0.59	26.83	27.65	0.970
0.64	28.10	28.70	0.979
0.70	30.63	33.05	0.927
0.74	35.13	38.63	0.909
0.80	46.42	53.86	0.862
0.84	66.98	73.05	0.917

[a]Ratio = photoelastic results/approximate solution.

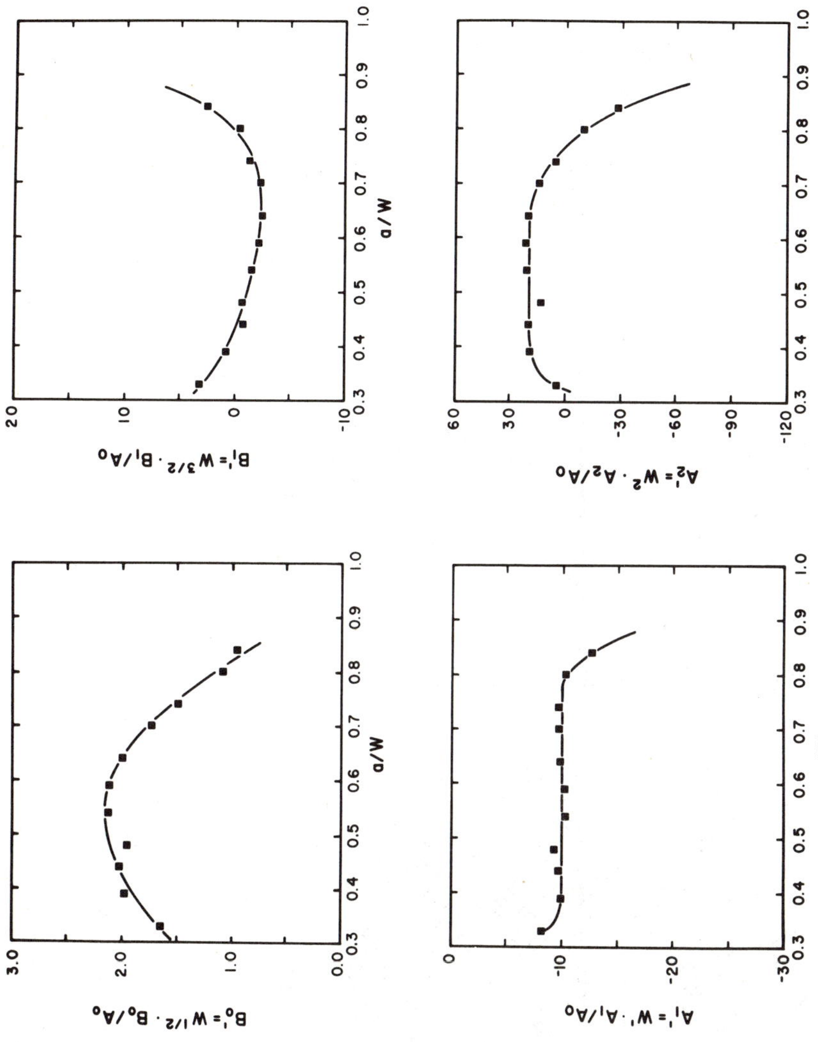

FIG. 9—*Variations with* a/W *of the first four nonsingular coefficients in the series representation of the crack-tip stress field.*

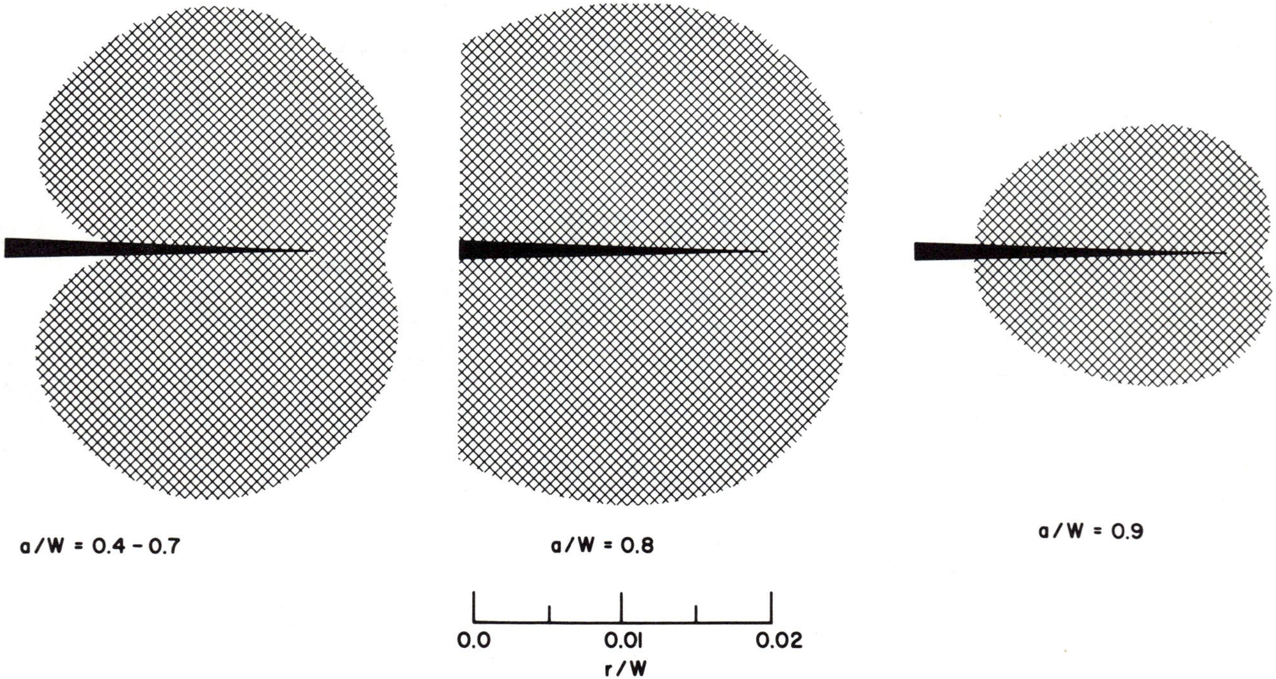

FIG. 10—*The changes with* a/W *in the size and shape of the singularity dominated zone in the planar chevron-notched geometry (based upon a nonsingular contribution to* σ_y *of 5%).*

It has been previously shown [*8*] that a comparison between the singular and nonsingular contributions to the crack opening stress σ_y, is a good quantitative measure of this singularity dominated zone. The changes with a/W in the size and shape of this singular zone are shown in Fig. 10, which is based on a 5% nonsingular contribution to the total stress magnitude. The minimum zone dimension, r_{min}, which occurs directly ahead of the crack tip has been selected as a characteristic dimension of the zone size, and Fig. 11 shows r_{min}/W as a function of a/W. The zone size is seen to be a constant (and rather small) percentage of the specimen width (0.5%) over much of the range of interest and is seen to get smaller rapidly beyond an a/W of 0.8.

Summary and Discussion

A new hybrid experimental-numerical (local collocation) technique for determining the geometric shape function in stress-intensity factor relations has been presented. The method uses experimental data from the region around the crack tip to determine the coefficients of the general series solution to the opening-mode crack problem. In the present case the method has been used with photoelastic data to calibrate the short-bar chevron-notched geometry. The results of this analysis are shown in Fig. 6 and Table 2 for the two-dimensional (through-crack) case.

The calibration of the three-dimensional (chevron-notched) geometry was obtained by applying the equivalent compliance change assumption proposed by Munz et al [*7*] for chevron-notched geometries. This assumption is equivalent to the constant strain energy release rate assumption for straight forward advance of the crack proposed by Freed and Krafft [*14*] for the side groove correction in plane specimens. In this sense the chevron-notched specimen can be viewed

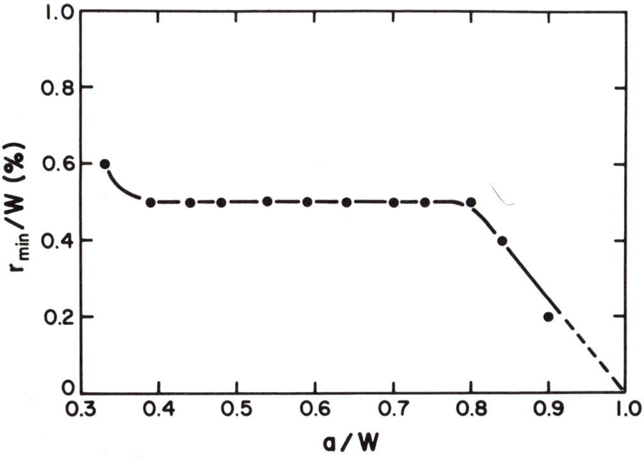

FIG. 11—*The singularity-dominated zone size,* r_{min}/W, *as a function of crack length,* a/W, *for the planar chevron-notched geometry.*

as a plane specimen with variable depth side grooving. Using this correction method, the geometric shape function for the chevron-notched geometry proposed by Barker [*9*] is shown in Fig. 8 and Table 3. It should be noted that the calibration of the three-dimensional geometry could have been performed directly using three-dimensional photoelastic models and the stress-freezing method [for example, Ref *15*]. Two-dimensional slices from these models could have then been analyzed using the local collocation method to obtain the shape function on each plane; however, the compliance equivalence approach appears to give results of sufficient accuracy so as to make a detailed three-dimensional photoelastic analysis unnecessary.

An additional advantage of the local collocation method over earlier methods for analyzing experimental data around a crack tip is the determination of additional stress field parameters. These parameters govern the size and shape of the region around the crack tip in which the near-field (that is, singular) stresses dominate the fracture process. The results of the present analysis indicate that the size of this zone for the planar chevron-notched geometry is of the same order as that previously obtained for other specimen geometries [*6,8*].

Acknowledgments

The work reported herein was supported by Sandia National Laboratories. The support and encouragement of Dr. J. J. Mecholsky is gratefully acknowledged. The authors also wish to thank Mr. R. E. Link for his assistance with the experiments and data analysis and the University of Maryland Computer Science Center for the computational facilities provided.

References

[*1*] Irwin, G. R., discussion of "The Dynamic Stress Distribution Surrounding a Running Crack—A Photoelastic Analysis," *Proceedings*, Society for Experimental Stress Analysis, Vol. XVI, 1958, pp. 93–96.

[*2*] Kobayashi, A. S., in *Experimental Techniques in Fracture Mechanics*, Monograph No. 1, Society for Experimental Stress Analysis, Iowa State University Press, Ames, Iowa, 1973, pp. 126–146.

[*3*] Sanford, R. J. and Dally, J. W., *Engineering Fracture Mechanics*, Vol. 11, 1979, pp. 621–633.

[*4*] Sanford, R. J., *Experimental Mechanics*, Vol. 20, 1980, pp. 192–197.

[*5*] Sanford, R. J. and Chona, R., *Proceedings*, SESA Annual Spring Meeting, Dearborn, Mich., May 1980, pp. 273–276.

[*6*] Whitman, G. D. and Bryan, R. H., "Heavy-Section Steel Technology Program Quarterly Progress Report for April-June 1982," NUREG/CR-2751/Volume 2 (ORNL/TM-8369/v2), 1982, pp. 47–53.

[*7*] Munz, D., Bubsey, R. E., and Srawley, J. E., *International Journal of Fracture*, Vol. 16, No. 4, 1980, pp. 359–374.

[*8*] Chona, R., Irwin, G. R. and Sanford, R. J. in *Fracture Mechanics: Fourteenth Symposium-Volume I: Theory and Analysis, ASTM STP 791*, J. C. Lewis and G. Sines, Eds., American Society for Testing and Materials, Philadelphia 1983, pp. I-3–I-23.

[*9*] Barker, L. M., *Engineering Fracture Mechanics*, Vol. 17, No. 4, 1983, pp. 289–312.

[*10*] Frocht, M. M., *Photoelasticity*, Vol. II, John Wiley, New York, 1948, p. 136.

[*11*] Sanford, R. J., *Mechanics Research Communications*, Vol. 6, 1979, pp. 289–294.

[12] Westergaard, H. M., *Transactions*, American Society of Mechanical Engineers, Vol. 61, 1939, pp. A49–A53.
[13] Newman, J. C., Jr., this publication, pp. 5–31.
[14] Freed, C. N. and Krafft, J. M., *Journal of Materials*, Vol. 1, No. 4, 1966, pp. 770–790.
[15] Schroedl, M. A. and Smith, C. W., *Fracture Analysis, ASTM STP 560*, American Society for Testing and Materials, Philadelphia, 1974, pp. 69–80.

Isa Bar-On,[1] Floyd R. Tuler,[1] and Itzhak Roman[2]

Comparison of Analytical and Experimental Stress-Intensity Coefficients for Chevron V-Notched Three-Point Bend Specimens

REFERENCE: Bar-On, I., Tuler, F. R., and Roman, I., "**Comparison of Analytical and Experimental Stress-Intensity Coefficients for Chevron V-Notched Three-Point Bend Specimens,** *Chevron-Notched Specimens: Testing and Stress Analysis, ASTM STP 855,* J. H. Underwood, S. W. Freiman, and F. I. Baratta, Eds., American Society for Testing and Materials, Philadelphia, 1984, pp. 98–113.

ABSTRACT: Chevron-notched specimens in a number of configurations have been recently used to determine plane-strain fracture toughness. Most researchers have employed tensile short-rod and short-bar specimens, while the three-point bend configuration has been used only to a limited extent despite its convenience for testing at ambient and high temperatures.

In the present study, both experimental and analytical K-calibrations for three-point bend chevron-notched specimens were carried out. Fracture toughness tests were performed in accordance with ASTM Standard of Test Method for Plane-Strain Fracture Toughness of Metallic Materials (E 399-81) to determine plane-strain fracture toughness, K_{Ic}, for polymethyl methacrylate (PMMA), 60/40 brass, 7075-T651 aluminum, and quenched and tempered 4140 steel. In parallel, three-point bend chevron-notched specimens of these materials were loaded to failure. The ratio $(K_{Ic}/P_{max}) B \cdot \sqrt{W}$ is the dimensionless quantity Y^*. This quantity was found to be 11.6 for the specimen geometry employed in the present study.

Three analytical models, which have been used successfully for the short-bar and short-rod configurations were used to calculate the stress-intensity coefficient and the critical crack length for the three-point bend configuration. The predicted stress intensity coefficients are 20 to 60% higher than the experimental value, and the predicted critical crack length is at least 20% lower than the directly measured critical crack length.

KEY WORDS: fracture toughness, chevron-notched specimen, three-point bend chevron-notch specimen, stress-intensity coefficient, critical crack length

The chevron-notched specimen has gained wide acceptance for fracture toughness testing since it was proposed by Barker [*1*]. The convenience of measuring maximum load (P_{max}) only without the need for fatigue precracking makes this

[1]Instructor and associate professor, respectively, Mechanical Engineering Department, Worcester Polytechnic Institute, Worcester, Mass. 01609.

[2]Senior lecturer, Materials Science Division, School of Applied Science and Technology, The Hebrew University of Jerusalem, Jerusalem, Israel.

approach attractive for fracture toughness testing of very brittle as well as metallic materials. As a result, the short-rod and short-bar specimens have been studied extensively, experimentally as well as analytically [2–6].

Chevron-notched bend bars in three-point bending have been investigated to a more limited extent, even though it seems that this specimen would be especially convenient for fracture toughness testing of brittle materials due to the ease of test setup and use at high temperature [8,9]. The stress-intensity coefficient has been determined experimentally as well as analytically for the short-bar, short-rod, and four-point bend specimen [4,7]. For the three-point bend specimen the stress-intensity coefficient has been determined experimentally only.

In this study, the stress-intensity coefficient Y^*[3] for specimens containing a chevron-notch loaded in three-point bending has been determined experimentally for four different materials. The result has been compared with Y^*-values which have been determined analytically by application of models which have been successfully used for the short-bar and short-rod specimens.

Background

The stress-intensity coefficient Y^* for the chevron-notched specimen can be determined in three ways:

1. Experimental determination of Y^* based on a comparison with standard K_{Ic}-values.
2. Analytical or semianalytical approach based on the compliance and the stress-intensity factor determined for specimens with straight cracks.
3. Full stress analysis, such as a three-dimensional finite element or three-dimensional boundary element analysis.

The first and third approaches should yield exact values of Y^*, while the second approach is basically an approximation which would need to be verified. The disadvantage of the first approach is that it must be repeated for each specimen geometry.

Barker has experimentally determined Y^*, showing that for a given geometry the critical crack length is material independent, assuming that the material exhibits a flat crack growth resistance curve [1]. The constant critical crack length permits the plane-strain fracture toughness to be determined for a given specimen size and geometry from the expression

$$K_{Ic}^* = P_{max} \cdot A \tag{1}$$

where A depends on the specimen and loading geometry only.

[3]An asterisk is used to designate the stress intensity coefficient and fracture toughness of the chevron-notched specimens.

For bend specimens of different sizes but with the same geometrical relationship

$$K_{Ic}^* = \frac{P_{max}}{B\sqrt{W}} Y^* \tag{2}$$

where B and W are the thickness and width of the specimen, respectively.

The stress-intensity coefficient Y^* is determined by measuring the standard fracture toughness K_{Ic} and the maximum load for three-point bend chevron specimens made from the same material. By repeating the experiments for different materials, this approach checks implicitly the assumption that the critical crack length is material independent. The critical crack length can be also determined independently by direct measurements.

Following Munz et al [*4,7*] the analytical or semianalytical approach is based on the compliance or stress intensity coefficient of a straight crack. The crack front of the specimen is characterized by the parameters shown in Fig. 1. After the crack has initiated at the apex of the triangle, for a given dimensionless crack length α the width of the crack is given by

$$b = B\,\frac{a - a_0}{a_1 - a_0} = B\,\frac{\alpha - \alpha_0}{\alpha_1 - \alpha_0} \tag{3}$$

Equating the necessary and available energy for crack propagation gives

$$K_{Ic}^* = \frac{P}{B\sqrt{W}} \left[\frac{1}{2}\frac{dC'}{d\alpha}\left(\frac{\alpha_1 - \alpha_0}{\alpha - \alpha_0}\right)\right]^{1/2} \tag{4}$$

where C' is the dimensionless compliance.

Thus

$$Y^* = \left[\frac{1}{2}\frac{dC'}{d\alpha}\left(\frac{\alpha_1 - \alpha_0}{\alpha - \alpha_0}\right)\right]^{1/2} \tag{5}$$

Straight Crack Assumption

To a first approximation, Munz et al assumed that $dC/d\alpha$ for a trapezoidal crack is the same as for a straight crack.

Thus

$$Y^* = Y\left[\frac{\alpha_1 - \alpha_0}{\alpha - \alpha_0}\right]^{1/2} \tag{6}$$

where Y is the stress-intensity coefficient for a straight crack for the specific geometry.

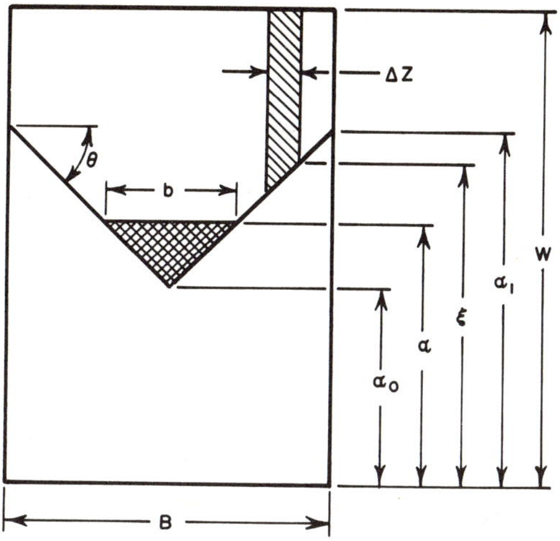

FIG. 1—*Notch geometry.*

Bluhm's Slice Model

The compliance calculation can be refined by use of Bluhm's slice model [*10*]. Here the compliance of the trapezoidal crack is the sum of the slice compliances, which are assumed to have a straight crack front. Using Munz's notation [*7*] the compliance of a specimen with a trapezoidal crack is given by

$$\frac{1}{C_{tr}} = \sum_{i=1}^{n} (C_S)_i = \left(\frac{\alpha - \alpha_0}{\alpha_1 - \alpha_0}\right) \frac{1}{C(\alpha)} + \frac{K}{n} \sum_{i=m+1}^{n} \frac{1}{C(\xi)_i} \tag{7}$$

where C_S is the compliance of a slice of thickness B/n, and K is the interlaminar shear correction factor. This interlaminar shear factor must be determined experimentally and thus to a certain extent accounts for the more complicated stress state resulting from the chevron notch.

Sakai's Modification of Bluhm's Model

Sakai and Yamasaki [*11*] have shown that for the compliance function to be continuous as α approaches α_1, the interlaminar shear factor K must be a function of α. According, Eq 7 can be rewritten as

$$\frac{1}{C_{tr}(\alpha)} = \left(\frac{\alpha - \alpha_0}{\alpha_1 - \alpha_0}\right) \frac{1}{C(\alpha)} + \frac{\alpha_1 - \alpha}{\alpha_1 - \alpha_0} \left\langle \frac{1}{C(\alpha)} \right\rangle \tag{8}$$

In this expression $\langle 1/C(\alpha)\rangle$ is defined as

$$\left\langle \frac{1}{C(\alpha)} \right\rangle = \frac{1}{\alpha_1 - \alpha} \int_{\alpha}^{\alpha_1} \frac{K(\xi)d\xi}{C(\xi)} \tag{9}$$

The two conditions that

$$\lim_{\alpha \to \alpha_1} C_{tr}(\alpha) = C(\alpha_1) \tag{10}$$

and

$$\lim_{\alpha \to \alpha_1} K_I(\alpha)|\text{nonstraight} = K_I(\alpha_1)|\text{straight through} \tag{11}$$

lead to the following definition of $K(\alpha)$

$$K(\alpha) \begin{array}{l} = K > 1.0 \\ = (K - 1)\dfrac{\alpha - \alpha_1}{\alpha^* - \alpha_1} + 1 \\ = 1.0 \end{array} \left\{ \begin{array}{l} \alpha_0 \leq \alpha \leq \alpha^* \\ \alpha^* \leq \alpha \leq \alpha_1 \\ \alpha \geq \alpha_1 \end{array} \right\} \tag{12}$$

where α^* defines a characteristic crack length beyond which the value of K changes towards 1.0.

In Sakai's study the K- and α^*-values were obtained by an approximate two-dimensional finite element analysis of the specimens. This analysis was performed for both the short-bar and the four-point bend specimens and also showed that K changes across the thickness of the specimen.

Materials and Experimental Procedure

Standard fracture toughness specimens and chevron-notched three-point bend specimens were cut from the same stock material and in the same directions (with one exception to be discussed later). Fracture toughness tests conforming to ASTM Test Method for Plane-Strain Fracture Toughness of Metallic Materials (E 399-78) were performed on four materials: PMMA, 60/40 brass, 7075-T651 aluminum, and quenched and tempered 4140 steel. A summary of materials, specimen configurations, and sizes is given in Table 1. For the hardened steel specimens the final notch extension was completed using electrical discharge machining after heat treatment.

The general configuration of the three-point bend specimens with the chevron notched is shown in Fig. 2. Stable crack propagation after reaching maximum load is influenced by the distance from the apex of the triangle to the edge of the specimen (a_0), the magnitude of the chevron angle, and the constraint that is achieved by the slot width of the notch. The specimen geometry chosen differs

TABLE 1—*Materials and specimen dimensions for standard fracture toughness specimens.*

Material	Source	No. of Speci-mens	Specimen Configuration	Dimensions, mm	Heat Treatment	Yield Strength, MPa
PMMA	plate 15-mm-thick isotropic	6	compact tension specimen	12.7 × 25.4 × 30.5		145[a]
60/40 brass	bar	4	3-point bend	10 × 10 × 55		380
7075-T651 aluminum	plate 25 mm thick	4	3-point L-T direction	10 × 20 × 110		565
4140	plate 30 mm thick	4	3-point L-S direction	12 × 24 × 120	oil quenched 845°C tempered at 300°C for ½ h	1570

[a]Use tensile strength, since yield strength is unavailable.

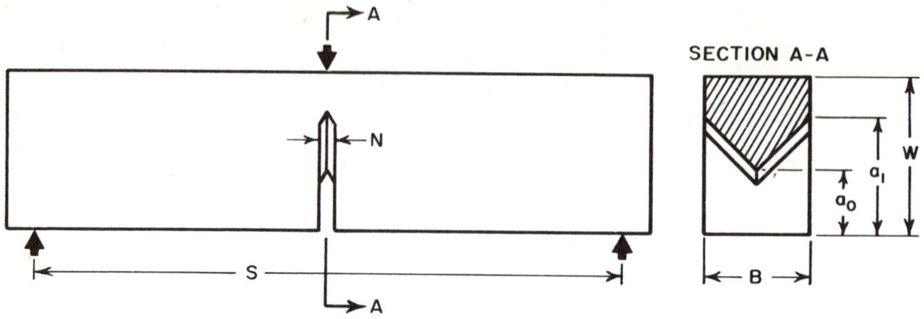

FIG. 2—*Three-point bend specimen.*

from the geometry proposed by Shih by the larger value of a_0. This promotes stable crack propagation after reaching maximum load. Interaction of the stress field arising from the loading pin with the stress field of the advancing crack was avoided by locating the base of the triangle far enough away from the loading point. A smooth load versus displacement curve was obtained for all materials used in this study with a specimen configuration having $a_0 = 0.42\ W$, $a_1 = 0.75\ W$, $B = \frac{2}{3}\ W$, and span to depth $S/W = 4$. The loading pin had a 12.7 mm diameter.

The materials and specimen sizes used in this study are summarized in Table 2. This table also contains the values for the quantity $2.5\ (K_{Ic}/\sigma_y)$,^{2} which is approximately 15 times the plane-stress plastic zone size. For three materials, PMMA, aluminum, and the steel this quantity is considerably smaller than the thickness of the specimen. Only for the brass specimen is this quantity somewhat larger than the thickness of the chevron-notched specimen used.

To further ensure plane-strain conditions the 1.0-mm-wide slot (1.5 mm wide for the largest specimens) had a 60° included angle pointed slot bottom, as recommended by Barker [*13*]. Barker showed that this slot bottom geometry provides excellent plane-strain constraint, regardless of the slot thickness. For the steel specimens the slot was 0.8 mm wide and had a rounded slot bottom.

The tests were performed on a 10-ton load capacity servocontrolled materials testing system and in a screw driven 5-ton load capacity machine. Displacement was measured using a double cantilever beam crack opening displacement gage with a 2 to 6 mm range. The standard K_{Ic} tests were performed in load control and the three-point bend chevron specimens were tested under stroke control.

Crack length was measured at maximum load on specimens made from PMMA, brass, and aluminum of various sizes. The fracture surface was marked by one of the following techniques: spraying penetrating paints into the crack; marking the fracture surface by unloading the specimen from maximum load and reloading; and cyclically loading the specimen after reaching maximum load. A summary of various techniques used is given in Table 3.

The interlaminar shear factor was determined by measuring the compliance of 7075-T651 aluminum specimens with $B = 13.3$ mm, $W = 20$ mm, $a_0 = 8.4$ mm, and $a_1 = 15$ mm. A slot was introduced using a 0.008-in.-diameter

TABLE 2—*Material and dimensions of chevron-notched specimens.*

Material	No. of Specimens	Dimensions, mm	$2.5(K_{Ic}/\sigma_y)^2$ mm
PMMA	5	6.7 × 10 × 55	1.2
	1	13.3 × 20 × 120	
PMMA[a]	3	6.7 × 10 × 55	
	3	16.6 × 40 × 180	
60/40 brass	5	6.7 × 10 × 55	8.0
7075-T651 aluminum	8	6.7 × 10 × 55	6.3
	3	13.3 × 20 × 55	
	1	25 × 40 × 180	
4140 steel (hardened)	2	16 × 24 × 110	1.3

[a]These specimens were cut from a different plate than the compact tension specimens.

wire saw and was extended gradually. The crack length was measured on both sides of the specimen after complete fracture. For each crack length, the load versus displacement was measured. To calculate the dimensionless quantity $C' = CEB$, Young's modulus for the 7075-T651 aluminum was determined on four tension specimens, cut from the same aluminum plate and in the same direction as the chevron-notched specimens. The value for the Young's modulus obtained was $E = 73.6$ kN/mm^2.

Results

The valid K_{Ic}-values obtained from the standard tests, P_{max}-values, and dimensions of the chevron specimens for the different materials are summarized in Table 4. The last column of Table 4 also contains the Y^*-values calculated from Eq 2 using the data given in the previous columns. Two different plates of PMMA were used for the chevron-notched specimens, but the standard K_{Ic}-value was determined for one material only. Nevertheless, the results from the second plate give a constant Y^*-value for two specimen sizes.

The thickness of some of the aluminum specimens might not have been sufficient to reach a true minimum fracture toughness value even though they fulfill the requirements of the ASTM Method E 399-81. The average value of Y^* is 12.00 for the four materials. Excluding the values for the aluminum specimens gives $Y^* = 11.6$.

An example of the critical crack length measurement is shown in Fig. 3. The measurements of the critical crack length for various materials are summarized in Table 3, giving $\alpha_c = 0.65$. Since in all cases the measurement was performed after the maximum load was reached, this may be an overestimate of the order of 5%. Thus the critical crack length is at least 0.62 W.

The effective thickness of the material at the critical crack length, b_c, can be calculated from

$$\frac{b_c}{B} = \frac{a_c - a_0}{a_1 - a_0}$$

TABLE 3—*Critical crack lengths for chevron-notched specimens.*

Material	Dimension, $B \times W$, mm	Measuring Method	a_c/W
Aluminum	25.4 × 10	penetrating paints	0.64
	13.3 × 20	penetrating paints and fatigue	
PMMA	13.3 × 20	penetrating paints	0.65
	6.7 × 10	unload	0.64
Brass	6.7 × 10	unload	0.65

Using the parameters of the three-point bend specimen of this study, b_c equals 0.61 B. For the short-rod specimen with $a_0 = 0.35\ W$, $a_1 = W$, and $a_c = 0.56$, the resulting b_c equals 0.32 B. Thus for the same specimen thickness the width of the crack front at maximum load is larger for the three-point bend geometry than for the short-rod or short-bar geometry.

Two parameters, Y^* and α_c, are available for comparison with the predictions of analytical or semianalytical models. Curves of Y^* versus α for the model assuming a straight crack are shown for the geometry used in this study in Fig. 4 and for Shih's geometry in Fig. 5.

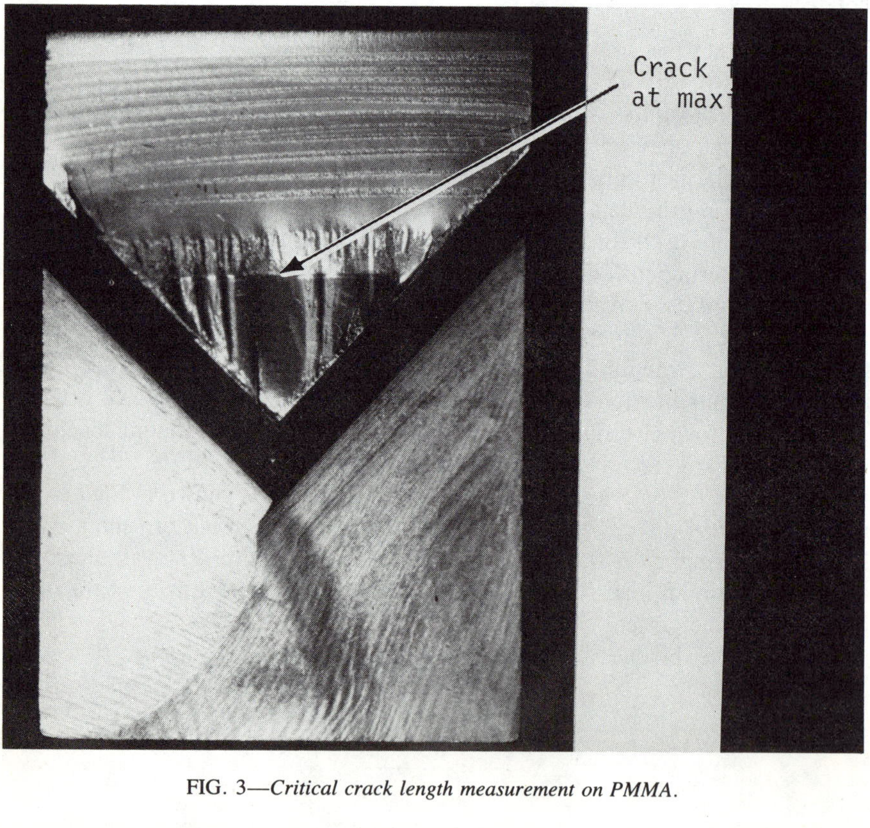

FIG. 3—*Critical crack length measurement on PMMA.*

TABLE 4—*Summary of results:* K_{Ic} *according to ASTM E 399-81 and data for chevron-notched specimens.*

Material	No. of Valid K_{Ic} Tests	K_{Ic}, MPa $m^{1/2}$	No. of Valid Chevron-Notched Tests	P_{max}, kN	Dimensions, $B \times W \times L$,[b] mm	Y^* Calculated
PMMA	3	0.971	5	55	6.7 × 10 × 55	11.8
			1		13.3 × 20 × 110	12.5
PMMA[a]		...	3	48	6.7 × 10 × 55	
		...	3	390	26.6 × 40 × 180	
60/40 brass	4	22.4	5	1 370	6.7 × 10 × 55	11.0
7075-T651 aluminum	4	28.3	6	1 462	6.7 × 10 × 55	12.9
			3	4 153	13.3 × 20 × 110	12.8
			1	12 164	25 × 40 × 180	11.6
4140 steel (hardened)	2	35.3	2	7 358	16 × 24 × 110	11.9

[a]These specimens were taken from a different plate.
[b]L = length of the specimen.

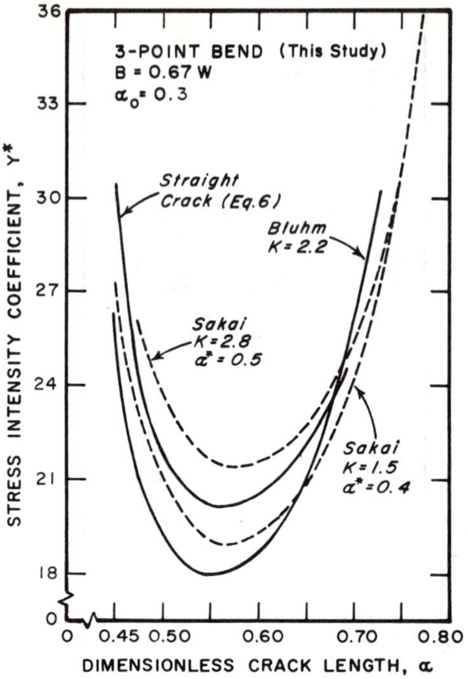

FIG. 4—*Stress-intensity coefficients predicted by three analytical models.*

Bluhm suggested a value for the interlaminar shear factor K of 1.45 to be used in the slice model for a span to depth ratio of 8 [*10*]. The K-value for our configuration was determined by comparing the measured and calculated compliances. Figure 6 shows the compliances calculated for various values of K and the experimentally determined values. Also it can be seen that as the crack length increases, the difference in the compliance as a function of K decreases. Figure 7 shows the stress intensity coefficients determined using Bluhm's model with 500 slices for $K = 1.4$ and $K = 2.2$. These results show that Y^* is not very sensitive to the exact value of K.

The analysis of Sakai et al uses two parameters: the shear factor K and the crack-length α^* beyond which the K-value varies as a function of crack length [*11*]. Since we did not perform a finite element analysis, we based our calculations on Sakai's results extrapolated to the three-point bend loading configuration. The extrapolation gives a shear factor value of $K = 2.8$. Sakai's model was used to calculate the stress intensity coefficient and the critical crack length using the following parameters: $K = 2.8$ and Bluhm's value, $K = 1.5$; $\alpha^* = 0.5$ and 0.65 for our geometry and $\alpha^* = 0.4$ and 0.5 for Shih's geometry. We also assumed that K varies with α in the same manner as for a short-bar specimen. The stress-intensity coefficients calculated using the three analytical models are plotted in Figs. 4 and 5. A summary of the minimum Y^* and the α_c-values are given in Table 5.

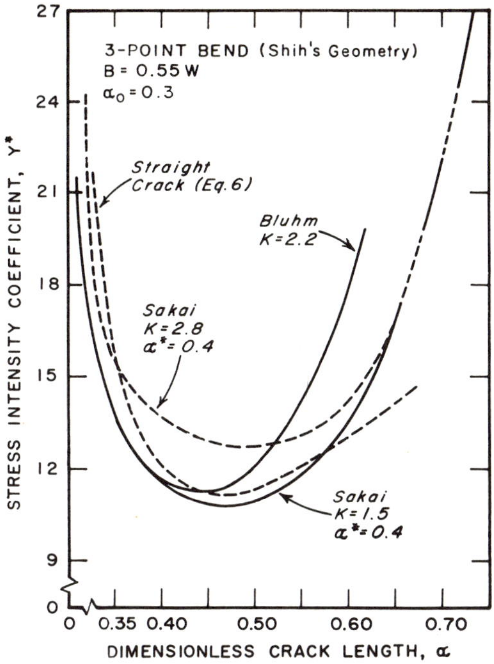

FIG. 5—*Stress-intensity coefficients predicted by three analytical models.*

Discussion

The experimentally determined values for the stress-intensity coefficient for the four materials in this study have a consistent value of $Y^* = 12.0$ (if we include the aluminum) and $Y^* = 11.6$ excluding the aluminum. The critical crack length measurements give $\alpha^* = 0.63$ to 0.65, independent of material.

Shih found $Y^* = 10.17$ for his specimen geometry, which is lower than the value determined in this work. This difference can be explained by the smaller value of α_0 for Shih's geometry. The critical crack length for Shih's geometry was determined from a photograph in Ref *8* to be $\alpha_c = 0.56$. All the analytical results that are based on the compliance of specimens with straight notches predict values of Y^* that are too high by at least 20% for Shih's geometry as well as for our geometry. Also, the predicted α_c-values are too low by more than 20%.

Similar discrepancies between experimental and analytical results can be seen in the work of Pook [*12*]. The K_{Ic}^* determined by the chevron-notched specimen are 30% higher than the K_{Ic}-values determined from standard fracture toughness tests.

Sakai and Yamasaki have shown that the value for the interlaminar shear is not only a function of the crack length but also of the crack width. Furthermore, they show that small changes in the shape of the compliance curve might result

TABLE 5—*Predicted Y^* and α_c values.*

		Geometry of This Study				Shih's Geometry		
		Y^*	a_c	Deviation from Y^*_{exp}		Y^*	a_c	Deviation from Y^*_{exp}
Straight crack assumption (Eq 6)		20.0	0.55	66%		12.9	0.48	27%
Bluhm's slice model								
$K = 1.5$		17.8	0.55	48%		12.2	0.44	20%
$K = 2.2$		17.9	0.55	49%		12.5	0.44	23%
Sakai's model								
$K = 1.5$	$\alpha^* = 0.5$	19.0	0.57	58%	$\alpha^* = 0.4$	12.2	0.46	20%
	$\alpha^* = 0.65$	18.6	0.55	55%	$\alpha^* = 0.5$	12.1	0.43	19%
$K = 2.8$	$\alpha^* = 0.5$	21.4	0.58	78%	$\alpha^* = 0.4$	14.2	0.49	40%
	$\alpha^* = 0.65$	19.15	0.54	60%	$\alpha^* = 0.5$	12.9	0.41	27%

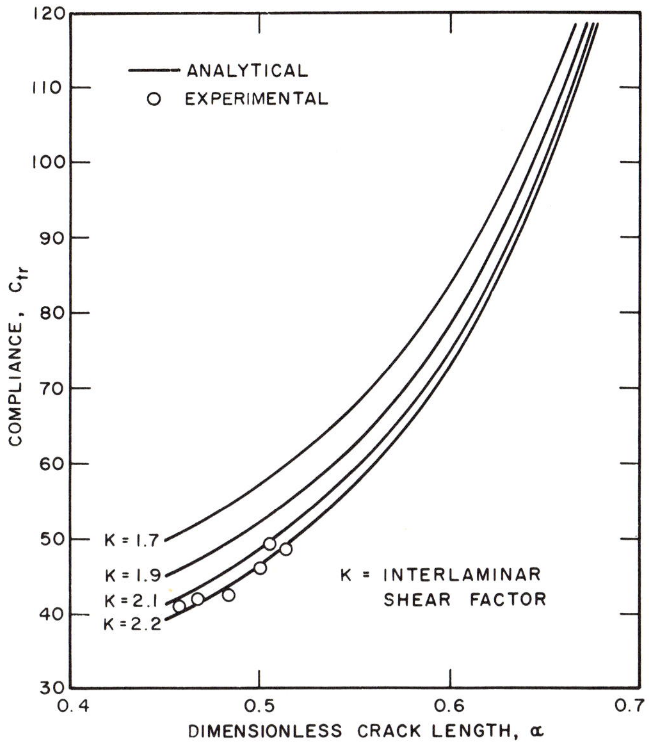

FIG. 6—*Comparison of experimentally and analytically determined compliance values.*

in substantial changes in the stress intensity coefficient curve. The sensitivity of the stress-intensity coefficient and critical crack length to the choice of stress parameters is apparent in Figs. 4 and 5. However, any reasonable variation of the stress parameters does not bring the experimental and analytical results into agreement.

Conclusions

1. The stress-intensity coefficient was determined independently for PMMA, 60/40 brass, 7075-T651 aluminum, and quenched and tempered 4140 steel. Y^* was determined to be 12.00 for the four materials and 11.6 excluding the aluminum, respectively.

2. The critical crack length was measured for PMMA, 60/40 brass, and 7075-T651 aluminum, giving $\alpha_c = 0.65$.

3. The Y^*-value predicted by the straight crack model, by Bluhm's model, and by Sakai's model is 20 to 70% higher than the value determined experimentally.

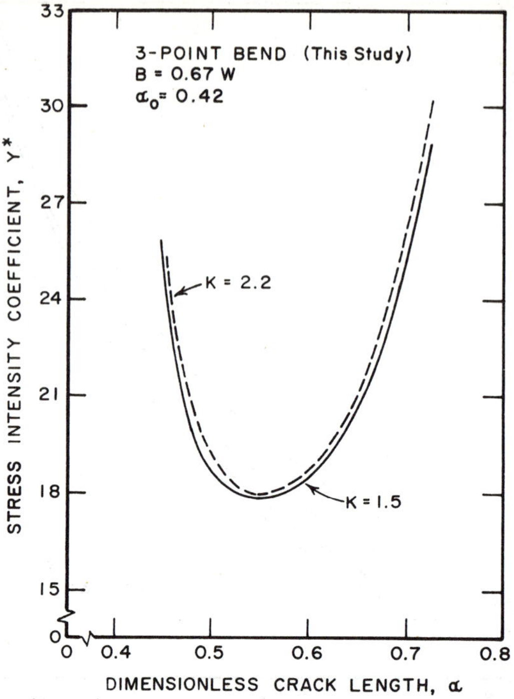

FIG. 7—*Stress-intensity coefficients using Bluhm's slice model.*

4. The predicted value for the critical crack length using the straight crack model, Bluhm's model, and Sakai's model is lower by 20%.

5. In view of the discrepancies of items 3 and 4, a three-dimensional stress analysis for the three-point bend chevron-notched specimen should be performed.

Acknowledgment

This work was performed at the Hebrew University of Jerusalem as part of the Ph.D. thesis of I. Bar-On.

References

[*1*] Barker, L. M., *Engineering Fracture Mechanics,* Vol. 9, 1977, pp. 361–369.
[*2*] Barker, L. M. and Baratta, F. I., *Journal of Testing and Evaluation,* 1980, Vol. 8, pp. 97–102.
[*3*] Munz, D., Bubsey, R. T., and Shannon, J. L., Jr. *Journal of Testing and Evaluation,* 1980, Vol. 8, pp. 103–107.
[*4*] Munz, D., Bubsey, R. T., and Srawley, J. E., *International Journal of Fracture,* 1980, Vol. 16, pp. 359–374.
[*5*] Munz, D., *Engineering Fracture Mechanics,* 1981, Vol. 15, pp. 231–236.
[*6*] Eschweiler, J. and Munz, D., *International Journal of Fracture,* 1982, Vol. 20, pp. R131–R133.

[7] Munz, D., Bubsey, R. T., and Shannon, J. L., Jr., *Journal of the American Ceramic Society,* 1980, Vol. 63, pp. 300–305.
[8] Shih, T. T., *Journal of Testing and Evaluation,* 1981, Vol. 9, pp. 50–55.
[9] Shih, T. T., *Engineering Fracture Mechanics,* 1981, Vol. 14, pp. 821–832.
[10] Bluhm, J. I., *Engineering Fracture Mechanics,* 1975, Vol. 7, pp. 593–604.
[11] Sakai, M. and Yamasaki, K., *Journal of the American Ceramic Society,* 1983, Vol. 66, pp. 371–375.
[12] Pook, L. P., *International Journal of Fracture,* 1983, Vol. 22, pp. R21–R23.
[13] Barker, L. M., *Fracture Mechanics, ASTM STP 743,* American Society for Testing and Materials, Philadelphia, 1981, pp. 456–475.

Test Method Development

Lynn M. Barker[1]

Specimen Size Effects in Short-Rod Fracture Toughness Measurements

REFERENCE: Barker, L. M., "**Specimen Size Effects in Short-Rod Fracture Toughness Measurements,**" *Chevron-Notched Specimens: Testing and Stress Analysis, ASTM STP 855*, J. H. Underwood, S. W. Freimen, and F. I. Baratta, Eds., American Society for Testing and Materials, Philadelphia, 1984, pp. 117–133.

ABSTRACT: Short-rod fracture toughness specimens of various sizes were used in toughness measurements on six different metals. The toughness values were observed to be independent of the specimen size, within experimental scatter, whenever the specimen diameter exceeded $1.25(K_{IcSR}/\sigma_{ys})^2$. The test records were analyzed by a linear elastic fracture mechanics (LEFM) method and also by an elastic-plastic method. When the specimen size was too small to yield a size-independent result, the LEFM analysis always tended toward conservative (smaller) toughness values, while the elastic-plastic analysis always tended toward nonconservative values. Thus, the two analysis techniques apparently establish upper and lower bounds for the toughness when the specimen is too small for a valid test. It was also found that the slots which form the chevron in the specimen must either be very thin or have sharply pointed slot bottoms ($\leq 60°$ included angle) to maintain good plane-strain constraint along the crack front.

KEY WORDS: fracture mechanics, crack propagation, chevron-notched specimen, specimen size criterion, metallic materials, data analysis, test specimen geometry

Properly designed chevron-notched specimens for fracture toughness testing have certain advantages such as the acquiring of a bona fide test crack during the toughness test without any fatigue precracking, even in very brittle materials. The first chevron-notched specimen tests were reported in 1966 by Tattersall and Tappin [*1*], who benefited from earlier work by Nakayama [*2,3*] on an asymmetrical specimen with half of a chevron notch. The first papers on chevron-notched toughness testing dealt with bend-type specimens and used a "work of fracture" data analysis method in which the toughness was calculated from the area under the load-displacement test record [*1–12*]. In 1972, however, Pook [*13*] recognized that just the peak load in a chevron-notched specimen test was sufficient to estimate the toughness. He suggested using the chevron-notched bend specimen for quality control purposes because of the economy and simplicity of the test.

[1]Distinguished member of the technical staff, Sandia National Laboratories, Albuquerque, N. Mex. 87185.

The recent introduction of the smaller "short-rod" specimen (Fig. 1) by Barker [*14*] has led to an increased interest in chevron-notched specimens [*15–56*]. Barker has attempted to develop short-rod and short-bar test methods which produce accurate measures of the fracture toughness [*24,32–34,40,41*], rather than only the quality control type estimates suggested by Pook. Currently a variety of chevron-notched specimens are being used and tested, including short rods and short bars of various configurations [*14–48*], as well as the older bend specimens [*49–56*].

There seems to be general agreement that properly designed and tested chevron notched specimens should provide good measurements of the plane-strain critical stress-intensity factor, provided the specimen conforms sufficiently well to the assumptions of linear elastic fracture mechanics (LEFM). For metals, the most important LEFM assumption is that there is negligible plasticity in the specimen. Since metals all have a zone of plasticity at the crack tip, the LEFM criterion is satisfied only when the specimen size is very large compared to that crack-tip plastic zone size which is characteristic of the material of which the specimen is made. In ASTM Standard Test Method for Plane-Strain Fracture Toughness of Metallic Materials (E 399), certain rules such as the requirement that the specimen thickness must exceed the quantity[2] $2.5(K_{Ic}/\sigma_{ys})^2$ have been established to assure that the test result is not degraded very much by nonconformance to LEFM criteria. However, insufficient test data have been generated on chevron-notched specimens to establish a comparable specimen size criterion.

The specimen size effect studies reported in this paper should help to establish a minimum size criterion for short-rod/short-bar specimens. A size criterion is suggested based on these studies, and a method is given for calculating both

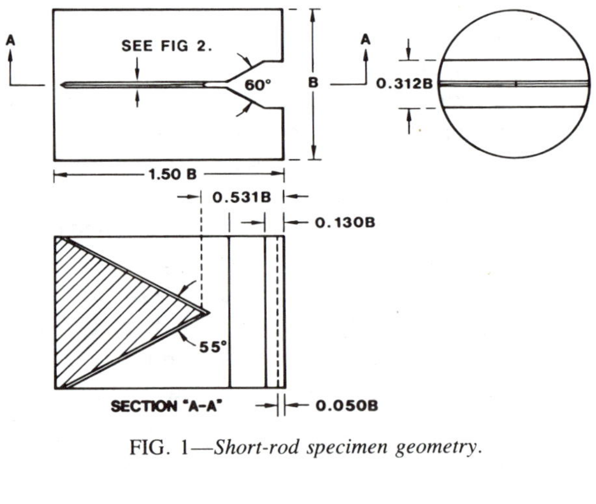

FIG. 1—*Short-rod specimen geometry.*

[2]Here, K_{Ic} is the plane-strain critical stress-intensity factor, and σ_{ys} is the tensile yield strength.

upper and lower bounds for the toughness when the specimen size is smaller than the minimum size criterion.

Experiments and Data Analysis

The six metals studied in the size effect investigation are listed in Table 1, along with crack orientations, heat treatments, chemistry, and estimated yield strengths. Four specimen sizes ranging from 6.35 to 50.8 mm diameter were tested. The specimens were machined approximately from the center of the original stock of material listed in Table 1. The test geometries were always similar to that shown in Fig. 1. Figure 2 shows the slot geometry constraints which were recommended as a result of a previous study of slot geometry effects [*33*]. The slots of the size effect test specimens followed these recommendations, except as noted in the text.

The toughness tests were performed on patented Fractometer II test machines and Fracjack test fixtures produced by Terra Tek Systems. The Fractometer II is a stand-alone test machine, whereas the Fracjack attaches to a tension test machine. In either case, the specimen is placed mouth-down onto grips which open the mouth of the specimen in the horizontal direction during the test. Gravity holds the specimen in place until the test is started. A three-pronged specimen mouth opening gage automatically slides into the mouth of the specimen as the specimen is placed on the grips. Alignment takes less than 10 s, and consists of centering the specimen on the grips so the mouth opening gage hangs free through slots in the grips, touching nothing except the inside of the specimen mouth. During the test, the flexing of the specimen is automatically matched by a rotation of the grips, so that the load line in the specimen mouth remains nearly constant. Any remaining load line variation is further compensated for in these test devices, as detailed in Ref *57*. Tests were typically accomplished in 1 to 2

TABLE 1—*Size effect study materials.*

Material	Form	Crack Orientation	Hardness, HRC	Estimated Yield Strength, MPa
2124-T851	76.2-mm plate	L-T	. . .	462
6061-T651	63.5-mm plate	L-T	. . .	276
7475-T7351	76.2-mm plate	L-T	. . .	407
4340[a,c]	50.8-mm rod	R-L	30	862
17-4PH[b,d]	50.8-mm rod	R-L	34	759
Ti-6A1-4V[e]	57.15-mm rod	R-L	33	910

[a]0.400C, 0.76Mn, 0.009P, 0.012S, 0.27Si, 1.67Ni, 0.81Cr, 0.24Mo, 0.14Cu.
[b]0.043C, 0.56Mn, 0.024P, 0.016S, 0.44Si, 4.29Ni, 15.82Cr, 0.3Mo, 3.15Cu, 0.28Nb, 0.01Ta.
[c]HT: 845°C 2 h, oil quench, 620°C 2 h, air cool.
[d]HT: H 1150 M.
[e]HT: 925°C 3 h, air cool, 590°C 4 h, water quench.

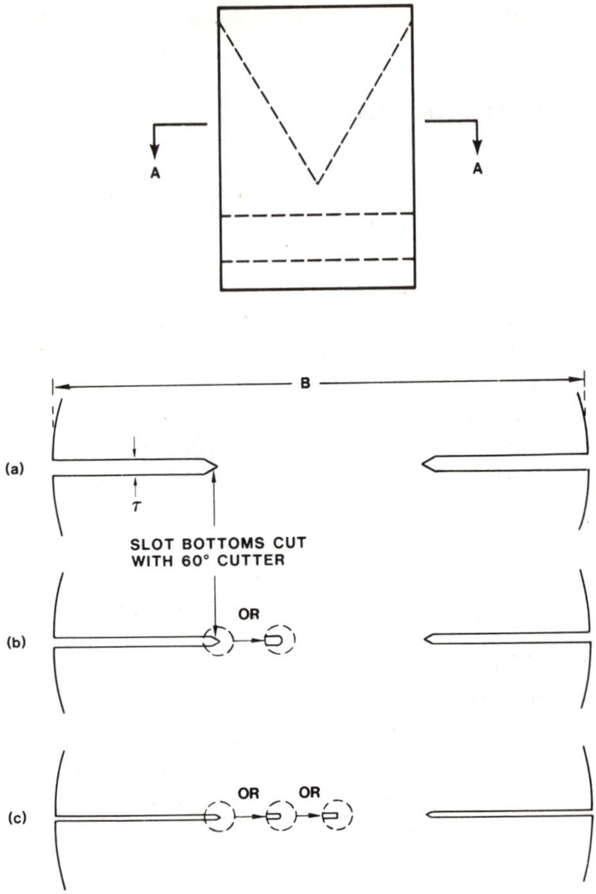

FIG. 2—*Enlarged scale drawings of cross sections of the center portions of short-rod specimens, showing recommended* [33] *chevron slot-bottom configurations. The scaled slot thickness,* τ/B, *should be* ≤ 0.03. *If* $0.02 < \tau/B < 0.03$, *the slot bottoms must be pointed with an included angle of 60° or less in order to maintain good plane-strain constraint* [33], *as shown in* (a). *If* $0.01 < \tau/B < 0.02$, *the slot bottoms can be either pointed with an angle of 60° or less, or rounded, as in* (b). *If* $0 < \tau/B < 0.01$, *any slot bottom configuration, including square, is acceptable* (c).

min each, using constant specimen mouth opening speeds. Unloadings during the test were done by reversing the grip displacement direction. Load-displacement records were made on an X-Y recorder.

It must be recognized that a minimum specimen size criterion will depend to some degree on the test and data analysis procedures. The current ASTM standard toughness test method E 399 could have relied purely on LEFM assumptions, in which case only the peak load in the test (along with the crack length) would ever be used to calculate the toughness. However, in this case, a larger minimum specimen size criterion would have been required for accurate results. Likewise, the purely LEFM approach which uses only the peak load in a short-rod or short-bar test would require a larger specimen size criterion than a somewhat more sophisticated test and data analysis technique. Since a more sophisticated technique is much easier to accept than a larger specimen size criterion, this paper presents the results of two data analysis methods which go beyond the purely LEFM approach of using only the peak load to calculate the toughness.

The first method involves only one basic change from the peak-load LEFM approach. Instead of using the peak load and assuming that the scaled crack length is always the same at the time of the peak load [*14*], a measure is obtained of the crack length by unloading compliance measurements which are made during the course of the test [*34*]. One or more actual load/crack-length pairs are then used to calculate the toughness. This method is actually the LEFM procedure for measuring the toughness at an arbitrary crack length, rather than the crack length at which the peak load is supposed to occur. Therefore, in this paper, the foregoing method will be referred to as the LEFM analysis.

The second method, called the elastic-plastic (EP) analysis, involves the above crack-length determination plus a measurement of the degree of plasticity, p, displayed by the specimen during the test. A theory for making and incorporating such a measurement into the toughness calculation appears in Ref *24*. For relatively small amounts of plasticity, that is, for p small compared to unity, the elastic-plastic equation for the plane-strain critical stress-intensity factor for a short-rod or short-bar specimen[3] is

$$K_{SR} = AF(1 + p)/B^{3/2} \qquad (1)$$

where

B = specimen diameter,
F = mouth opening load required to advance the crack, and
A = dimensionless function of the scaled crack length which has a minimum value of 22.0 for the specimen configurations of this study [*42*].

The value of the plasticity, p, was obtained for each specimen by making autographic plots of the load versus the specimen mouth opening displacement during the test. Two unloading-reloading cycles were plotted at different crack lengths, which allowed a graphical determination of p, as indicated in Fig. 3. The unloading-reloading cycles also facilitate the crack length determination. A detailed explanation of the EP data analysis method is given in Ref *34*.

The results of applying these analysis methods to the short-rod specimen size

[3]The K_{SR} and K_{IcSR} nomenclatures are used to distinguish short-rod/short-bar fracture toughness measurements from toughness measurements made according to ASTM Method E 399, for which the ASTM has reserved the symbol K_{Ic}. In this paper, K_{SR} is analogous to K_Q in E 399 tests, and K_{IcSR} is analogous to K_{Ic} in the sense that it is felt to be an accurate toughness measurement, essentially independent of specimen size effects. Comparison tests [*32*] have shown good agreement between K_{Ic} and K_{IcSR}.

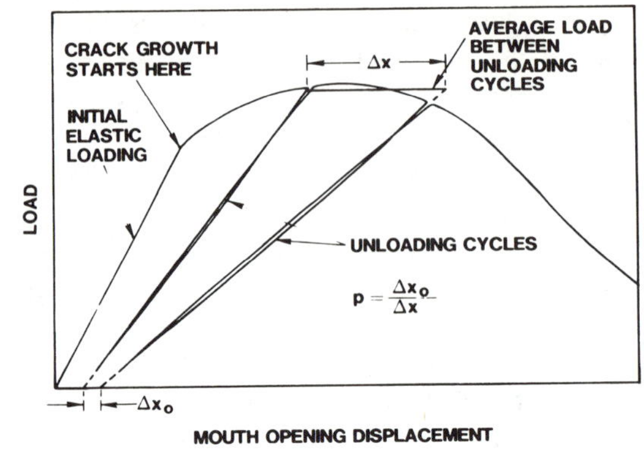

FIG. 3—*Schematic of a short-rod test record illustrating the method of determining the plasticity*, $p = \Delta x_o/\Delta x$. *A more complete discussion of* p-*measurements is given in Ref* 34.

effect tests are presented in the next section. Based on prior experience, the materials and specimen size range were chosen such that in most cases the smallest specimens were *not* expected to provide accurate measures of the toughness. No comparisons of short-rod toughness measurements with ASTM E 399 toughness measurements are reported, since no E 399 tests were made of the materials used in this study. Some short-rod versus E 399 comparisons have been reported in Ref *32*, and a more extensive comparison study is currently being undertaken by ASTM [*58*]. The purpose of the present study was solely to help identify the minimum specimen size criterion for valid short-rod fracture toughness measurements.

Results

The size effect test results are presented in graphs of apparent toughness, K_{SR}, versus the specimen diameter, B, in the following discussions of each material tested. The diameter is indicative of the specimen size since the other dimensions were scaled in proportion to the diameter. The data points obtained by the use of the EP theory, that is, using Eq 1, are shown with error bars which show the data scatter. A number in parentheses gives the number of tests represented by each data point. The same experiments were also analyzed according to LEFM assumptions, that is, Eq 1 with p set equal to zero. The data points so obtained are shown as open circles on the graphs. The LEFM data scatter was always similar to that shown for the EP analysis.

With sufficient data and sufficiently well-controlled material and experiments, the value of the apparent toughness, K_{SR}, will approach the plane-strain critical stress intensity factor, K_{IcSR}, as the specimen size is increased. The combined results of the size effect tests conducted to date, including those of this study, have been used to suggest the minimum specimen size criterion

$$B \geqslant 1.25(K_{IcSR}/\sigma_{ys})^2 \qquad (2)$$

In each size effect data figure of the following individual material discussions, the "validity line" corresponding to this criterion is shown.

2124-T851 Aluminum

This material had a small crack-tip plastic zone size, and it was expected that the toughness measurements would probably be independent of specimen size for all four sizes tested. However, a moderate but consistent increase in apparent toughness with decreasing specimen size was initially measured by both LEFM and EP analysis methods. An examination of the crack surfaces showed that the surface texture was not constant, but was visibly more coarse along a band a few millimetres wide running longitudinally through the centers of the specimens. It was reasoned that if the coarseness was indicative of a higher toughness band at the center of the parent plate of material, the apparent size effect would be explained, because the smaller the specimen, the tougher would be the material (on the average) sensed by the test crack at the time of the toughness measurement. To determine the validity of this hypothesis, seven short-rod specimens, each 12.7 mm diameter, were subsequently machined from the same parent 76-mm-thick plate of material and tested to determine toughness values. The centers of the seven specimens were taken at seven equally spaced locations through the thickness of the parent plate so as to obtain a profile of the toughness through the thickness of the plate.

The test results are plotted in Fig. 4, which shows that the toughness is indeed

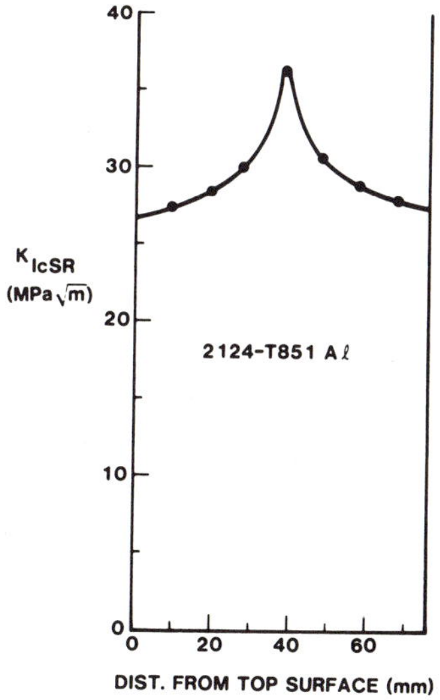

FIG. 4—*Measured toughness variation through the thickness of a 76 mm thick plate of 2124-T851 aluminum alloy. The crack surface had a visibly rougher texture in a narrow band at the center of the plate where the toughness spike was found. Each toughness measurement represents an average toughness over a crack-front width of about 4 mm.*

sharply peaked at the center of the plate. Thus, the apparent size effect was not real but was an artifact of the toughness spike at the center of the parent plate of material. Note that the central toughness is about 30% higher than the toughness close to the surfaces of the original plate. Such variations in toughness have been observed previously in various aluminum alloys [*43,59,60*] and have been shown [*32*] to be responsible for apparent variations between toughness values measured by the short rod method and toughness measurements done according to the ASTM E 399 standard.

Having measured the toughness profile of the plate of 2124-T851 aluminum, and assuming that the specimens were all centered on the toughness peak (as the visual observation of the crack surface roughness indicated), one can make a reasonable estimate of how much the toughness variation affected the size effect tests. If one assumes that the specimen measures the average toughness over the center one third of the specimen diameter (the breadth of the crack front), the toughness measurements to be expected from the profile of Fig. 4 would be about 33.0, 34.8, 36.2, and 36.5 MPa$\sqrt{m}$ for the specimen sizes of 50.8, 25.4, 12.7, and 6.35 mm, respectively.

Thus, relative to the 50.8-mm specimen, the toughness variation affected the 25.4, 12.7, and 6.35-mm test results by about 5.5, 9.7, and 10.6%, respectively. These corrections were therefore applied, and the corrected data appear as shown in Fig. 5, where very good specimen size independence is indicated, especially to the right of the validity line.

6061-T651 Aluminum

The original tests of the 50.8-mm specimens showed considerable nonplane-strain effects in the form of shear lips and an initially high, then decreasing load-displacement test record, as had been seen in some of the tests for nonplane-

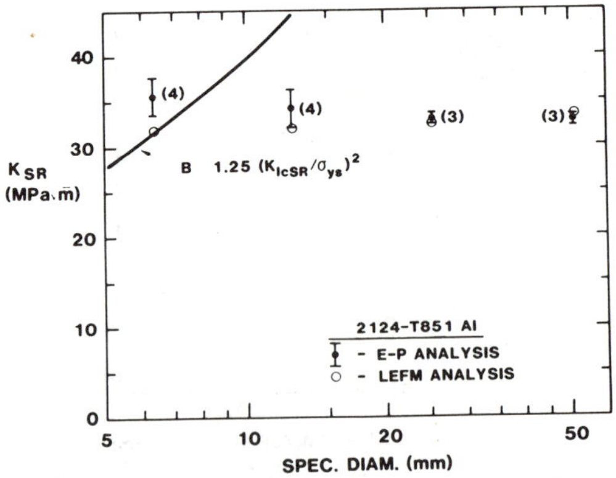

FIG. 5—*Size effect test results for 2124-T851 aluminum corrected for variation of toughness through the thickness of the parent plate. The solid curve is the suggested validity criterion resulting from this study.*

strain effects in Ref *33*. This occurred in spite of the fact that the chevron slot bottoms were machined to a point with a 90° included angle. The 12.7-mm specimens also showed the same effect, but to a lesser degree. The 25.4-mm specimens showed good plane-strain constraint because the slot thickness to specimen diameter ratio was only half as large as in the 12.7-mm specimens.

Since sufficient material was left of the original plate, the 50.8 and 12.7-mm specimens were re-machined, this time with the 60° or sharper included angles at the slot bottoms as used in Ref *33*. The test records for these specimens showed acceptable plane-strain constraint, that is, no shear lips and a load-displacement curve similar to that of very brittle materials. This confirmed that 60° pointed slot bottoms are satisfactory, while 90° pointed slot bottoms are not. The results of the size effect tests are shown in Fig. 6, where good specimen size independence is evident for the specimen diameters greater than $1.25(K_{IcSR}/\sigma_{ys})^2$. Because of excessively large plasticities, that is, because the specimens bent open without much crack growth, both the LEFM and EP treatments of the data from the 6.35-mm specimens were obviously invalid. The EP data for these specimens are off the top of the graph in Fig. 6.

7475-T7351 Aluminum

The size effect test results are shown in Fig. 7. Good specimen size independence is shown for specimen diameters greater than $1.25(K_{IcSR}/\sigma_{ys})^2$ for both the elastic-plastic and LEFM analyses. The smallest size specimens gave obviously invalid results by bending open without much crack growth. The EP data for these specimens are off the top of the graph in Fig. 7.

There are two factors which undoubtedly contributed to the high elastic-plastic toughness values obtained for the 12.7-mm 7475-T7351 aluminum specimens of Fig. 7, in addition to the fact that they fall well to the left of the validity line. The first is that the cracks in the 12.7-mm specimens always tunnelled more

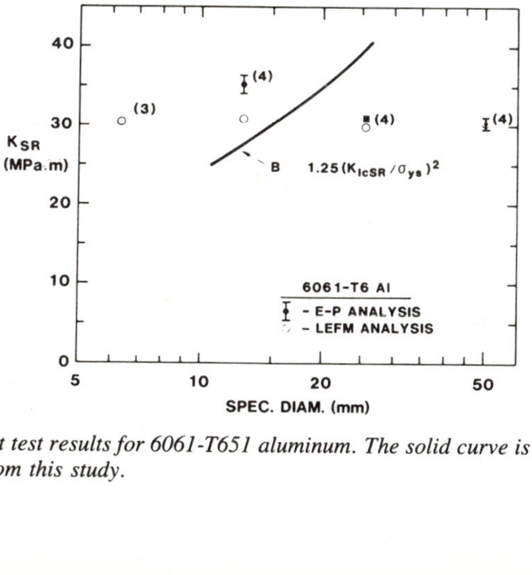

FIG. 6—*Size effect test results for 6061-T651 aluminum. The solid curve is the suggested validity criterion resulting from this study.*

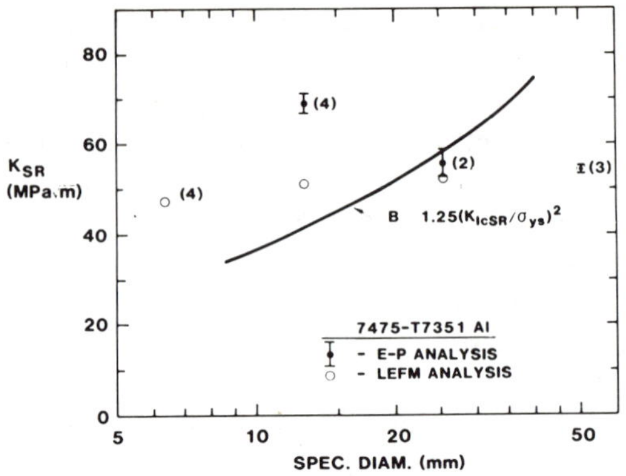

FIG. 7—*Size effect test results for 7475-T7351 aluminum. The solid curve is the suggested validity criterion resulting from this study.*

into one side of the specimen than the other, instead of following the intended crack plane. The cracks remained planar in the larger specimens. Thus, the 12.7-mm specimens created a larger crack surface area in proportion to their size, which probably required an increased load, thus increasing the apparent toughness over what it would have been for a planar crack in both the elastic-plastic and LEFM analyses. The second contributing factor to the high-toughness values concerned the chevron slots. Their thickness was about 3% of the specimen diameter, and, due to an error in machining, the slot bottoms were rounded instead of the recommended pointed geometry for 3% slot thickness (Fig. 2). It was found in Ref *33* that the nonplane-strain constraint associated with this slot configuration causes a 2 to 5% increase in the apparent toughness. This effect also influenced both the elastic-plastic and the LEFM results. Without these effects, therefore, the 12.7-mm data points would be somewhat lower, which would cause the elastic-plastic data to show less size dependence, but the LEFM data to show more size dependence.

4340 Steel

Size effect test results are shown in Fig. 8. Although the material gave a rather large data scatter, the average elastic-plastic toughness was nearly the same for all three sizes tested. The LEFM data show some decrease in toughness with decreasing specimen size, even to the right of the validity line. Only one specimen of the smallest size (6.35 mm diameter) was tested, but it gave a plasticity factor, p, which was much too large (about 0.6), thus invalidating the test. The LEFM value for the 6.35-mm test is about 20% too small, as can be seen from Fig. 8.

17-4 Stainless Steel

The size effect test results are shown in Fig. 9. The larger specimens had a small crack jump behavior, whereas the 6.35-mm specimens and one of the

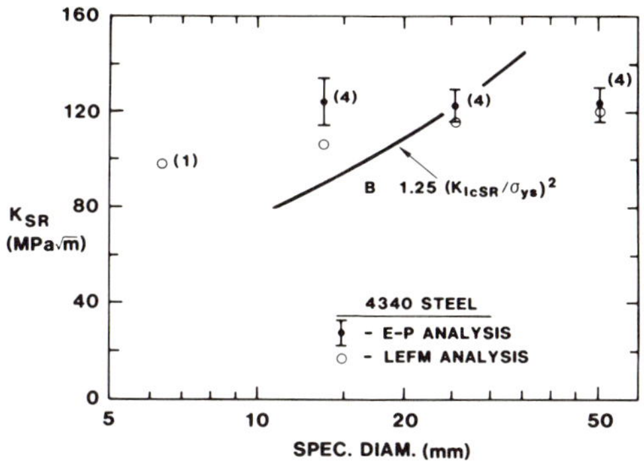

FIG. 8—*Size effect test results for 4340 steel. The solid curve is the suggested validity criterion resulting from this study.*

12.7-mm specimens were classified as smooth crack growth cases. Since the theory for the EP analysis is based on the assumption of smooth crack growth, only the LEFM analysis was done for the larger specimens. The results seem to be reasonably independent of specimen size and give no clear indication of an increase or a decrease in measured toughness with decreasing specimen size. However, as with all the other tests, the EP analysis gives higher toughness measurements than the LEFM analysis at small specimen sizes.

Ti-6A1-4V Titanium

The size effect test results are shown in Fig. 10. All four specimen sizes lie well to the right of the validity line, and little difference was found between the EP and the LEFM analyses of the data. Negative *p*-values, due to residual stresses in the largest test specimens [*40*], caused the EP-values to be smaller than the LEFM results.

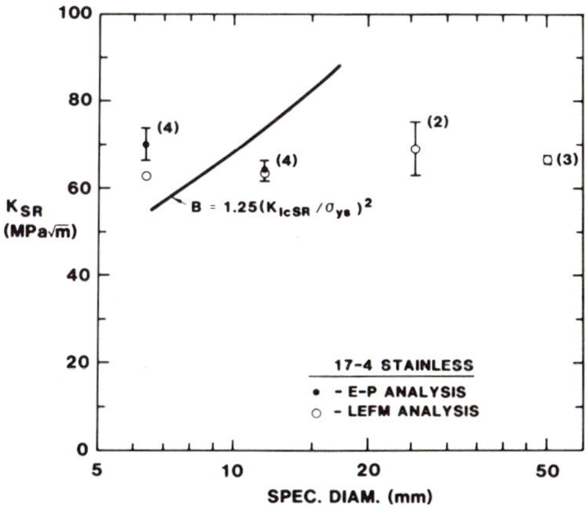

FIG. 9—*Size effect test results for 17-4 stainless. The solid curve is the suggested validity criterion resulting from this study.*

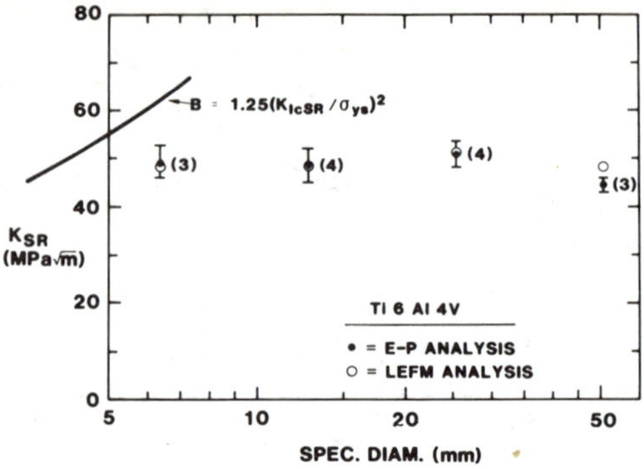

FIG. 10—*Size effect test results for Ti-6A1-4V. The solid curve is the suggested validity criterion resulting from this study.*

Discussion

Based on the tests of 6061-T6 aluminum reported in Ref *24*, it was expected that a specimen size validity criterion analogous to that in ASTM E 399 would apply to short-rod fracture toughness testing, and that the size requirement would be of the order of $B \geqslant 1.0(K_{IcSR}/\sigma_{ys})^2$. Since the data from the Ti-6A1-4V tests do not span this specimen size region, they do not contribute much to the determination of a minimum acceptable specimen size criterion.

The five remaining materials did provide data on both sides of the expected minimum size criterion. The *EP analyses* of the three aluminums begin to show an apparent increase in toughness at the smallest specimen sizes, whereas the 4340 steel and 17-4 stainless steel results seem to indicate no size effect down to the smallest size tested. The *LEFM analyses* of the three aluminums show remarkably good size independence. Only the smallest specimens of the 7475-T7351 alloy show a significant deviation from the large-specimen toughness, and the deviation is toward a conservative (smaller) toughness value. The LEFM results for the 17-4 stainless also seem to be size independent well to the left of the tentative validity line, but the 4340 data to the left of the validity line deviate significantly toward conservative values.

If the suggested test validity criterion given by Eq 2 were written into a standard short-rod test method, all of the measurements which fall to the left of the curves in Figs. 5 through 10 would be discarded because of possibly nonnegligible specimen size effects on the test results. This would leave only data which are essentially independent of the specimen size. The coefficient 1.25 in Eq 2 is exactly half of the coefficient used in the ASTM E 399 standard for determining the minimum ASTM specimen thickness. A short-rod specimen which just satisfies Eq 2 has only 3% of the material of a minimum size ASTM compact toughness specimen.

The question naturally arises as to why the EP analysis often begins to give toughness values which are too high as soon as the plasticity, p, becomes nonnegligible. In the derivation of the elastic-plastic data treatment, it is assumed that the work per unit area required to create the new crack surfaces remains constant as the specimen size is decreased. It may be that for some materials this assumption becomes invalid as soon as p becomes nonnegligible.

A plot of the plasticity, p, as a function of the specimen size scaled by the material's plastic zone size parameter, $(K_{IcSR}/\sigma_{ys})^2$ appears in Fig. 11, which shows the data of the present size effect study for which $p > 0.01$. Each dot on the graph represents the average of the several measurements taken on a given material and specimen size. It is apparent that the data are well-described by a straight line on the log-log plot. The equation of the line is

$$p = C\beta^n \tag{3}$$

where

$\beta = B/(K_{IcSR}/\sigma_{ys})^2$,
$C = 0.075$, and
$n = -1.6$.

This relation between p and $B/(K_{IcSR}/\sigma_{ys})^2$ can be used to estimate the value of K_{IcSR} and to calculate the specimen size required for a valid test if an invalid result has been obtained because the specimen was too small.

The reasonably well-defined relation between p and $B/(K_{IcSR}/\sigma_{ys})^2$ can be adversely affected by macroscopic residual stresses in the test specimen [*40*]. The residual stress contribution to the value of p in a test could not be expected to bear any relation to $B/(K_{IcSR}/\sigma_{ys})^2$, and residual stresses would therefore produce some scatter in the relation shown in Fig. 11.

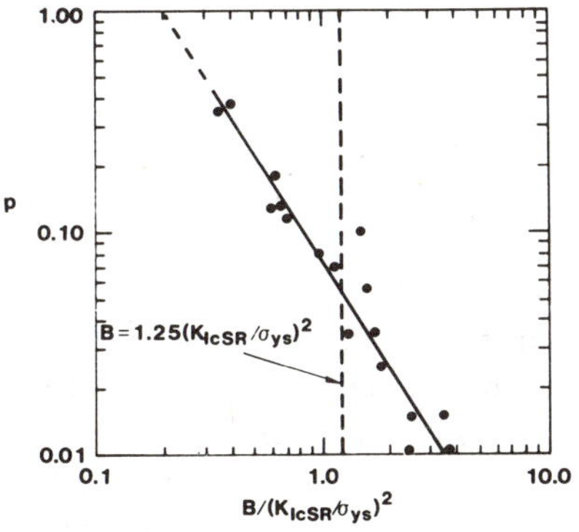

FIG. 11—*Variation of the plasticity,* p, *as a function of the scaled specimen size. Each dot represents the average of several tests on a given material using a given specimen size. The solid line is a best fit of the data. The dashed line is the suggested specimen size validity criterion resulting from this study.*

The present size effect test results are characterized by the following features:

1. According to the EP analysis, the specimens which were too small to satisfy the validity criterion (Eq 2) always showed a toughness *as large or larger* than the valid-specimen toughness.
2. According to the LEFM analysis, the specimens which were too small to satisfy the validity criterion (Eq 2) always showed a toughness *as small or smaller* than the valid-specimen toughness.

These observations are consistent with previously reported data [*24,61*]. They are also consistent with the theoretically expected small-specimen limits of the EP and LEFM analyses. Thus, if the value of p is due primarily to plasticity as opposed to residual stress effects [*40*], it appears that short-rod tests can provide both upper and lower bounds on the toughness, even when the specimen is too small for a valid measurement. The upper limit is found by using the p-factor in Eq 1, and the lower limit is found by the same equation with p set equal to zero. This is much better information than simply a statement that the test result is invalid.

Although the p-values due to plasticity apparently cannot be very large in a valid toughness test, the measurement of p should nevertheless be mandatory because of the additional information which such a measurement provides. First, if p is small, a valid test is indicated. Second, if p is too large but falls close to the data of Fig. 11, the test is invalid, but upper and lower bounds can be placed on the toughness, as discussed previously. Third, if p does not lie reasonably close to the data of Fig. 11, the value of p is probably influenced by the presence of macroscopic residual stresses in the test specimen. In this case, since the toughness equation which corrects for macroscopic residual stress effects is identical [*40*] to Eq 1, the short-rod test may provide both a valid toughness measurement and a warning of residual stresses in the test specimen.

It should be noted that all six materials of this paper had a smooth crack growth behavior, except for the small crack jumps in the larger 17-4 stainless specimens. However, a number of materials do display a marked crack jump behavior [*34*]. Inasmuch as the elastic-plastic theory is based on the assumption of smooth crack growth, upper and lower bounds on the toughness from EP and LEFM analyses may not be available in tests of crack-jump materials, since the EP analysis is invalid from the start. Also, because of the basically different test behavior of crack jump materials, it is not obvious that the validity line deduced from smooth crack growth materials will also apply to crack jump materials. Nevertheless, from prior tests on crack jump materials [*61*], and in lieu of further information, and for the sake of consistency, it appears advisable to apply the $B \geqslant 1.25(K_{IcSR}/\sigma_{ys})^2$ criterion to crack jump materials as well, as least for the present.

An unexpected but important finding to emerge from this study was that the slot bottoms with included angles of 90° did *not* provide good plane-strain constraint, whereas those with 60° included angles did. Thus, for valid tests of short-rod specimens, the importance of proper design of the specimen slot bottoms (Fig. 2) cannot be overemphasized.

Conclusions

1. Two unloading-reloading cycles should be performed in each short-rod fracture toughness test for crack length determination and for measurement of plasticity or residual stress effects or both. The unloading-reloading cycles are important test diagnostics for improving accuracy.

2. As a result of this series of size effect tests, a minimum short-rod specimen size criterion of $B \geqslant 1.25(K_{IcSR}/\sigma_{ys})^2$ is suggested. Short-rod specimens of this size have only 3% of the mass of a minimum-size ASTM specimen.

3. When a short-rod specimen is smaller than the minimum size criterion, it appears that upper and lower bounds can still be found for K_{IcSR} by the EP and the LEFM analysis techniques, respectively, provided that residual stresses do not contribute significantly to the value of p. The upper bound is calculated using the measured value of p in Eq 1, and the lower bound is calculated by arbitrarily setting p equal to zero in Eq 1.

4. Nonplane-strain effects can become a source of error in short-rod/short-bar testing if the chevron slots are not sufficiently thin or if the slot bottoms are not sufficiently sharp.

Acknowledgment

The author is indebted to Randy Guest for his contributions in the experimental portion of this work and for helpful consultations.

This work was supported in part by the National Science Foundation under NSF Award No. DAR-7820993, and in part by the U.S. Department of Energy under Contract DE-AC04-76-DP00789. Sandia National Laboratories is a U.S. Department of Energy Facility.

References

[1] Tattersall, H. G. and Tappin, G., *Journal of Materials Science,* Vol. 1, 1966, pp. 296–301.
[2] Nakayama, J., *Japan Journal of Applied Physics*, Vol. 3, 1964, p. 422.
[3] Nakayama, J., *Journal of the American Ceramic Society,* Vol. 48, 1965, p. 583.
[4] Davidge, R. W. and Tappin, G., *Journal of Materials Science*, Vol. 3, 1968, p. 165.
[5] Harris, B. and DeFarran, E. M., *Journal of Materials Science*, Vol. 4, 1969, pp. 1023–1026.
[6] Coppola, J. A. and Bradt, R. C., *Journal of the American Ceramic Society*, Vol. 55, 1972, pp. 455–60.
[7] Kuszyk, J. A. and Bradt, R. C., *Journal of the American Ceramic Society*, Vol. 56, 1973, pp. 420–23.
[8] Lynch, J. F. and Bradt, R. C., *Journal of the American Ceramic Society*, Vol. 56, 1973, pp. 228–29.
[9] Kessler, J. B., Ritter, J. E., Jr., and Rice, R. W. in *Surfaces and Interfaces of Glass and Interfaces of Glass and Ceramics*, V. D. Frechette, W. C. Lacourse, and V. L. Burdick, Eds., Plenum, New York, 1974, pp. 529–544.
[10] Simpson, L. A., *Journal of the American Ceramic Society*, Vol. 56, 1973, p. 7.

[*11*] Bluhm, J. I., *Engineering Fracture Mechanics*, Vol. 7, 1975, pp. 593–604.
[*12*] Bluhm, J. I. in *Fracture 1977*, Vol. 3, pp. 409–417, D. M. R. Taplin, Ed., University of Waterloo Press, Waterloo, Ont. Canada, 1977.
[*13*] Pook, L. P., *International Journal of Fracture Mechanics*, Vol. 8, 1972, p. 103.
[*14*] Barker, L. M., *Engineering Fracture Mechanics*, Vol. 9, 1977, p. 361.
[*15*] Barker, L. M. in *Proceedings of the ICM-II Conference*, Boston, 1976, p. 1547.
[*16*] Barker, L. M. and Leslie, W. C., *Fracture 1977*, Vol. 2, D. M. R. Taplin, Ed., Waterloo, Canada, 1977, p. 305.
[*17*] Barker, L. M. in *Fracture Mechanics of Ceramics*, Vol. 3, R. C. Bradt, D. P. H. Hasselman, and F. F. Lange, Eds., Plenum, New York, 1978, p. 483.
[*18*] Koblitz, F. F., Luna, V. R., Glenn, J. F., and DeVries, K. L., *Organic Coatings and Plastics Chemistry*, Vol. 38, 1978, p. 322.
[*19*] Farb, J. E., "The Fracture Toughness of Cobalt Cemented Tungsten Carbides," M.S. thesis, Department of Materials Science and Engineering, University of Utah, Salt Lake City, 1978.
[*20*] Smyth, R. R., Bensten, L., and Youngblood, G. E., "Corrosion Effects in a Fused-Cast $Mg0{:}MgA1_20_4$ Refractory," 81st Annual Meeting of the American Ceramic Society, Cincinnati, April 1979.
[*21*] Marion, R. and Cuthrell, R. E., "Effect of Various Liquids on Crack Propagation in Glass," 81st Annual Meeting of the American Ceramic Society, Cincinnati, April 1979.
[*22*] Drake, E. F. and Krawitz, A. D., "Fatigue Damage and Crack Propagation in a WC-Cemented Carbide Composition," 7th NSF Hard Materials Workshop, Illinois Institute of Technology, June 1979.
[*23*] Koblitz, F. F., Luna, V. R., Glenn, J. F., De Vries, K. L., and Draughn, R. A., *Polymer Engineering Science*, Vol. 19, 1979, p. 607.
[*24*] Barker, L. M., *International Journal of Fracture*, Vol. 15, 1979, pp. 515–536.
[*25*] Barker, L. M. in *Fracture Mechanics Applied to Brittle Materials, ASTM STP 678*, American Society for Testing and Materials, Philadelphia, 1979, pp. 73–82.
[*26*] Guest, R. V. in *Proceedings Third Engineering Mechanics Specialty Conference*, American Society of Civil Engineers, 1979, pp. 413–416.
[*27*] Munz, D., Bubsey, R. T., and Shannon, J. L., Jr., *Journal of Testing and Evaluation*, Vol. 8, 1980, pp. 103–107.
[*28*] Munz, D., Bubsey, R. T., and Srawley, J. E., *International Journal of Fracture*, Vol. 16, 1980, pp. 359–374.
[*29*] Beech, J. F. and Ingraffea, A. R., *International Journal of Fracture*, Vol. 18, No. 3, March 1982, pp. 217–229.
[*30*] Brown, K. R., "Evaluation of the Resource Enterprises (Fractometer II) Fracture Toughness Test on Aluminum Alloys," Report No. ZCFT-80-25-KRB, Kaiser Aluminum and Chemical Corp., Pleasanton, Calif., 1980.
[*31*] Bubsey, R. T., Munz, D., Pierce, W. S., and Shannon, J. L., Jr., *International Journal of Fracture*, Vol. 18, Feb. 1982, pp. 125–133.
[*32*] Barker, L. M. and Baratta, F. I., *Journal of Testing and Evaluation*, Vol. 8, 1980, pp. 97–102.
[*33*] Barker, L. M. in *Fracture Mechanics, ASTM STP 743*, American Society for Testing and Materials, 1981, pp. 456–475.
[*34*] Barker, L. M., "Data Analysis Methods for Short Rod and Short Bar Fracture Toughness Tests of Metallic Materials," Terra Tek Report TR 80-12, 1980.
[*35*] Cutler, R. A., Goodman, E. C., Guest, R. V., Hendrickson, R. R., and Leslie, W. C., "Elevated Temperature Fracture Toughness and Fatigue Testing of Steels for Geothermal Applications," Terra Tek Report TR 80-87, 1980.
[*36*] Guest, R. V., "Fracture Toughness Testing Using Short Rod Specimens, Illustrated by Controlled Temperature Tests on M-2 High Speed Tool Steel," Symposium on New Developments in the Processing and Properties of High Speed Tool Steels, AIME Annual Meeting, Las Vegas, Nev., Feb. 1980.
[*37*] Shannon, J. L., Jr., Bubsey, R. T., Munz, D., and Pierce, W. S. in *Advances in Fracture Research ICF5*, D. Francois, Ed., Vol. 2, 1981, p. 1127.
[*38*] Costin, L.S. and Wawersik, W. R., "Creep Healing of Fractures in Rock Salt," Report SAND-80-0392, Sandia National Laboratories, 1980.

[39] Costin, L. S. in *Fracture Mechanics for Ceramics, Rocks, and Concrete, ASTM STP 745*, S. W. Freiman and E. R. Fuller, Jr., Eds., American Society for Testing and Materials, 1981, p. 169.
[40] Barker, L. M. in *Advances in Fracture Research ICF5*, D. Francois, Ed., Vol. 5, 1981, p. 2563.
[41] Barker, L. M. in *Wear and Fracture Prevention*, Proceedings of an ASM conference held May 21–22, 1980, Peoria, Ill., American Society for Metals, Metals Park, Ohio, 1981.
[42] Barker, L. M., *Engineering Fracture Mechanics*, Vol. 17, No. 4, 1983, pp. 289–312.
[43] Brown, K. R., "An Evaluation of the Chevron Notched Short-Bar Specimen for Fracture Toughness Testing in Aluminum Alloys," Kaiser Aluminum and Chemical Corporation, to be published.
[44] Munz, D., *Engineering Fracture Mechanics*, Vol. 15, 1981, p. 231.
[45] Coyle, R. T. and McFadden, J. D. O., "Stress Corrosion of a High-Transmittance Glass for Solar Applications," Proceedings of the Electrochemical Society Symposia on Materials for Batteries and Solar Energy, held in Denver, Colo., Oct. 1981.
[46] Brown, K. R. in *Strength of Metals and Alloys*, R. C. Gifkins, Ed., Vol. 2, Pergamon Press, New York, 1982, pp. 765–771.
[47] Shannon, J. L., Jr., Bubsey, R. T., Pierce, W. S., and Munz, D., *International Journal of Fracture*, Vol. 19, No. 3, July 1982, pp. R55–R58.
[48] Cutler, R. A. and Leslie, W. C., *Journal of Testing and Evaluation*, Vol. 11, Jan. 1983, pp. 3–15.
[49] Shih, T. T. and Opoku, J., *Engineering Fracture Mechanics*, Vol. 12, 1979, p. 479.
[50] Munz, D., Bubsey, R. T., and Shannon, J. L., Jr., *Journal of the American Ceramic Society*, Vol. 63, 1980, p. 300.
[51] Munz, D., Shannon, J. L., Jr., and Bubsey, R. T., "Fracture Toughness Calculation from Maximum Load in Four Point Bend Tests of Chevron Notch Specimens," to be published in *International Journal of Fracture*.
[52] Shih, T. T., *Engineering Fracture Mechanics*, Vol. 14, 1981, p. 821.
[53] Shih, T. T., *Journal of Testing and Evaluation*, Vol. 9, 1981, p. 50.
[54] Munz, D., Himsolt, G., and Eschweiler, J. in *Fracture Mechanics for Ceramics, Rock, and Concrete, ASTM STP 745*, S. W. Freiman and E. R. Fuller, Jr., Eds., American Society for Testing and Materials, 1981, pp. 69–84.
[55] Wu Shang-Xian, *Engineering Fracture Mechanics*, Vol. 19, 1984, p. 221.
[56] Wu Shang-Xian, "An Investigation of Chevron Notch Three Point Bend Fracture Toughness Specimen," *International Journal of Fracture*, to be published.
[57] Barker, L. M. and Guest, R. V., U. S. Patent No. 4,198,870, 22 April 1980.
[58] ASTM Task Group E24.01.04, Dr. Kevin Brown, chairman.
[59] DeJong, H. F., "Thickness Direction Inhomogeneity of Mechanical Properties and Fracture Toughness as Observed in Aluminum 7075-T651 Plate Material," *Engineering Fracture Mechanics*, Vol. 13, 1980, pp. 175–192.
[60] Barker, L. M., "Fracture Toughness Measurements of Aluminum by the Short Rod/Short Bar Method," Terra Tek in-house report, April 1980.
[61] Barker, L. M., "Evaluation of a New Fracture Toughness Measuring Technique, and Adaptation of the Technique to Use Ultra-Small Specimens," Final Report, Contract No. DAAG 46-78-C-0040, Report TR-79-53, Army Materials and Mechanics Research Center, 1979.

R. Thomas Coyle[1] *and Marshall L. Buhl, Jr.*[1]

A Computer-Assisted Technique for Measuring K_I-V Relationships

REFERENCE: Coyle, R. T. and Buhl, M. L., Jr., "**A Computer-Assisted Technique for Measuring K_I-V Relationships,**" *Chevron-Notched Specimens: Testing and Stress Analysis, ASTM STP 855,* J. H. Underwood, S. W. Freiman, and F. I. Baratta, Eds., American Society for Testing and Materials, Philadelphia, 1984, pp. 134–151.

ABSTRACT: We developed a computer-assisted data collection technique for flatjack-loaded, short-rod specimens to readily obtain crack velocity V as a function of stress intensity factor K_I. The technique facilitates the study of stress corrosion in glass and ceramic materials and allows us to determine the crack velocity at which the critical stress intensity factor is evaluated. The experimental apparatus included (1) a stepper motor to automate pressurization of the flatjack and (2) a digital oscilloscope to acquire information about the flatjack pressure and sample displacement at preselected time intervals.

We conducted fracture experiments using samples of soda-lime-silicate float glass and alumino-borosilicate glass (Corning 7809 Solar Glass) in an environmental chamber at 25°C and 30% relative humidity. We found that the critical stress-intensity factor for float glass, extrapolated to a crack velocity of 0.1 m/s, was 0.77 MPa $m^{1/2}$ compared to 0.81 MPa $m^{1/2}$ for the solar glass. Comparison of K_I versus V data for the two glasses showed that the coefficient n in the empirical relationship $V = AK_I^n$ is 27.5 for the solar glass and 18 for float glass.

KEY WORDS: glass, fracture, stress corrosion, computer, fracture toughness, stress intensity, crack velocity, short-rod specimen

Nomenclature

A	Depth of short rod sample
A_F	Flatjack calibration constant, 8.26
B	Width of short rod sample
c	Relative compliance of sample
c_o	Initial relative compliance of sample
D	Relative displacement
$f(R)$	Correction factor ($K_I f(R) = L_S$)
K_I	Stress-intensity factor
K_{Ic}	Critical stress-intensity factor

[1]Senior materials scientist and senior computer scientist, respectively, Solar Energy Research Institute, Golden, Colo. 80401.

l Length of crack
l_o Initial slot length to "V" tip
L Length of short rod sample
L_S Sample load
L_{SC} Peak value in a fast fracture experiment
P Mercury pressure in flatjack
R Compliance ratio, c_o/c
V Crack velocity
θ ½ angle of "V" tip
ΔR Change in compliance ratio
Δt Time interval required for ΔR
Λ Dimensionless crack length (l/B)

Many current and projected solar energy applications will require the use of brittle materials. For instance, glass is used in mirrors to redirect solar radiation and in glazings and envelopes to transmit incoming solar radiation while blocking outgoing thermal energy. Brittle, high-temperature ceramic materials are required in photothermal receivers to convert concentrated solar radiation to high-temperature thermal energy. Ceramic materials may also be used in heat exchangers compatible with various high-temperature heat transfer fluids and in tanks for storing solar thermal energy as sensible heat in high-temperature liquids. However, the brittle nature of these materials makes them subject to fracture by the growth of cracks when they experience tensile stresses caused by thermal shock, thermal gradients, differential thermal expansion, and high pressures in the heat transfer fluids.

The purpose of this work was to develop a computer-assisted technique for measuring the fracture behavior of these brittle materials. We developed a technique for testing short-rod specimens [*1*] loaded with a flatjack [2] along with a data acquisition technique to evaluate fracture parameters from experimental data.

The experiments in this work compared the fracture behavior of a high solar transmittance-glass composition (Corning Glass Works Code 7809 aluminoborosilicate glass [*3,4*]) with that of soda-lime-silicate float glass, which is currently the most widely used composition of glass for mirrors and glazings. The fracture properties studied were the fracture toughness [*5*] or critical stress-intensity factor K_{Ic}, and the stress-corrosion susceptibility [*6–9*], which can be determined from stress intensity factor K_I versus crack velocity V information.

Experimental Procedure

Glass Preparation

The short-rod specimens were cut from 15 by 15 by 1.4-cm slabs of layered 7809 and float glass and from 30 by 30 by 1.3-cm slabs of unlayered float glass—1.3-cm-thick unlayered sheets of 7809 glass were unavailable for making

TABLE 1—*Composition of Corning 7809 and float glass.*

	Float	7809
SiO_2	72	66
Na_2O	14.3	9
K_2O	0.3	5
CaO	8.2	2
MgO	3.5	. . .
Al_2O_3	1.3	9
Fe_2O_3	0.1	0.02
SO_3	0.3	. . .
B_2O_3	. . .	8
TiO_2	. . .	0.5
As_2O_3	. . .	0.2

specimens. Table 1 gives an approximate composition for the commercial soda-lime-silicate float glass [*10*] and the composition for the 7809 glass [*4*]. In the layered slabs, we thermally fused five 15 by 15 by 0.3-cm sheets together, after cleaning them with pumice, rinsing, and drying them at 40°C. The glass stack was fused with carbon blocks on the top and bottom of the stack. We fused the 7809 glass at 740°C for 3 h and annealed it from 575 to 510°C at −2°C/h. The soda-lime-silicate float glass was fused at 685°C for 3 h and annealed from 560 to 490°C at −2°C/h.

Preparation of Short-Rod Specimens

We had the slabs cut into pieces with these specifications: $A = 11.05 \pm 0.04$ mm, $B = 12.70 \pm 0.05$ mm, $L = 19.05 \pm 0.075$ mm, as shown in Fig. 1. We used a diamond saw to slot pieces to make the short-rod fracture specimens[2]

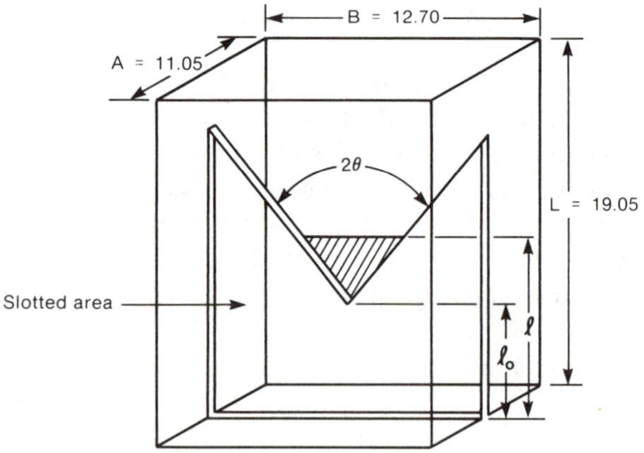

FIG. 1—*Dimensions of the short-rod fracture toughness specimen with rectangular cross section.*

[2]These rectangular samples are sometimes called short bars with the term short rod sometimes reserved for cylindrical samples, although the two types are mechanically equivalent [*11*].

with the V shown in the figure. The slots were cut to meet these specifications: $l_o = 6.35 \pm 0.07$ mm, $2\theta = 58°C \pm 0.5°C$, and the slot was centered within ± 0.025 mm. The layers of the samples were parallel to the slot cut, and the sample was designed so the slot was in the center of the third layer (away from the region where the adjacent sheets fused). Directly after cutting, we cleaned each sample ultrasonically in deionized water for 60 s and in acetone for 60 s followed by another ultrasonic cleaning cycle. The slotted short-rod specimens were then annealed according to the schedule indicated earlier. We then stored the samples immediately in a desiccator for at least one hour to ensure dryness before testing. Fused silica short-rod samples used for compliance ratio measurements were annealed at $-2°C/h$ from 1155 to 1050°C.

Computer-Assisted Fracture Apparatus

Figure 2 shows a schematic of the computer-assisted fracture apparatus. The apparatus consists of a thin bladder containing mercury, called a flatjack (supplied by TerraTek, Inc.) over which the V, cut in the sample, sets. This bladder uniformly loads the sample over the 12.7-mm sample width and to the height of 6.1 mm, near the V. We can manually increase or decrease the pressure of the mercury in the flatjack bladder or use a stepper motor with a micrometer-driven piston connected to the flatjack bladder by a mercury-filled pressure tube. A pressure transducer senses the bladder pressure, and a mouth opening displacement gage (supplied by TerraTek, Inc.) senses the sample displacement to within 50 nm. Using a digital oscilloscope, we took pressure and displacement data at preselected time intervals ranging from 5 μs to 200 s.

Since the temperature and relative humidity affect the stress corrosion susceptibility, these were controlled to 25 ±0.5°C and 30% ±5% relative humidity in the environmental chamber shown in Fig. 2. We monitored the temperature

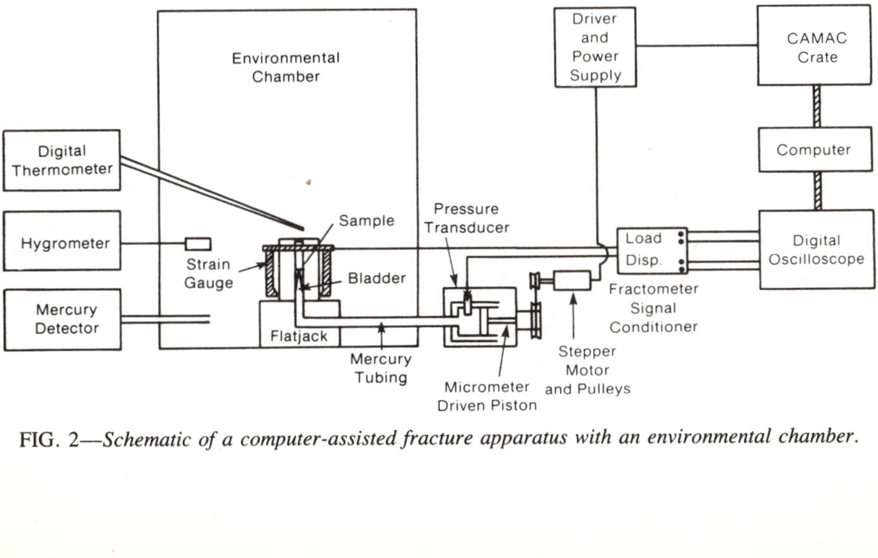

FIG. 2—*Schematic of a computer-assisted fracture apparatus with an environmental chamber.*

with a digital thermometer and the relative humidity with a hygrometer. We used a mercury detector to check for any leaks in the flatjack.

Computer Control and Data Acquisition Systems

A driver with a 24-V power supply runs the stepper motor in Fig. 2, which is activated by a pulse train generator in a CAMAC crate controlled by a Hewlett Packard (HP) 85 computer through a GP-IB crate controller. The original intent was to get pressure and displacement data through a multi-channel A/D converter in the CAMAC crate, but the HP 85 and GP-IB controller were too slow (one conversion every 0.1 s) without in-crate data storage. We used a two-channel digital storage oscilloscope with a GP-IB interface instead. The computer controlled the stepper motor and digital oscilloscope and analyzed the data stored in the oscilloscope. Computer accessories needed for the system included a 16 000-byte memory expansion chip, a matrix ROM, an advanced programming ROM, and a GP-IB interface board.

Sample Introduction Procedure

Before placing the sample on the flatjack, we coated it with a thin film of rubber adhesive where it contacts the strain gage. The adhesive prevents the strain gage from slipping on the sample. We always mounted the strain gage on the sample at its lowest possible position so we could reproduce the mounting position. We adjusted the position of the strain gage at the bottom of the sample by gently manipulating the strain gage with a needlepoint probe until we observed a minimum reading on the displacement output. Before each test, we zeroed and calibrated the fracture analysis apparatus. To establish zero gage pressure (atmospheric pressure) of the mercury in the flatjack, we opened an atmospheric vent. Using electronic calibration, we set the range of the pressure scale. The short-rod samples were stored in a desiccator until they were used. We then equilibrated them for at least 30 min in an environmental chamber at 25 ±0.5°C and 30 ±5% relative humidity before testing.

Computer Assisted Method for Determining K_I and V

We designed a computer program, FRACT, to enable the digital control and data acquisition equipment (shown in Fig. 2) to determine the fracture parameters K_I and V. This program allows one to choose between two types of experiments. One is a fast fracture experiment where crack velocities above 10^{-5} m/s are measured. The second is a slow fracture experiment where the crack velocity is below 10^{-4} m/s.

The logic flow diagram for FRACT is shown in Fig. 3. The program can automatically cause crack growth in a sample until the crack length is suitable for the start of an experiment—this is sometimes called the "pop-in" of the crack. It can also maintain any chosen K_I by sampling the flatjack pressure and

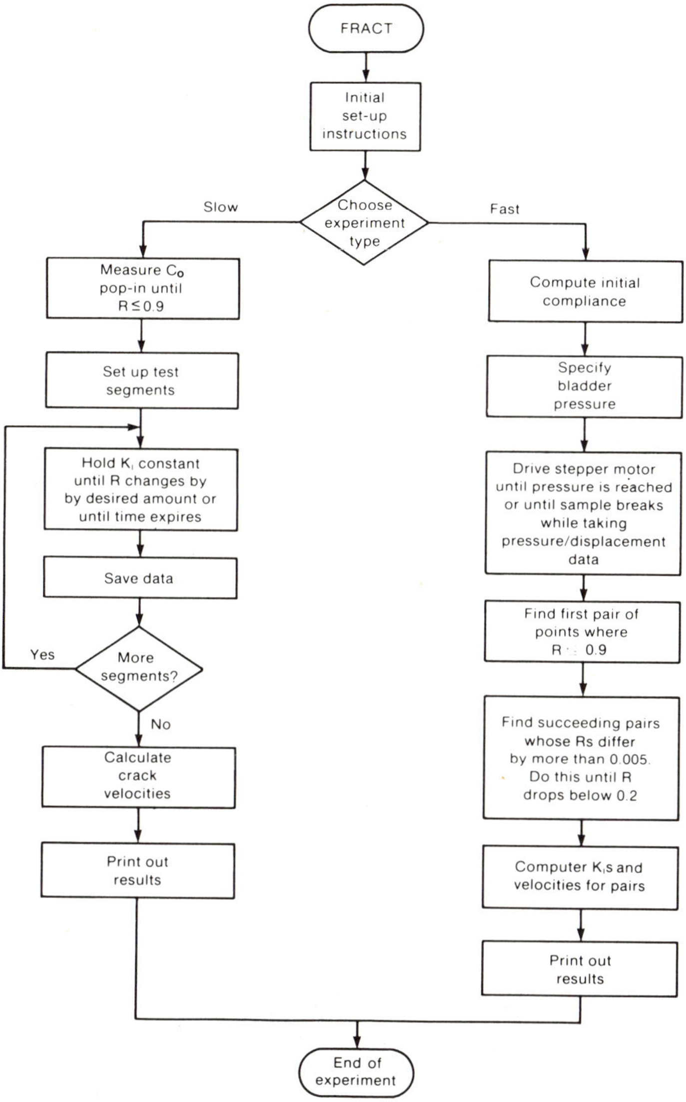

FIG. 3—*Logic flow diagram for the FRACT computer program.*

the crack length every second and then adjusting the pressure to maintain K_I constant to within about 0.5%. From information on the change in crack length with time, FRACT can calculate V.

Fast-Fracture Experiment

Typical load-displacement test records for computer-assisted, fast-fracture experiments are shown in Fig. 4 where an experiment was conducted to measure

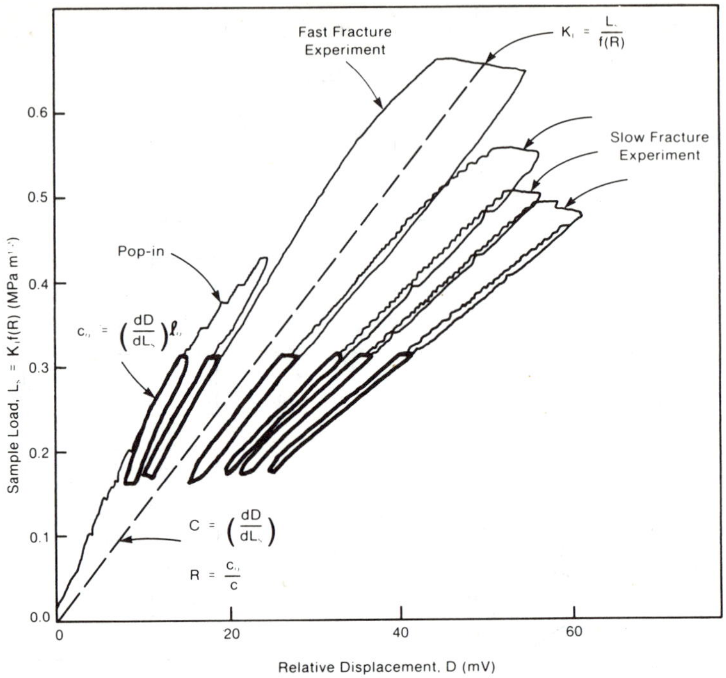

FIG. 4—*Test record for a layered 7809 sample tested with the digital control and data acquisition system. (The figure shows the measurement of* c_o, *popin, a fast-fracture experiment and three in a series of slow-fracture experiments. The dashed line is a hypothetical unloading-reloading curve.*)

the K_I-V relationship at large V and in Fig. 5 where we conducted an experiment to determine the fracture toughness K_{Ic}.

The strain gage measures in millivolts the relative displacement D of the sample, the abscissa in these figures. The sample load L_S, the ordinate in the figures, is expressed in terms of $K_I f(R)$ where K_I is the crack opening stress intensity factor [*12*], and $f(R)$ is a correction factor that depends on the compliance ratio R. The factor R is defined by the equation $R = c_o/c$ where c_o is the initial relative compliance of the sample. The reciprocal slope of the initial loading curve in the figures is c_o, where the slot length (see Fig. 1) is l_o before the crack has grown into the V of the sample. The relative compliance c increases as the crack length l (see Fig. 1) increases. We can determine a value for c from each point along the test record by drawing a line (such as the dashed line shown in the figures) back to the origin and measuring its reciprocal slope. This is the reciprocal slope we would obtain if the sample were unloaded and reloaded from that point on the test record. We determined the values of $f(R)$ by modifying the Barker and Guest [*13*] relationship obtained using the flatjack-loaded short-rod specimen—see the Results and Discussion section for more detail.

The fractometer electronically generates the value for the sample load L_s from the pressure transducer signal using the expression $L_s = A_F P B^{1/2}$ where P is the

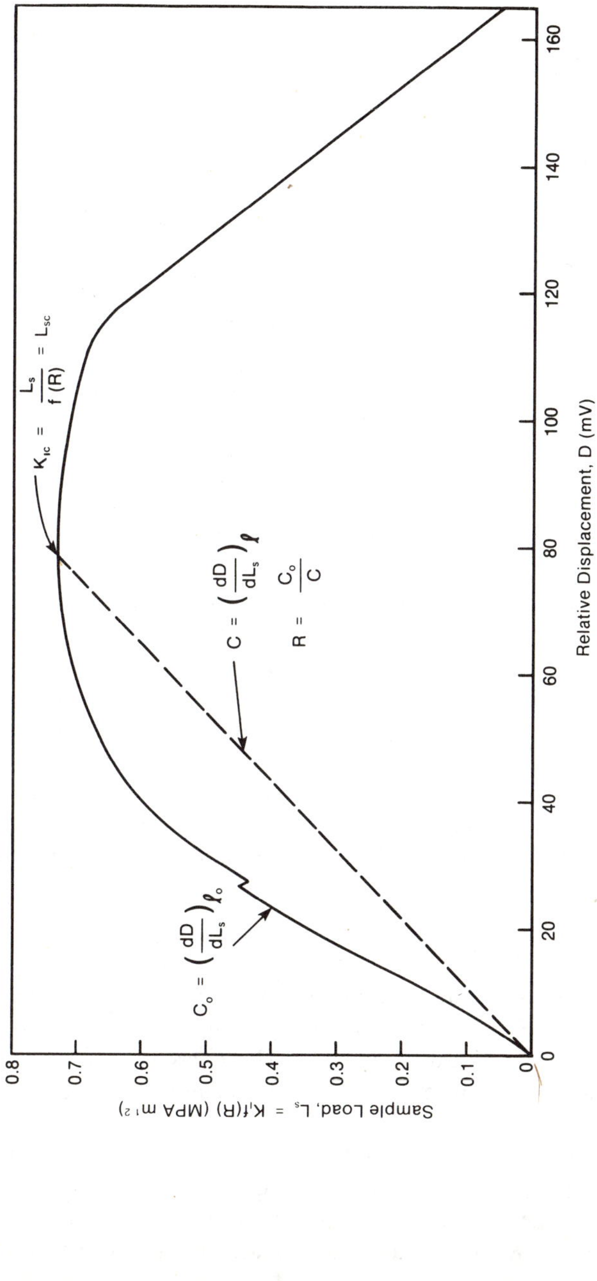

FIG. 5—*Typical fast-fracture test record for measuring* K_{Ic}. *(The figure shows the initial relative compliance* c_o *at crack length* l_o *and the relative compliance of crack length* l *for a hypothetical unloading and reloading curve, dashed line.)*

mercury pressure, B is the width of the sample (0.0127 m), and A_F is the calibration constant (8.26) for a flatjack-loaded, short-rod sample [*13*]. To determine a value for K_I at the crack tip of a sample, one divides the value of L_S on the test record by the appropriate value for $f(R)$.

The FRACT program generates K_I-V information in a fast fracture experiment as shown in Fig. 4, by first being used in the slow experiment mode to measure c_o and to achieve pop-in, and then by being restarted and branched into the fast mode. We determine c_o by calculating the least squares straight line through about 35 L_S and D points, one point every 50 ms, taken as L_S goes from 0.2 to 0.3 (see Fig. 4). This measurement of c_o is repeated any desired number of times to improve its accuracy, resulting in a heavy trace on the test record as seen in the figure. In the same manner c is determined after pop-in, and the starting value of R is calculated from c_o/c.

The sample is then loaded into the range of rapid V, while the data acquisition system takes L_S and D points at preselected intervals ranging from 1 ms to 0.2 s. At the conclusion of the fast-fracture experiment, FRACT searches for points that have compliance ratios between 0.2 to 0.9 and that are separated by a ΔR of 0.005 or more. For each pair of R-values, FRACT generates an average value of R and an average L_S, calculates $f(R)$, and uses the latter two to calculate K_I. The crack velocity V is determined from the relationship [*14*]

$$V = B(\Delta R)(d\Lambda/dR)/\Delta t \quad (1)$$

where ΔR is the change in compliance ratio, Δt is the time interval required for ΔR, and $d\Lambda/dR$ is the slope of the dimensionless crack length versus compliance ratio curve at R. The dimensionless crack length Λ is defined by l/B (see Fig. 1). We determined the values for the derivative from a compliance calibration done by Barker and Guest on fused silica using line loading [*15*]. We found their calibration to be sufficiently accurate for flatjack loading as seen in the Results and Discussion section. Only data derived from values of R within the range from 0.25 to 0.80 were considered valid in these experiments, as this is the range of validity for $f(R)$ and for the dimensionless crack length derivative [*12,14*].

Slow-Fracture Experiments

Both slow- and fast-fracture experiments can be done on the same specimen as shown in Fig. 4. FRACT can successively run up to 20 segments—three are shown in Fig. 4—each with a separate K_I. The pressure in the flatjack is automatically adjusted, as seen in the third slow-fracture segment, to maintain K_I constant as the crack grows in the specimen. Each segment is continued until a prechosen time allotment has expired or the value of R has changed by the amount preselected for that segment. The value of R is determined and printed out at chosen time intervals during a segment. After the segments have all been completed, values for K_I and the associated V are printed out for each segment.

In the slow- and fast-fracture experiments, the program will detect a specimen that has fractured completely. When this occurs during the experiment, the mercury gage pressure is reduced to zero, and the results up to that point are printed out.

Results and Discussion

Correction Factor f(R) *for a Flatjack-Loaded Short Rod*

We use $f(R)$ as a necessary calibration in deriving the stress-intensity factor K_I from the sample load L_S and the relative displacement D in a flatjack-loaded, short-rod specimen. We used the digital control and data acquisition system to verify the $f(R)$ expression derived from the Barker and Guest work [*13*] for this specimen geometry and method of loading. Their expression is

$$(1 + s)L_{SC}^2 - (L_S + 2rs/c_o D)L_{SC} + r^2 s/c_o^2 D^2 = 0 \qquad (2)$$

where $r = 0.5$, $s = -0.6$, and L_{SC} is the peak value in a fast-fracture experiment where the crack velocity is maintained at a high value (approximately 2×10^{-2} m/s) throughout the fracture process so K_{Ic} can be determined. For the short-rod geometry, L_{SC} is equal to K_{Ic} [*1*]. This is shown at the peak of the test record in Fig. 5.

The value of R at any hypothetical unloading-reloading curve, such as the dashed line in the figure, can be written $R = L_S c_o/D$. Substituting this expression and the values of r and s into Eq 1 and multiplying the resulting expression by 20 gives

$$8\, L_{SC}^2 - 20\, L_S L_{SC} + 12\, L_{SC} L_S/R - 3\, L_S^2/R^2 = 0 \qquad (3)$$

Since the experiments used in generating Eq 2 were done as in Fig. 5 where the critical stress-intensity factor K_{Ic} was applied throughout the fracture process, each value of L_S along the top of the test record must yield a value of K_{Ic} by being divided by a factor f that varies with R. The relationship is $K_{Ic} = L_{SC} = L_S/f(R)$. Substituting this into Eq 3 yields

$$3\, f^2 + R(20\, R - 12)f - 8\, R^2 = 0 \qquad (4)$$

The solution of this quadratic gives the relationship

$$f(R) = R/6\, [((20\, R - 12)^2 + 96)^{1/2} - (20\, R - 12)] \qquad (5)$$

This equation was used in FRACT to calculate K_I from L_S and R, or to load a specimen to the proper L_S to achieve a desired K_I.

The results of experiments to verify Eq 5 are shown in Fig. 6, where log V is plotted versus R. In these experiments, FRACT was programmed to maintain a constant K_I over a wide range of R using Eq 5 and to measure crack velocities periodically across the range. If Eq 5 is accurate, V should be independent of R. Figure 6 shows that for the experiment with $K_I = 0.482$ MPa m$^{1/2}$, V increases by a factor of two over the range of R from 0.85 to 0.3. A similar line fits the experiment with $K_I = 0.503$ MPa m$^{1/2}$ over the range of 0.8 to 0.4, although the points at $R < 0.4$ indicate that the variation in V may be less than a factor of two.

The crack velocity for these experiments was between 10^{-6} m/s and 10^{-7} m/s. As seen by inspecting the fracture data in Fig. 7, the scatter in V is typically a factor of 5 to 10 in the vicinity of $V = 10^{-5}$ m/s to 10^{-8} m/s. Thus, we did not attempt to correct Eq 5 in this work since we only expected minor reductions in the scatter of the data and since the required correction was not clearly enough established by the data in Fig. 6. This work indicates that the error in V caused by inaccuracy in Eq 5 is small compared to the typical scatter in the K_I versus V data.

Λ(R) *for a Flatjack Loaded Short Rod*

To determine V using the technique in this work requires a knowledge of the dimensionless crack length Λ as a function of the compliance ratio R. This yields the derivative $d\Lambda/dR$ used in calculating V as discussed earlier. Previous work [*6*] used the compliance calibration of Barker and Guest [*15*] for this derivative.

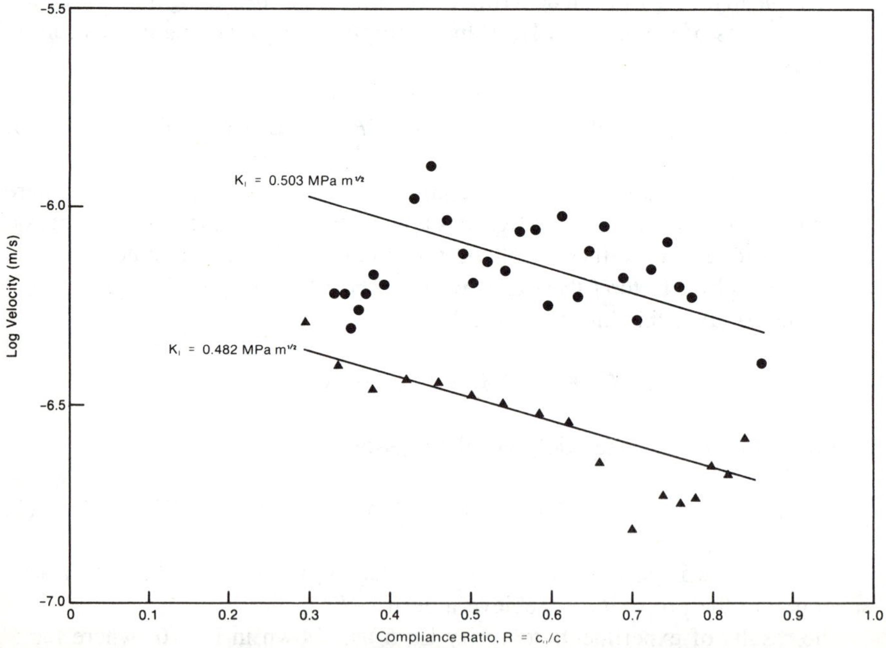

FIG. 6—*Crack velocity versus compliance ratio at* K_I = *0.503 and* K_I = 0.482.

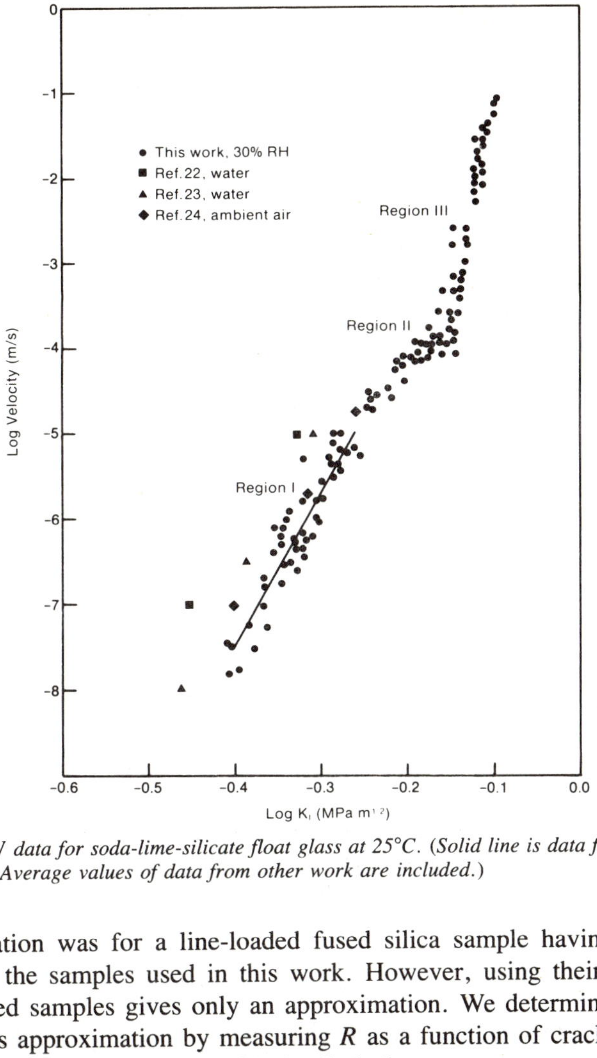

FIG. 7—K_I-V *data for soda-lime-silicate float glass at 25°C. (Solid line is data from Coyle and McFadden* [6]. *Average values of data from other work are included.)*

Their calibration was for a line-loaded fused silica sample having the same geometry as the samples used in this work. However, using their results for flatjack-loaded samples gives only an approximation. We determined the suitability of this approximation by measuring R as a function of crack length for a series of glass samples loaded with the flatjack.

Figure 8 shows the comparison of $\Lambda(R)$ between flatjack- and line-loaded samples. The line-loaded results of Barker and Guest fall below the results determined in this work for the flatjack-loaded samples, but the values nearly merge at higher values for the dimensionless crack length Λ. The difference in the derivative $d\Lambda/dR$ for the two types of loading is insignificant at the lower values of R and is less than 10% at $R = 0.8$, the highest value of R used in this work. The 10% error in $d\Lambda/dR$ leads to a 10% error in V as seen in Eq 1. This is insignificant in view of the factor of 5 to 10 scatter in V that is encountered in fracture experiments.

Also, note that the flatjack-loaded results shown in Fig. 8 include layered and

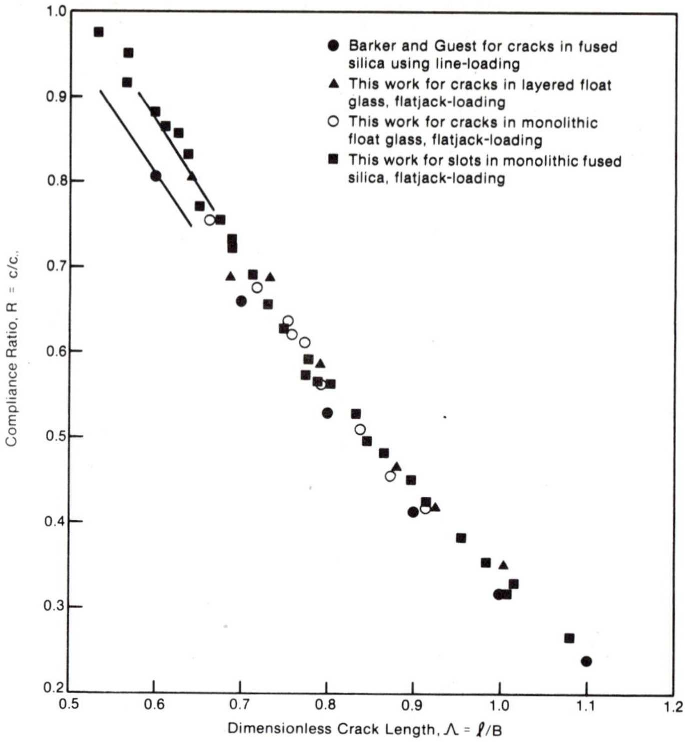

FIG. 8—*Crack length versus compliance ratio data for flatjack and line-loaded glass samples.*

monolithic samples where we measured crack arrest lines and compliance ratios as well as monolithic samples that had cracks simulated by a 0.38-mm-width saw cut. The $\Lambda(R)$ information for these various types of specimens and cracks was in good agreement.

Crack Velocity Versus Stress-Intensity Factor Measurements

We measured V as a function of K_I using the digital control and data acquisition system and used specimens made of layered soda-lime-silicate float glass and layered 7809 alumino-borosilicate glass. Figure 7 shows the results for float glass and Fig. 9 shows the results for 7809 glass, where the dashed line in Fig. 9 shows the average values for the data in Fig. 7. The results in these figures are presented as log V versus log K_I, and the solid lines in the two figures are the results of previous measurements for the two glasses using a manually operated flatjack-loaded, short-rod technique.

Figures 7 and 9 show that data obtained in this work on both glasses compares well with data obtained in Coyle and McFadden for Region I [*16*] of the K_I-V diagram. Table 2 shows the results of fitting the Region I data in the figures to

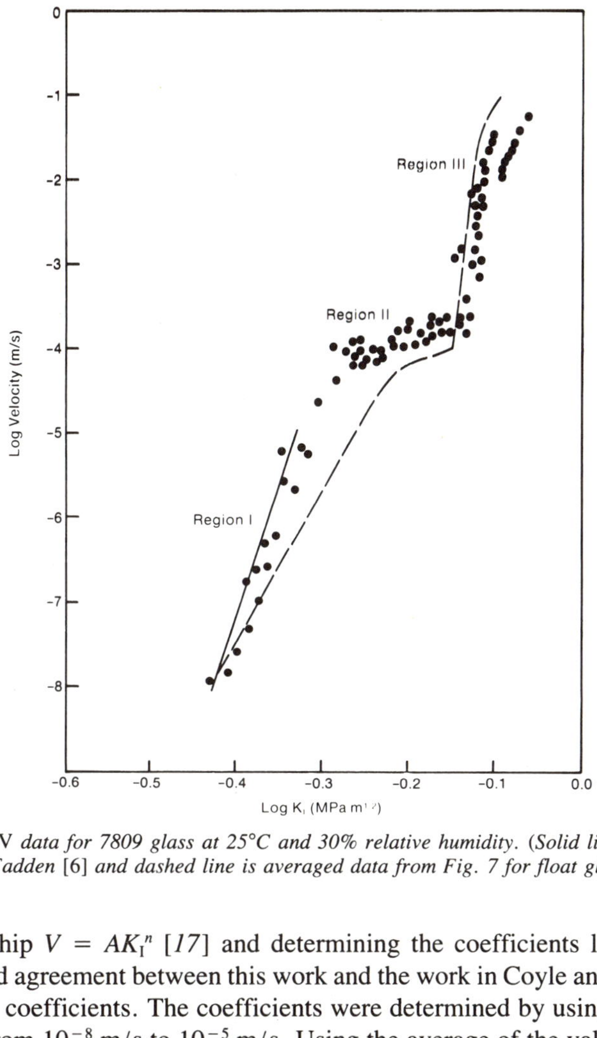

FIG. 9—K_I-V *data for 7809 glass at 25°C and 30% relative humidity.* (*Solid line is data from Coyle and McFadden* [6] *and dashed line is averaged data from Fig. 7 for float glass.*)

the relationship $V = AK_I^n$ [*17*] and determining the coefficients log A and n. There is good agreement between this work and the work in Coyle and McFadden [*6*] for these coefficients. The coefficients were determined by using data in the range of V from 10^{-8} m/s to 10^{-5} m/s. Using the average of the values reported in the table gives an n of 27.5 and 18 for 7809 and float glass, respectively. In previous work on soda-lime-silicate glasses that were similar in composition to float glass, values of about 16 were reported for n when the measurements were done in a moist environment [*5*].

Although this computer-assisted technique should be ideal for measuring very slow crack velocities, we encountered difficulties. We could not successfully conduct experiments with V less than 10^{-8} m/s because of what appeared to be strain gage slippage or drift. For a V of 10^{-8} m/s, this slippage resulted in an apparent change in R similar in the worst case but usually about ten times less than the change in R associated with crack growth. Further, the slippage or drift caused increasing values of R or apparent crack healing. Thus, it would be necessary to immobilize the strain gage or institute another method for measuring the sample displacement to measure V below 10^{-8} m/s.

We also obtained Region III data for both of these glasses. The high velocity Region III data in the figures allows us to accurately determine K_{Ic}. The velocity of 0.1 m/s is cited [*5*] as being used for determining K_{Ic}. One can readily obtain K_I at this velocity from the data in the figures by extrapolation. However, in these figures the slope of the data decreases sharply above approximately 10^{-2} m/s. We think this is caused by a difference in mercury pressure between the flatjack and the pressure transducer, which are separated by tubing through which mercury flows, as seen in Fig. 2. This could result from pressurizing the mercury system too rapidly so a significant pressure difference exists; or such a pressure difference could arise by having a rapid drop in flatjack pressure caused by a large crack velocity.

A maximum turning rate of ¼ turn per second is commonly used when the flatjack-loaded, short-rod technique is used to determine K_{Ic} [*18*]. In our experiments at this turning rate, we found that V was about 2×10^{-2} m/s at the R of 0.5 where the K_{Ic} determination is made. This V is about the same as seen in Figs. 7 and 9 where the slope decreases sharply. Thus, the extrapolation to 0.1 m/s should be made with the steeply sloping Region III data.

Extrapolation to 0.1 m/s using these data in the figures results in a K_{Ic} of about 0.81 MPa $m^{1/2}$ for 7809 glass and about 0.77 MPa $m^{1/2}$ for the float glass. As seen in Table 2, these values of K_{Ic} are higher by about 0.06 MPa $m^{1/2}$ for both glasses than data from the previous work in Coyle and McFadden [*6*], which indicates that the K_{Ic} determinations in that work were done at velocities considerably below 0.1 m/s, estimated in that work to be about 2×10^{-3} m/s.

The data in the figures also show the Region II data for the two glasses. The float glass in Fig. 7 shows a very narrow range for Region II contrasted to the broad range for 7809 glass seen in Fig. 9. It is this broad Region II for the 7809 glass compared with the float glass that makes for a higher V for 7809 glass at intermediate K_I, even though n and K_{Ic} for 7809 indicate that the crack growth velocity in 7809 should be less than in float glass. However, as seen in Fig. 9, for crack velocities below about 10^{-8} m/s 7809 can support higher K_I for the

TABLE 2—K_{Ic} *and Coefficients from* $V = A K_I^n$ *for fracture data on 7809 glass and float glass at 25°C and 30% relative humidity.*

	K_{Ic}/MPa $m^{1/2}$	log A/m/s	n
7809, this work	0.81 ±0.03(20)	4.0 ±0.8(15)[a]	28.5 ±2.2(15)[a]
7809 [*6*]	0.75 ±0.01(9)[a]	3.5 ±0.8(31)	26.6 ±2.2(31)
Float, this work	0.77 ±0.03(30)	−0.2 ±0.4(47)	18.2 ±1.3(47)
Float [*6*]	0.71 ±0.02(8)	−0.3 ±0.4(38)	17.9 ±1.1(38)

[a]The standard deviation is given after each value and the number of datum points is given in parentheses.

same *V* than float glass if the common practice is followed of extrapolating the Region I behavior to low *V* [*19*]. As a cautionary note, however, this practice is misleading in some cases because of the presence of a stress corrosion limit in some glasses [*20–22*]. For instance, Weiderhorn and Bolz [*22*] have made K_I-V measurements on soda-lime-silicate glass in water at 25°C that indicate a stress corrosion limit below 0.3 MPa $m^{1/2}$ and 10^{-10} m/s. Thus, for a valid comparison between the two glasses below 10^{-8} m/s, one needs to obtain data at very low *V*.

There is good agreement between the K_I-V results obtained using this short-rod technique and results obtained using the double cantilever beam (DCB) technique as shown in Fig. 7 for experiments on similar soda-lime-silicate glasses. The results of Weiderhorn and Bolz [*22*] were obtained in water and, as seen in the figure, are displaced to the left of our data by about 0.07 MPa $m^{1/2}$, about the same separation observed by Weiderhorn [*17*] between K_I-V data measured in water versus data measured at 30% relative humidity as in our work. The data of Freiman et al [*23*] also measured by the DCB technique in water are also displaced to the left of our data as seen in the figure. Another set of DCB data shown in the figure was obtained by Champomier [*24*] in ambient air (this is probably similar to our conditions of 25°C and 30% relative humidity); it is in good agreement with our measurements.

The K_{Ic}-value obtained for float glass in this work, 0.77 MPa $m^{1/2}$, was about in the middle of the range of values (0.71 to 0.85 MPa $m^{1/2}$) for soda-lime-silicate glasses given in the recent review by Freiman [*5*]. The K_{Ic} obtained by using this short-rod technique also compares well with recent values of 0.75 and 0.74 measured by Champomier [*24*] using double torsion (DT) and DCB, respectively, 0.76 measured by Pletka et al [*25*] using DT and 0.75 measured by Shinkai et al [*26*] using a microflaw technique. However, a strict comparison of these values is difficult in that the only one where an associated crack velocity was reported was for the DT experiments of Champomier where *V* was 0.4 m/s compared to 0.1 m/s in our work.

Conclusions

We developed a computer-assisted control and data acquisition system for conducting flatjack-loaded, short-rod fracture tests and wrote a software package called FRACT to operate it. Results using this system were in good agreement with previous Region I results on 7809 glass and float glass. Calibrations required for determining K_I and *V* were also evaluated using this system and were found to be accurate.

The critical stress intensity factor at a crack velocity of 0.1 m/s was found to be 0.81 ± 0.03 MPa $m^{1/2}$ and 0.77 ± 0.03 MPa $m^{1/2}$ for 7809 glass and float glass, respectively. K_{Ic} information from other studies on the float glass agreed well with the results of this work.

Crack velocity versus stress intensity measurements were made for 7809 glass and float glass. Crack velocity determinations done in this work above 10^{-8} m/s agreed well with previous work on float glass while those below 10^{-8} m/s were of questionable accuracy because of strain gage slippage or drift. The stress corrosion coefficient, n, was found to be 27.5 ± 2.2 for the 7809 glass and 18.0 ± 1.3 for the float glass.

Acknowledgments

We acknowledge the encouragement and support of A. W. Czanderna and D. K. Benson of SERI and helpful discussions with L. M. Barker of Sandia National Laboratory, Albuquerque. This work was supported by the U.S. Department of Energy under contract No. EG-77-C-01-4042.

References

[*1*] Barker, L. M., *Engineering Fracture Mechanics,* Vol. 9, 1977, p. 361.
[*2*] Barker, L. M., *Fracture Mechanics of Ceramics,* Vol. 3, R. C. Bradt, D. P. H. Hasselman, and F. F. Lange, Eds., Plenum Press, New York, 1978, pp. 481–494.
[*3*] Coyle, R. T. and Livingston, Rob, *Evaluation and Market Testing of Corning Code 7809 Thin Solar Glass,* SERI/TR-733-1230, Solar Energy Research Institute, Golden, Colo., 1981.
[*4*] Coyle, R. T., Lind, M. A., Shelby, J. E., Vitko, J., and Shoemaker, A. F., *Journal of Non-Crystalline Solids,* Vols. 38 and 39, 1980, pp. 233–238.
[*5*] Freiman, S. W. in *Glass Science and Technology,* D. R. Uhlmann and N. J. Kreidl, Eds., Academic Press, New York, 1980.
[*6*] Coyle, R. T. and McFadden, J. D. O., *Proceedings,* Electrochemical Society Symposia on Materials for Batteries and Solar Energy, Denver, Colo., Oct. 1981.
[*7*] Doremus, R. H., *Glass Science,* Wiley, New York, 1973.
[*8*] Weiderhorn, S. M., *Journal of the American Ceramic Society,* Vol. 55, No. 2, 1972, p. 81.
[*9*] Charles, R. J., *Journal of Applied Physics,* Vol. 29, No. 11, 1958, pp. 1549–1560.
[*10*] Kingery, W. D., Bowen, H. K., and Uhlmann, D. R., *Introduction to Ceramics,* 2nd Edition, Wiley, New York, 1976.
[*11*] Barker, L. M. in *Fracture Mechanics Applied to Brittle Material, ASTM STP 678,* American Society for Testing and Materials, Philadelphia, Aug. 1979, pp. 73–82.
[*12*] Lawn, B. R. and Wilshaw, T. R., *Fracture of Brittle Solids,* Cambridge University Press, New York, 1975.
[*13*] Barker, L. M. and Guest, R. V., *Evaluation of the Short Rod Method for Measuring the Fracture Toughness of Beryllium,* Report No. TR79-1, TerraTek, Inc., Salt Lake City, Utah, 1979.
[*14*] Barker, L. M., Sandia National Laboratory, Albuquerque, N. Mex., private communication.
[*15*] Barker, L. M. and Guest, R. V., *Compliance Calibration of the Short Rod Fracture Toughness Specimen,* Report No. TR78-20, TerraTek, Inc., Salt Lake City, Utah, 1978.
[*16*] Davidge, R. W., *Mechanical Behavior of Ceramics,* Cambridge University Press, New York, 1979.
[*17*] Weiderhorn, S. M. in *Fracture Mechanics of Ceramics,* Vol. 2, R. C. Bradt, D. P. H. Hasselman, and F. F. Lange, Eds., Plenum Press, New York, 1974, pp. 613–646.
[*18*] *The Fractometer System 4201 Owners Manual,* TerraTek, Inc., Salt Lake City, Utah, 1979.
[*19*] Ritter, J. E. Jr., *Fracture Mechanics of Ceramics,* Vol. 4, R. C. Bradt, D. P. H. Hasselman, and F. F. Lange, Eds., Plenum Press, New York, 1978, pp. 667–686.
[*20*] Charles, R. J. and Hillig, W. B. in *Symposium on Mechanical Strength and Ways of Improving It,* Union Scientifique Continentale du Verre, Charleroi, Belgium, 1962, p. 511.
[*21*] Hillig, W. B. and Charles, R. J. in *High Strength Materials,* V. F. Zackey, Ed., Wiley, New York, 1965, p. 682.
[*22*] Weiderhorn, S. M. and Bolz, L. H., *Journal of the American Ceramic Society,* Vol. 53, 1970, p. 543.

[23] Freiman, S. W., Gonzalez, A. C., and Weiderhorn, S. M., "Lifetime Predictions for Solar Glasses," to be published in the American Ceramic Society Bulletin.
[24] Champomier, F. P. in *Fracture Mechanics Applied to Brittle Materials, ASTM STP 678,* S. W. Freiman, Ed., American Society for Testing and Materials, Philadelphia, 1979, p. 60.
[25] Pletka, B. J., Fuller, E. D., Jr., and Koepke, B. G. in *Fracture Mechanics Applied to Brittle Materials, ASTM STP 678,* S. W. Freiman, Ed., American Society for Testing and Materials, Philadelphia, 1979, p. 19.
[26] Shinkai, N., Bradt, R. C., and Rindone, G. E., *Journal of the American Ceramic Society,* Vol. 64, 1981, p. 426.

Anthony R. Ingraffea,[1] *Kirk L. Gunsallus,*[2] *John F. Beech,*[2] *and Priscilla P. Nelson*[3]

A Short-Rod Based System for Fracture Toughness Testing of Rock

REFERENCE: Ingraffea, A. R., Gunsallus, K. L., Beech, J. F. and Nelson, P. P., **"A Short-Rod Based System for Fracture Toughness Testing of Rock."** *Chevron-Notched Specimens: Testing and Stress Analysis, ASTM STP 855,* J. H. Underwood, S. W. Freiman, and F. I. Baratta, Eds., American Society for Testing and Materials, Philadelphia, 1984, pp. 152–166.

ABSTRACT: A system for fracture toughness determination of rock using a short-rod specimen is described and evaluated. The calibration of the specimen by finite-element analysis for two loading cases is reviewed. A description of the preparation procedure is given showing the practicality of the testing system. Verification of the system is accomplished through testing of polystyrene, Indiana limestone, and Westerly granite. The results from these tests compare favorably to fracture toughnesses measured by other more expensive and time consuming methods.

KEY WORDS: calibration, fracture toughness, testing, rock, short rod

Fracture toughness measurements have been made on a wide range of rock types [*1*]. Toughness values are being used in a variety of applications including massive hydraulic fracturing [*2*], stability analyses [*3*], explosive fracturing [*4*], and fundamental studies into the theoretical strength of rock [*5,6*].

As a measure of energy of comminution, fracture toughness might be used for more sensitive predictions of a tunnel boring machine (TBM) performance than are possible with other index measures in current use. This paper describes part of an ongoing analytical/experimental program aimed at evaluation of K_{Ic} in such a role. The program includes efforts to:

1. Provide a firm analytical basis for a test specimen geometry.
2. Develop simple, inexpensive testing equipment and techniques.
3. Evaluate the accuracy of the above by testing rock and other materials and comparing the measured fracture toughnesses with those obtained by other methods.

[1]Associate professor and manager, Experimental Research, Cornell University, Ithaca, N. Y. 14853.

[2]Graduate students, Cornell University, Ithaca, N. Y. 14853.

[3]Assistant professor, University of Texas at Austin, Austin, Tex. 78712.

4. Measure fracture toughness of a substantial number of different rocks from ongoing TBM projects.
5. Compare fracture toughness results with other strength measures.
6. Compare all test results with observed TBM performance.

Presented here are descriptions and results of the first three of these tasks. Their objective is the proof testing of an accurate, economical system for fracture toughness testing of rock. Rock material samples are most often available as core. Currently used specimens such as the single-edge-notch bend (SENB) and the compact tension (CT) require extensive and accurate machining operations to be prepared from core. They also are wasteful of sometimes scarce material, and often are substandard in size because of core size limitations. Sometimes the length of available core pieces is insufficient to prepare a specimen of this type. Other proposed geometries, such as the hollow, pressurized cylinder, can be prepared from core. However, such a geometry requires wire-sawed notches, internal jacketing, and probably would be unsuitable for rock containing weak discontinuities.

The large number of tests that would be required to characterize reasonably the natural variation of toughness, for example, along many miles of a tunnel passing through a number of lithologic units argues against not only time-consuming and expensive specimen preparation, but also slow and costly testing procedures. Again, currently used specimens can require sophisticated loading frames or pressurization equipment or both, electronic equipment for load and crack length measurement and data reduction, and considerable amount of man-time for the testing and reduction process. Such specimens and testing systems were originally intended for quality control or proof testing of manmade materials or both, and their expense and accuracy cannot often be justified for a practical application to rock.

Rather, what is required is a practical, core-based system which wastes little material, has an easily and quickly prepared specimen geometry, is amenable to simple, mechanical loading, accommodates a wide range of rock types, and yet yields accuracy at least comparable to that obtainable in the research laboratory using a more complex and expensive test procedure. The short rod has potential to be the specimen geometry in such a system.

Preliminary results of an evaluation of the system with respect to TBM performance prediction, the last three of the above tasks, have been reported in Ingraffea et al [*7*].

Short-Rod Description and Analysis

The short-rod geometry used in the present testing and analysis (Fig. 1) is the same as that studied by Beech and Ingraffea [*8*]. The purpose of their three-dimensional finite element analyses was to provide a rigorous calibration of the short-rod specimen. Barker [*9*] showed that the critical stress-intensity factor

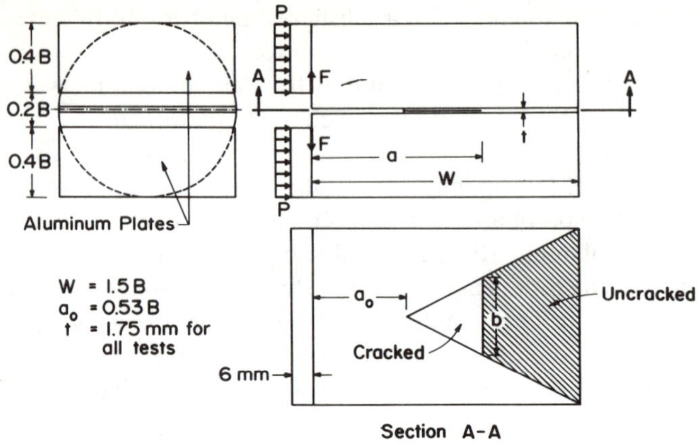

FIG. 1—*Geometry of short-rod specimen used in the present study.*

could be determined from short-rod tests using the following equation

$$K_{Ic} = AF_c/[B^{3/2}(1 - \nu^2)^{1/2}] \tag{1}$$

where

B = specimen diameter,
ν = Poisson's ratio,
F_c = load at crack instability, and
A = calibration constant which is a function only of geometry.

Note that, as indicated in Eq 1, short-rod testing does not require a determination of crack length.

Beech and Ingraffea [8] used the compliance calibration technique to determine the average stress-intensity factor along the crack front as a function of crack length. Their results are shown in Fig. 2. The average stress-intensity factor is computed using

$$K_I = \frac{F}{B\sqrt{W}} Y_F = \left[\frac{F^2E'}{2b}\frac{dC}{da}\right]^{1/2} \tag{2}$$

where

$E' = E/(1 - \nu^2)$, for plane strain,
E = Young's modulus,
Y_F = normalized average stress-intensity factor due to splitting force only,
F = splitting force,
W = specimen length (see Fig. 1),
b = crack front length (see Fig. 1), and
dC/da = rate of change of compliance, C, with crack length, a.

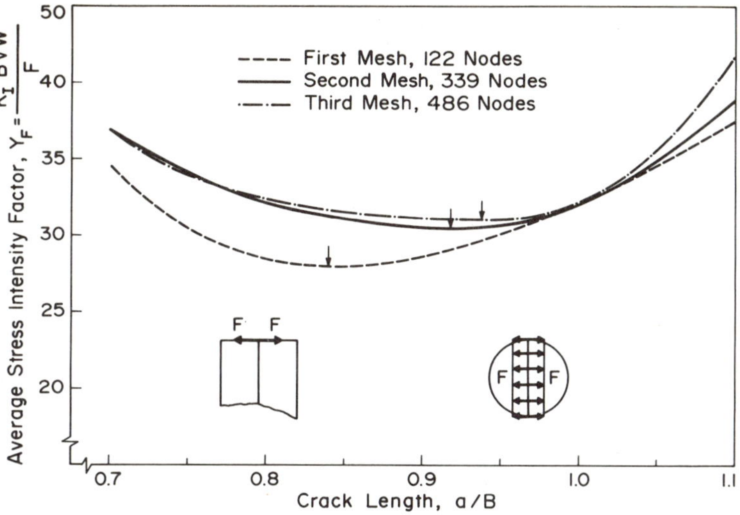

FIG. 2—*Stress-intensity factors for splitting force. Arrow denotes* $Y_{F_{min}}$.

The minimum value of this curve corresponds to $a = a_c$, $F_c = F_{max}$, and $K_I = K_{Ic}$. Beech and Ingraffea [*8*] determined the calibration constant, *A*, by evaluating the right side of Eq 2 at $a = a_c$ and setting it equal to Eq 1. However, it was shown that a further mesh refinement would provide a more accurate value of *A*. Such an analysis was performed in the present investigation, and a value of 25.2 was obtained for *A*. Updated values of Y_F are plotted in Fig. 2. The shape of the crack front and the critical crack length predicted by the finite-element analyses have been confirmed in tests on transparent polystyrene specimens.

As reported next, premature failure often developed in association with specimen discontinuities. To inhibit this mode of failure, an axial load was applied, as shown in Fig. 3. An axial load, distributed as shown, results in an additional

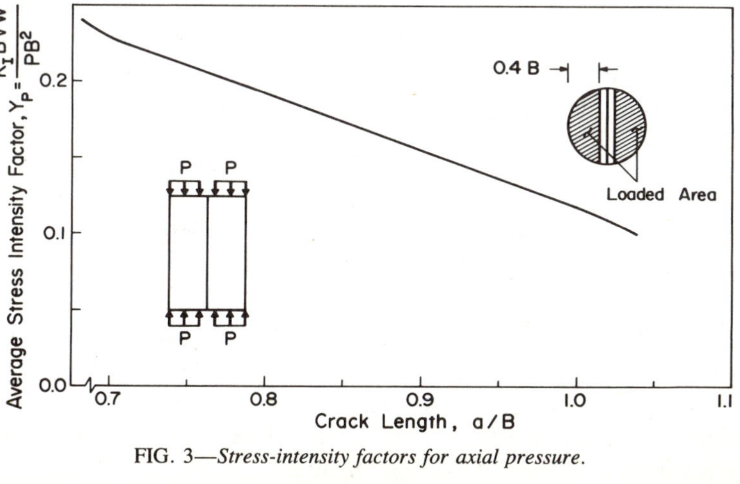

FIG. 3—*Stress-intensity factors for axial pressure.*

stress intensity along the crack front. This means a smaller value of F_c is required to reach instability, and Eq 1 is no longer valid. The effect of axial loading was analyzed using the same compliance calibration technique mentioned above. The results of this study are shown in Fig. 3.

The resultant critical stress intensity factor is obtained from superposition by

$$K_{Ic} = [(F_c Y_F + PB^2 Y_P)/B\sqrt{W}]_{min} \quad (3)$$

where

P = applied axial pressure and
Y_P = normalized average stress-intensity factor due to axial pressure only.

Test Program

Specimen Preparation

The short-rod specimens were prepared from core drilled from block samples at Cornell or obtained during preliminary subsurface investigations for various tunneling projects. All of the core had a nominal diameter, B, of approximately 54 mm. The actual diameters of the specimens ranged from 47.2 to 54.9 mm. Although some of the core exhibited marked irregularities, the diameters of the specimens used in this program varied less than 1% along their length, typically, by about 0.3%.

Three operations were used to prepare a short-rod specimen from rock cores. First, the specimen is cut to a nominal length of 1.5B using a standard water cooled, rock cut-off saw. The actual length of specimens was found to vary from 1.47B to 1.53B. The ends are made parallel by proper adjustment of the core guide prior to cutting. Second, the diametral cuts are made. The specimen is held at the proper angle for each of two cuts necessary to produce the chevron notch by a fixture that allows no core rotation about its axis between cuts. The resulting saw cut kerf was approximately 2 mm. Lastly, aluminum end plates, of approximately 57 by 13 by 6 mm, are epoxied to the top surfaces to act as loading lines for the splitting force. The end plates must be parallel and equidistant from the central crack. This was accomplished with a parallel-sided spacer bar which has a central stem which rests in the saw cut. The plates are abutted against the spacer bar when epoxied. The use of plates as opposed to a groove was chosen for load transfer because no special machinery is required for their use. Preparation time is such that a technician with minimal training could prepare 20 to 30 specimens a day from rock cores.

Test Procedure

The splitting force is applied with a specially constructed mechanical testing apparatus. The loading device is hand held and operated. Turning the actuating knob creates the splitting force which is read directly on an integral force gage. Gage accuracy is ±1% of full-scale reading.

The general test procedure consists of inserting the jaws of the loading device between the end plates, turning the knob, and recording the failure load which is given by a following needle on the gage. Figure 4 shows the loading device in place on a specimen, *left*, and a properly fractured specimen, *right*.

Preliminary testing of rock from some of the TBM work sites, however, showed that application of the splitting force sometimes caused horizontal shearing failure in specimens with weak bedding planes when these planes were close to perpendicular to the expected fracture plane, Fig. 5, *lower right*.

The method used to eliminate this phenomenon was an application of an axial pressure, *P*, prior to testing. This pressure serves to give the bedding planes greater shearing resistance. The clamps used to apply the pressure were calibrated with a presettable torque wrench, load cell, and voltmeter. Accordingly, a known torque on the clamps produced a known pressure. The modification to the general testing procedure for axial loading involved applying and tightening the clamps to give the desired axial pressure (equal on both sides) prior to insertion of the

FIG. 4—*Testing system with splitting force only.*

jaws of the loading device. As before, the failure load is recorded. Figure 5 shows the clamps attached to a specimen ready for testing.

Verification

Although the accuracy of the short-rod testing procedure has been independently confirmed [*10*] with tests on metals, additional verification tests on a plastic and two rock types were performed in the present investigation.

Polystyrene Specimens

Five tests were performed on 26-mm-diameter polystyrene short-rod specimens. These produced an average K_{Ic} of 2720 MN $m^{-3/2}$ (s = 133 MN $m^{-3/2}$). This value agrees well with the value of 2690 $\pm$ 50 MN $m^{-3/2}$ reported by Krenz et al [*11*] for polystyrene with the same molecular weight and made by the same manufacturer.

Indiana Limestone Specimens

A total of 34 tests were performed on short-rod specimens of Indiana limestone. A detailed petrographic description of Indiana (also known as Salem) limestone

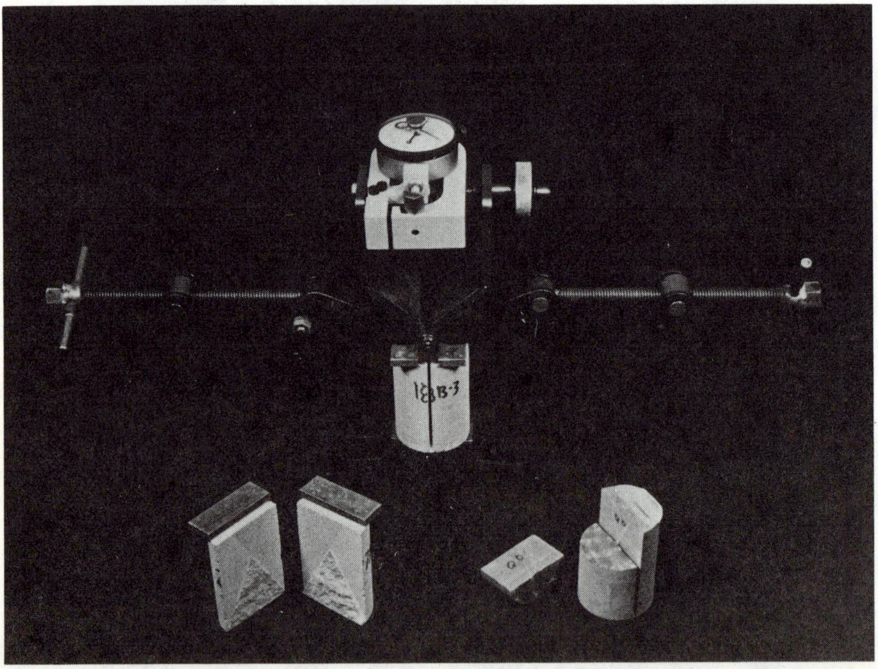

FIG. 5—*Testing system with splitting force and axial pressure.*

can be found in Ref *12*. Cores were taken from beams previously used in three-point bend (SENB) determinations of K_{Ic} [*13*]. Tests were performed at three applied axial pressures, *P*. Results of the tests are shown in Table 1. These results indicate that the average conditional toughness, K_Q, determined from the axially loaded specimens differs from that of the others by, at most, 5%. The consistent decrease in average K_Q with increasing axial pressure for Indiana limestone is probably due to the fact that the applied pressures are not insignificant relative to the strength of this rock. Some reported strengths are 43.9 MPa uniaxial compression (ASTM D 2264-67) and 5.2 MPa direct tension (ASTM D 2936-71) [*12*]. Consequently, some weakening of the specimens probably occurs under the application of the axial load. The good agreement of the K_Q results for the specimens with and without axial load tends to support the axial load analytical calibration. For this rock, axial loading contributes about 30% of the total stress-intensity factor at instability and cannot be ignored.

Figure 6 shows the comparison between SENB and short-rod results for Indiana limestone. The single value of K_Q from each SENB test is compared to the average and range of 4 to 6 short-rod tests on specimens made from the SENB specimen. The short-rod values tend to be higher than those from the SENB tests. A possible reason for this discrepancy is that the length of natural crack was longer in the short-rod specimens (about 19 versus about 6 mm). The trend of increasing K_Q with increasing natural crack length in rock has previously been reported [*6,14*]. Also, the K_Q determination for the SENB tests was made from the initial loading after precracking. Fracture toughness has been shown to increase upon subsequent loadings [*14*].

It should be noted, however, that the values of K_Q determined here by short-rod testing fall well within the range of toughness values previously reported [*1*].

Westerly Granite Specimens

A total of 27 K_Q determinations were made from short-rod specimens of Westerly granite. Again, a detailed petrographic description of this rock can be

TABLE 1—*Variation of mean* K_Q *(kPa* $\sqrt{m}$*) with applied axial load.*

	Axial Pressure, *P*(MPa)		
Rock Type[a]	0	5.56	8.35
Indiana limestone	1025	1008	975
s	93	24	63
n	22	3	9
Westerly granite	2277	...	2255
s	187	...	8
n	17	...	10

[a]s = standard deviation. n = number of samples.

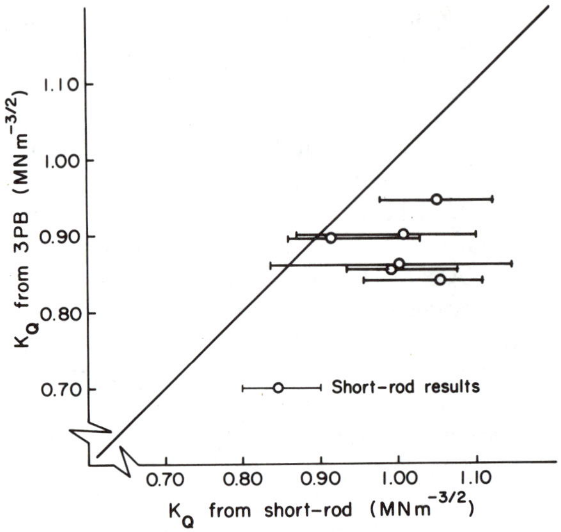

FIG. 6—*Comparison of Indiana limestone toughnesses measured by SENB* [13] *with present short-rod tests.*

found in Krech et al [*12*]. A summary of results is shown in Table 1. Again the results with and without axial pressure agree well with each other, this time to within 1%. In contrast to the situation with the Indiana limestone, the applied pressure used here is less than 4% of the uniaxial compressive strength of Westerly granite as measured by Krech et al [*12*] and noticeable weakening of the specimens by axial loading is not expected.

Schmidt and Lutz [*14*] have done extensive fracture toughness testing of Westerly granite using 3PB and compact tension specimens with differing total crack lengths. The results of the short-rod toughness determinations are compared with their results in Fig. 7. It is seen that the present short-rod values agree well with those of Schmidt and Lutz [*14*]. It should be noted that the crack length used to plot the short-rod results is the theoretical a_c for the specimen diameter used. The crack lengths used by Schmidt and Lutz [*14*] were computed from measured compliances.

Discussion

The variation in K_Q-values determined by the short-rod test is relatively small compared to that of other properties of the rocks tested [*12*]. The coefficients of variation for the data shown in Table 1 range from a high of 9% to a low of only 4% for the larger sample sizes. Obviously, some of this variation is due to material inhomogeneity. However, two other causes of variation are apparent. The first is due to a material characteristic, anisotropy, while the second derives from random experimental error.

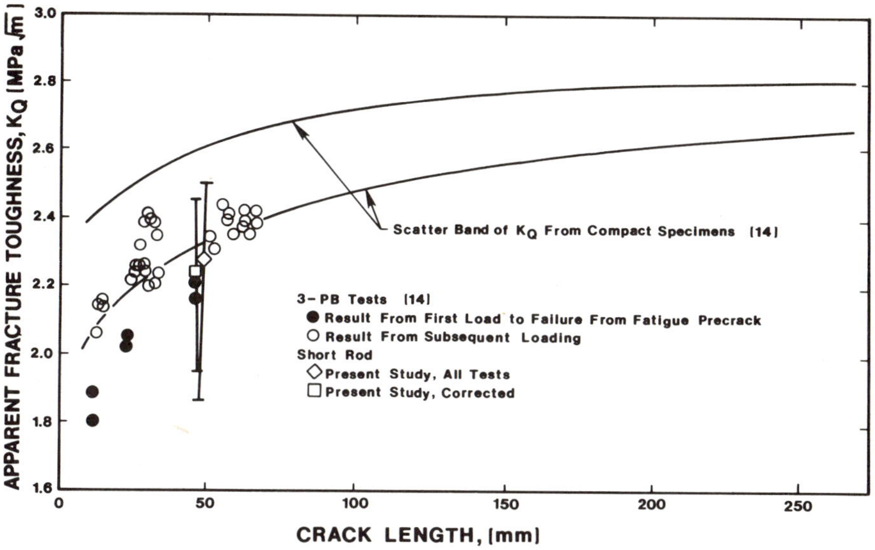

FIG. 7—*Comparison of Westerly granite toughnesses measured by SENB and CT* [14] *with present short-rod tests including those corrected for* a_o/B *effect.*

Effect of Anisotropy

While the Westerly granite is isotropic, it has previously been determined [*15*] that Indiana limestone is slightly orthotropic in toughness. Thus, toughness should vary with orientation of fracture plane. Initially, no attempts were made to keep the azimuth of the fracture plane in the present tests constant, although the crack was always propagated perpendicular to bedding. Subsequent tests on nine specimens with the same fracture plane orientation resulted in a decrease in the standard deviation of approximately 40% under these conditions, as shown in Table 2.

Effect of Geometry Variation

A source of random experimental error is inaccuracy in the length a_o (Fig. 1) created in each specimen. Post-failure examination of the rock specimens in-

TABLE 2—*Effect on mean* K_Q *(kPa* $\sqrt{m}$*) of fracture plane azimuth in Indiana limestone.*

Rock Type[a]	A Random Azimuth	B Constant Azimuth	A + B
$\bar{X}$	995	1066	1025
s	102	60	93
n	13	9	22

[a] s = standard deviation. n = number of samples.

dicated that the specimen preparation conditions allowed a variation of a_o/B from 0.50 to 0.57. Figure 8 shows the experimental variation of K_Q (from specimens tested without axial load) with measured a_o/B for all the Westerly granite and the nine Indiana limestone specimens tested with constant fracture plane orientation. It is seen that as measured a_o/B increases, K_Q decreases for both the granite and limestone although the trend is much clearer for the former than the latter. Using a least squares, linear regression fit of the data in Fig. 8, one observes that a 10% increase in a_o/B (from 0.50 to 0.55) causes 11 and 14% decreases in K_Q for the limestone and granite, respectively. This trend can be explained by noting that the calibration constant used in evaluating K_Q is fairly sensitive to a_o/B. Bubsey et al [*10*] have shown experimentally that $Y_{F\min}$ varies directly with a_o/B, all other geometrical parameters being the same. This observation has been confirmed numerically by Ingraffea et al [*16*] for two particular values of a_o/B. Their results are shown in Fig. 9. If it is assumed that the variation of $Y_{F\min}$ with a_o/B is linear, and, if this linear variation is extrapolated slightly

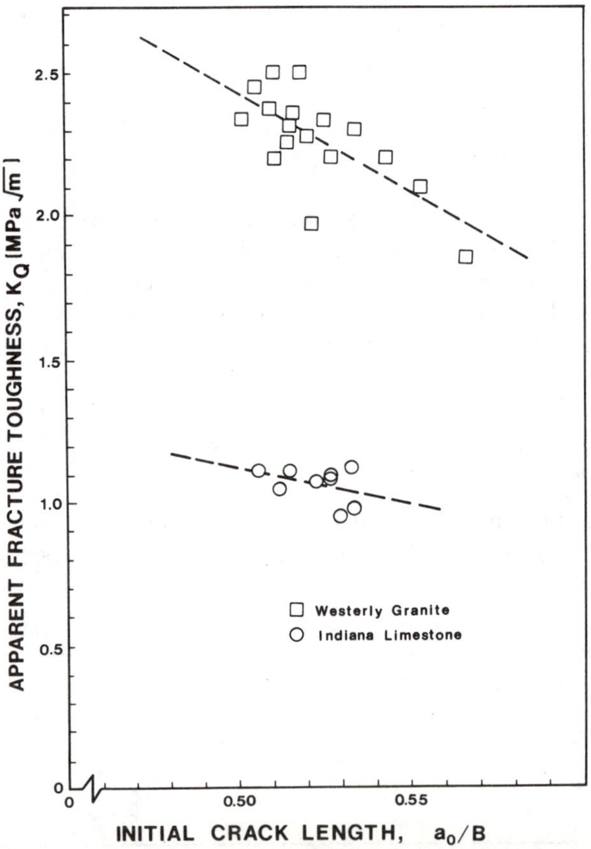

FIG. 8—*Variation of* K_Q *with measured initial crack length.*

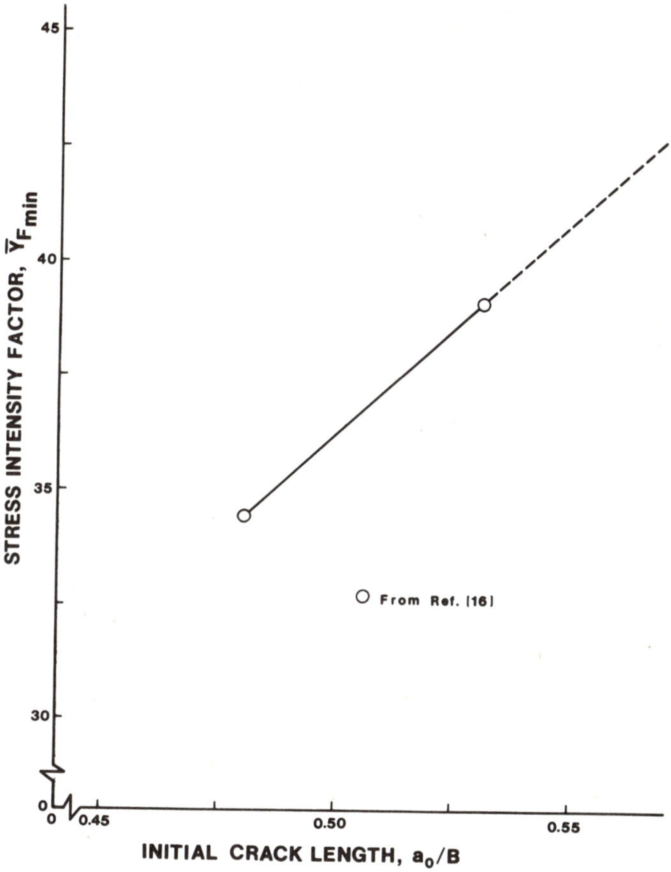

FIG. 9—*Variation of calibration constant,* $Y_{F_{min}}$ *(plane strain assumption), with initial crack length.*

as shown in Fig. 9, a correction factor for the experimental results can be obtained. Based on the curve shown in Fig. 9, corrected values of apparent toughness are given by

$$K_Q^* = K_Q[2.36(a_o/B) - 0.25] \tag{4}$$

where

K_Q = apparent toughness measured with the assumption that $a_o/B = 0.53$ and

K_Q^* = corrected value using actual, measured value of a_o/B.

The correction given by Eq 4 was applied to the data shown in Fig. 8 with the results shown in Table 3 and Figs. 7 and 10. It is seen that substantial

TABLE 3—*Correction to mean* K_Q *(kPa* $\sqrt{m}$*) for* a_o/B *effect, Eq 4.*

Rock Type	Before Correction	After Correction
Indiana limestone[a]	1066	1048
s	60	55
Westerly granite[b]	2277	2242
s	187	126

[a]n = 9.
[b]n = 17.

reduction in standard deviation results from this approximate correction for specimen preparation error. At the present time, this correction is available only for specimens tested without axial load. The effect of a_o/B variation on the Y_P-values is the subject of future numerical investigation. When a generally applicable and more accurate correction factor becomes available, the necessity to prepare specimens very accurately, a difficult and usually unjustified requirement with rock, will be precluded.

Effect of Specimen Size

A final point of discussion is the use of the apparent toughness designation, K_Q, rather than K_{Ic} for the results reported here. All present tests have been

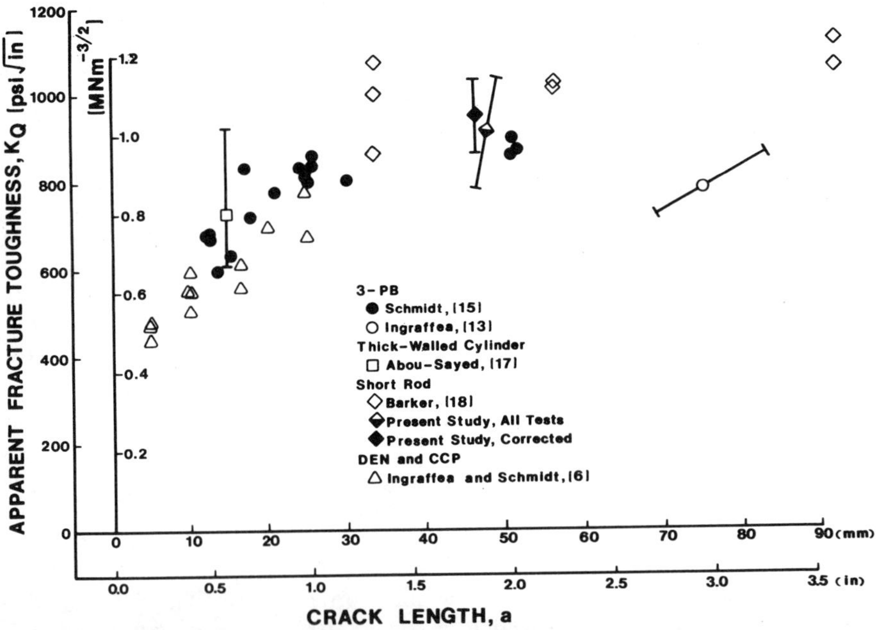

FIG. 10—*Comparison of previous results for Indiana limestone toughnesses with present short-rod tests including those corrected for* a_o/B *effect.*

performed on nominally the same size specimens. Figure 7 shows that the present value of mean K_Q for Westerly granite is somewhat less than the K_{Ic}-value reported by Schmidt and Lutz [*14*] who used two different geometries in a wide range of sizes. Figure 10 summarizes toughness data on Indiana limestone from a number of workers using five specimen geometries. Again it can be seen that present short-rod results are slightly lower than the K_Q-values measured on the largest specimens. It is likely, therefore, that the present results are lower bounds on the K_{Ic} of these rocks. Of course, this does not diminish the value of the short-rod geometry or testing system reported here. It means that a larger diameter core should be used. Alternatively, a new short-rod geometry with larger W/B or shorter a_o/B could be calibrated and used with the same core diameter used here.

Conclusions

A short-rod based system for fracture toughness testing of rock has been described. Several conclusions about the system can be made:

1. A firm analytic basis for the short-rod test is available.
2. Preparation of the specimen is relatively quick and simple and does not require extensive machining operations.
3. Testing does not require sophisticated equipment.
4. A straightforward modification to the general testing procedure permits fracture toughness determination of rocks with weak bedding planes. The modification consists in applying a known axial load to the specimen before application of the splitting force. The additional stress intensity induced by the axial load has been computed and verified by testing.
5. Short-rod system results with and without axial load for polystyrene, Indiana limestone, and Westerly granite agree well with data from other testing configurations.
6. The apparent fracture toughness was observed to vary inversely with initial crack length. Analysis predicts this trend. A simple correction factor was derived which can be used to adjust results for the actual initial crack length in each specimen.
7. The short-rod testing system is an accurate and efficient method to make fracture toughness determinations of a large number of samples of rock.

Acknowledgments

The authors express thanks to Professors Fred Kulhawy and Thomas O'Rourke for their continued enthusiastic support of this research program, and to Mr. Francis Fong for helping prepare specimens.

References

[*1*] Ouchterlony, F., "Review of Fracture Toughness Testing of Rock," Swedish Detonic Research Foundation, DS1980:15, Stockholm, 1980.

[2] Wiles, T. D. and Roegiers, J. C., ''Modeling Hydraulic Fractures Under Varied In Situ Conditions Using a Displacement Discontinuity Approach,'' Petroleum Society of CIM Paper No. 82-33-70 (preprint) presented at the 33rd Annual Technical Meeting of the Petroleum Society of CIM, Calgary, Alta., June 1982.
[3] Ingraffea, A. R., *Proceedings of the 20th U.S. Symposium on Rock Mechanics,* Austin, Tex., 1979, pp. 153–169.
[4] Warpinski, N. R., Schmidt, R. A., Cooper, P. W., Walling, H. C., and Northrop, D. A., *Proceedings of the 20th U.S. Symposium on Rock Mechanics,* Austin, Tex., 4–6 June 1979, pp. 143–152.
[5] Holzhausen, G. R. and Johnson, A. M., *International Journal of Rock Mechanics and Mining Sciences and Geomechanics Abstracts,* Vol. 16, No. 3, June 1979, pp. 163–177.
[6] Ingraffea, A. R. and Schmidt, R. A., *Proceedings of the 19th U.S. Symposium on Rock Mechanics,* Stateline, Nev., 1978, pp. 247–253.
[7] Ingraffea, A. R., Gunsallus, K. L., Beech, J. F., and Nelson, P., *Proceedings of the 23rd U.S. Symposium on Rock Mechanics,* Berkeley, Calif., Aug. 1982, pp. 463–470.
[8] Beech, J. F. and Ingraffea, A. R., *International Journal of Fracture,* Vol. 18, No. 3, March 1982, pp. 217–229.
[9] Barker, L. M., *Engineering Fracture Mechanics,* Vol. 9, No. 2, 1977, pp. 361–369.
[10] Bubsey, R. T., Munz, D., Pierce, W. S., and Shannon, J. L., Jr., *International Journal of Fracture,* Vol. 18, No. 2, Feb. 1982, pp. 125–133.
[11] Krenz, H. G., Ast, D. G., and Kramer, E. J., *Journal of Materials Science,* Vol. 11, No. 12, Dec. 1976, pp. 2198–2210.
[12] Krech, W. W., Henderson, F. A., and Hjelmstad, K. E., ''A Standard Rock Suite for Rapid Excavation Research,'' U.S. Bureau of Mines Report of Investigations 7865, U.S. Department of Interior, Washington, D.C., 1974.
[13] Ingraffea, A. R., *Proceedings, 22nd U.S. Symposium on Rock Mechanics,* Cambridge, Mass., 1981, pp. 186–191.
[14] Schmidt, R. A. and Lutz, T. J. in *Fracture Mechanics Applied to Brittle Materials, ASTM STP 678,* American Society for Testing and Materials, Philadelphia, 1979, pp. 166–182.
[15] Schmidt, R. A., *Experimental Mechanics,* Vol. 16, No. 5, 1976, pp. 161–167.
[16] Ingraffea, A. R., Perucchio, R., Han, T.-Y., and Gerstle, W. H., this publication, pp. 49–68.
[17] Abou-Sayed, A. S. and Simonson, E. R., *Proceedings of the 18th U.S. Symposium on Rock Mechanics,* 1977, pp. 2A3-1, 2A3-8.
[18] Barker, L. M., ''K_{Ic} Measurements Using Short-Rod Specimens—The Elastic-Plastic Case,'' Report 77-91R, Terra Tek, Oct. 1977.

Leon Chuck,[1] *Edwin R. Fuller, Jr.,*[1] *and Stephen W. Freiman*[1]

Chevron-Notch Bend Testing in Glass: Some Experimental Problems

REFERENCE: Chuck, L., Fuller, E. R. Jr., and Freiman, S. W., **"Chevron-Notch Bend Testing in Glass: Some Experimental Problems,"** *Chevron-Notched Specimens: Testing and Stress Analysis, ASTM STP 855,* J. H. Underwood, S. W. Freiman, and F. I. Baratta, Eds., American Society for Testing and Materials, Philadelphia, 1984, pp. 167–175.

ABSTRACT: This study describes experimental difficulties in the use of the chevron-notch bend test to determine the plane-strain fracture toughness, K_{Ic}, for brittle materials. Four-point flexure tests were performed on soda-lime-silica glass and vitreous silica in both "wet" and "dry" environments and at various loading rates. Results show that an inability to produce stable crack growth in the chevron-notch bend test can lead to serious overestimates of K_{Ic} as well as to significant scatter in the data. It is also shown that water enhanced crack growth can reduce the measured value of K_{Ic}.

KEY WORDS: fracture, glass, chevron notch, bend test, fracture toughness, crack growth

Because of a number of factors (including among others: simplicity of loading under extreme conditions, for example, at elevated temperatures or in reactive environments; reproducibility of results; and small amount of material required per specimen), the chevron-notch bend test (CNBT) has been suggested as an excellent technique for measuring plane-strain fracture toughness,[2] K_{Ic}, of brittle materials, such as glass, ceramics, and concrete. Another advantage of this configuration is that in principle there is no need for a separate crack initiation step, since a crack is presumed to grow stably from the apex of the chevron. K_{Ic} is typically determined by loading the specimen at a predetermined displacement rate in a test machine; K_{Ic} is calculated from existing expressions for the stress-intensity factor, K_I, in terms of the maximum load observed and the specimen geometry.

[1]Mechanical engineer, physicist, and ceramic engineer, respectively, Inorganic Materials Division, National Bureau of Standards, Washington, D.C. 20234.

[2]Throughout this paper we will use the designation K_{Ic} to denote the plane-strain fracture toughness. We concede that this value is not necessarily the value obtained by the operational definition embodied in ASTM procedure E-399, but rather corresponds to a literature definition of "critical" fracture toughness for brittle materials which exhibit time-dependent crack growth. Namely, it corresponds to a (critical) point on the curve of crack growth rate versus stress-intensity factor for the material. As these curves are generally very steep for crack speeds $\gtrsim 0.1$ m/s, the exact point has not proven to be a critical issue.

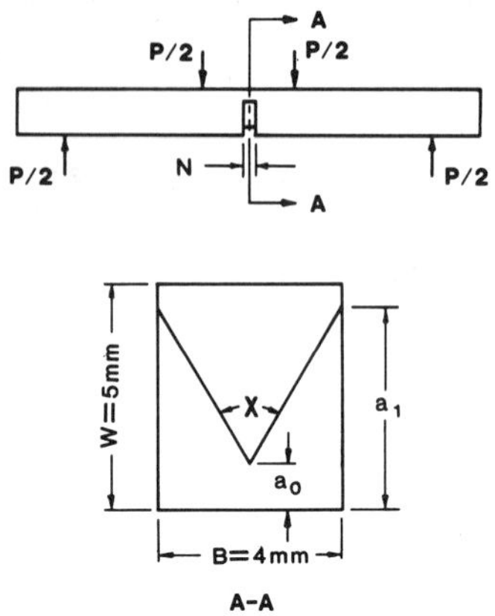

FIG. 1—*Nominal dimensions and nomenclature for the chevron-notched bend specimens used in this study.*

Munz et al [1] have used four-point bend specimens with both straight-through and chevron notches to determine fracture toughness of an aluminum oxide (Al_2O_3) ceramic. For the chevron-notched specimens, they varied the chevron parameters a_o and a_1 (Fig. 1) from 0.07 to 0.37 W and from 0.6 to 1.0 W, respectively. The measured K_{Ic} data were independent of variations in a_o as would be expected, but showed an unexpected dependence on a_1. The authors suggested that this variation was possibly due to inaccuracies in the expression which was used to calculate K_{Ic}. Yet using a similar analysis, except now for chevron-notched short-bar specimens, Munz et al [2] found that the calculated K_{Ic}-values for a hot-pressed silicon nitride were independent of these variations, while those for the same Al_2O_3 had a similar dependence on a_1 as for the four-point bend specimens. In discussing this result, the authors raised the possibility that this dependence was due to a rising crack growth resistance curve (R-curve) behavior for this Al_2O_3, namely, that the resistance to fracture increases with increasing crack extension, or crack area.[3]

Considering these variations in measured K_{Ic}-values with chevron geometry, the original objective of the present study was twofold. The first was to determine the effects, if any, of variations in chevron-notched geometry on measured K_{Ic}-values for a glass, which being homogeneous and isotropic down to the molecular level, cannot produce rising R-curve behavior.[4] The second objective was to determine whether moisture-induced (time-dependent) subcritical crack growth

[3]This influence of a rising R-curve on the K_{Ic}-value measured with chevron-notched specimens is discussed in more detail by two other papers in this volume: one by Shannon and Munz [3] for Al_2O_3; and another by Krause and Fuller [4] for polymer concrete materials.

[4]A rising fracture resistance with increasing crack extension can be attributed to the development of nonelastic processes in the crack-tip region (such as microcracking, phase transformations, or grain interlock), that result from microstructural inhomogeneities in the material [3,4]. Such influences in glass, occurring at the molecular level, would be unmeasurable.

is an important factor in the determination of K_{Ic}-values with the CNBT. We shall see that the unpredictable behavior of the test made it impossible to carry out the first objective and limited the data that we could obtain toward determining the second objective.

Experimental Procedure

Chevron-notched-bend specimens were machined from both soda-lime-silica glass and vitreous silica. The specimen configuration is shown in Fig. 1. The dimensions B, W, and a_o were kept constant at nominal values of 4, 5, and 1 mm, respectively. Three values of the angle X were used, 60°, 90°, and 120°, to give corresponding values of a_1 of 0.893 W (4.46 mm), 0.600 W (3.00 mm), and 0.431 W (2.15 mm). These configurations were chosen because of the observations of Munz et al [*1,2*] that measured values of K_{Ic} for Al_2O_3 generally varied with a_1, but were not as sensitive to a_o. The machined notch width gap N was 0.5 mm. All specimens were annealed after machining. Testing was performed in four-point bending (inner span of 10 mm, outer span of 40 mm) on a Universal testing machine. Testing was conducted at various machine displacement rates and in both "wet" and "dry" environments.[5] K_{Ic} was calculated from the maximum load observed on the chart recorder and the specimen dimensions using an analysis developed by Munz et al [*1,5*]. This analysis uses an assumption, dubbed the straight-through crack assumption (STCA), which assumes that the change of specimen compliance with crack length for the chevron-notched specimen is the same as that for a straight-through notched specimen, so that the stress-intensity factor expression is only modified by a factor incorporating the increasing crack area with crack extension. Although this analysis is purportedly not as accurate as the Bluhm slice analysis [*6*], it is considerably simpler to use. Since one of our objectives was an experimental calibration of the stress-intensity factor, we initially opted for the simpler STCA with which to compare our results. The Srawley and Gross [*7*] analysis for pure bending was used for the straight-through-crack component of the calculation.

Results

One of the primary advantages claimed for the CNBT is the initiation of a sharp crack during loading, eliminating the need to precrack the specimen. Evidence for the presence of a stably growing crack is purported to be a smooth reduction in specimen compliance giving rise to a gradual turnover of the loading curve near the maximum load. We will show that in our test configuration the occurrence of such a compliance change could not be predicted before a test was performed. When the smooth turnover in the loading curve did not occur, the K_{Ic}-value calculated from the maximum load was much higher than val-

[5]Nominally "dry" environments were achieved with bottled gases of either nitrogen or argon; whereas "wet" environments were typified either by testing in water or by a mixture of a dry gas and a gas bubbled through water.

TABLE 1—K_{Ic} *from chevron notch bend tests.*[a]

Material	Specimen Condition	No. of Specimens	Test Environment	Loading Rate, N/s	Type of Load Curve[b]	Calculated[c] K_{Ic}, MPa $m^{1/2}$
Soda-lime glass	annealed	7	dry N_2	0.017 and 0.17	1	1.51 ± 0.18
Soda-lime glass	annealed	7	H_2O	0.017	1	0.92 ± 0.07
Soda-lime glass	annealed	3	H_2O	0.017	2	0.63 ± 0.01
Soda-lime glass	abraded 320 grit SiC	3	dry N_2	0.017	1	0.73 ± 0.05
Soda-lime glass	abraded 320 grit SiC	4	dry N_2	0.017	2 and 3	0.72 ± 0.06
Vitreous SiO_2	razor precrack	2	argon	0.85	1	0.98 ± 0.07
Vitreous SiO_2	razor precrack	4	argon	0.85	2 and 3	0.84 ± 0.03
Vitreous SiO_2	razor precrack	5	argon	8.5	1	>0.90
Vitreous SiO_2	razor precrack	3	argon; 50% rh	0.85	1	>1.0
Vitreous SiO_2	razor precrack	1	argon; 50% rh	0.85	2	0.85
Vitreous SiO_2	razor precrack	1	argon; 50% rh	0.85	3	0.76

[a]All specimens had X = 60° (Fig. 1).
[b]See Fig. 2.
[c]Calculated from expression of Munz et al [*1*].

ues reported in the literature for the same material. Test results are listed in Table 1.

Soda-Lime-Silica Glass

The first specimens tested were cut from soda-lime-silica glass and had the configuration shown in Fig. 1, with $a_o/W = 0.2$ and $X = 60°$. Initial tests were performed on as-annealed bars in dry nitrogen (N_2) gas at machine displacement rates of either 1×10^{-2} or 1×10^{-3} mm/min, corresponding to initial loading rates of 0.17 and 0.017 N/s, respectively. None of the seven specimens tested in this way showed any evidence of a decreasing slope (compliance change) in the loading curve. Specimens broke suddenly with a precipitous drop in load. (Type 1 failure in Fig. 2). The value of K_{Ic} and standard deviation calculated from these seven specimens was 1.51 ± 0.18 MPa $m^{1/2}$ as compared to an accepted value of 0.75 MPa $m^{1/2}$ for this glass [*8*]. No difference in measured values could be detected between specimens tested at the two loading rates.

Several as-annealed soda-lime-silica specimens were also tested in water (H_2O) at a displacement rate of 1×10^{-3} mm/min. Of the 10 specimens tested, seven failed suddenly, yielding a K_{Ic} of 0.92 ± 0.07 MPa $m^{1/2}$, while three showed a Type 2 failure. K_{Ic}'s for the latter were calculated from the maximum load on the smooth portion of the curve following the load drop. These values were 0.63 ± 0.01 MPa $m^{1/2}$, significantly lower than the K_{Ic} for this glass obtained under dry conditions.

It was thought that the annealing treatment may have removed much of the surface damage at the chevron tip, making crack initiation at the tip more difficult. For this reason, we abraded the tip of the chevron in a number of specimens with 320 grit silicon carbide (SiC) paper, and retested in dry N_2 gas at a displacement rate of 1×10^{-3} mm/min. Of the seven specimens tested, three failed suddenly in a Type 1 pattern as before, but now yielding a K_{Ic}-value of 0.73 ± 0.05 MPa $m^{1/2}$, while the other four loading curves showed either a Type 2 or a Type 3 pattern. The K_{Ic} calculated for the latter four specimens was 0.72 ± 0.06 MPa $m^{1/2}$. Both of these values are close to the accepted value of 0.75 MPa $m^{1/2}$. The independence of the calculated K_{Ic}-value on the formation

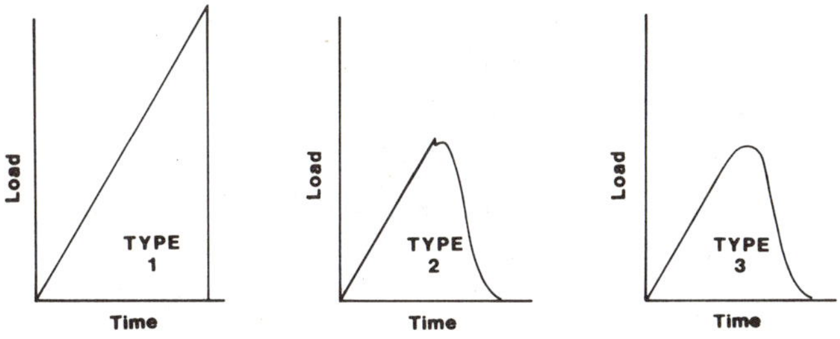

FIG. 2—*Schematic of the three types of load versus time curves which were observed.*

of a stable crack may be due to the extent of the damage put into the specimen by the SiC abrasive paper.

Vitreous Silica

Because of our experience with the annealed soda-lime-silica specimens, an attempt was made to ensure the presence of cracks in the vitreous silica (SiO_2) specimens, by carefully tapping on the tip of the chevron with a razor blade until a small crack could be observed using a low power optical microscope. As we will see, even this procedure did not ensure a valid test result.

Six SiO_2 specimens were tested in dry argon gas at a displacement rate of 5×10^{-2} mm/min (loading rate of 0.85 N/s). Of these, four gave ''valid'' tests (either Type 2 or Type 3 curves); K_{Ic} calculated from these tests was 0.84 ± 0.03 MPa $m^{1/2}$, slightly higher than the 0.79 MPa $m^{1/2}$ typically reported for vitreous SiO_2 [*8*]. There was no discernible difference between the results obtained from the Type 2 or Type 3 curves. The two specimens which failed abruptly gave a value of $K_{Ic} = 0.98 \pm 0.07$ Mpa $m^{1/2}$. Attempts to test at a more rapid loading rate, that is, 5×10^{-1} mm/min, resulted in all invalid tests (Type 1 curves) and K_{Ic}'s higher than 0.9 MPa $m^{1/2}$.

Since one of the goals of this study was to determine possible effects of moisture on the results of chevron notch tests, experiments were also conducted in a mixture of wet argon (that is, argon which had been bubbled through water) and dry argon to yield a gas of 50% relative humidity. Of the five specimens tested at a rate of 5×10^{-2} mm/min (0.85 N/s) under these conditions, three broke abruptly ($K_{Ic} > 1$ MPa $m^{1/2}$); one specimen broke in a Type 2 mode ($K_{Ic} = 0.85$ MPa $m^{1/2}$) and one broke in a Type 3 mode ($K_{Ic} = 0.76$ MPa $m^{1/2}$). We believe that the difference in the latter two values is significant and is due to the fact that when the crack is able to grow stably from the outset (Type 3 curve), it will grow more extensively in the presence of water. This extra crack extension from environmental crack growth results in a longer ''real'' crack length at a given applied load than that assumed by only the time-independent growth inherent in the chevron analysis. Thus, the point at which the crack goes unstable occurs at a lower applied load, leading to a lower value of the calculated K_{Ic} parameter.

Part of the initial objective of this study was to determine possible effects of notch geometry on K_{Ic}. However, no valid tests (as evidenced by prior stable crack extension) were obtained from specimens with either 90 or 120° angles.

Discussion

One of the primary observations stemming from this work is that despite predictions to the contrary, the presence of a chevron notch in a bend specimen does not guarantee the formation of a stably growing crack at low loads. A criterion for validity is the appearance on the loading curve of a smooth change in specimen compliance which is indicative of the stable growth of a crack. In contrast, a sharp drop in the load usually indicates dynamic popin, and may take place at loads which would yield K_{Ic}-values significantly higher than that measured on the same material by other techniques. This behavior is interpreted to mean that no sharp crack occurred in the chevron until a load was reached such that the crack popped in to a position beyond the point at which it is stable.

In the present study unstable fractures invariably occurred in annealed spec-

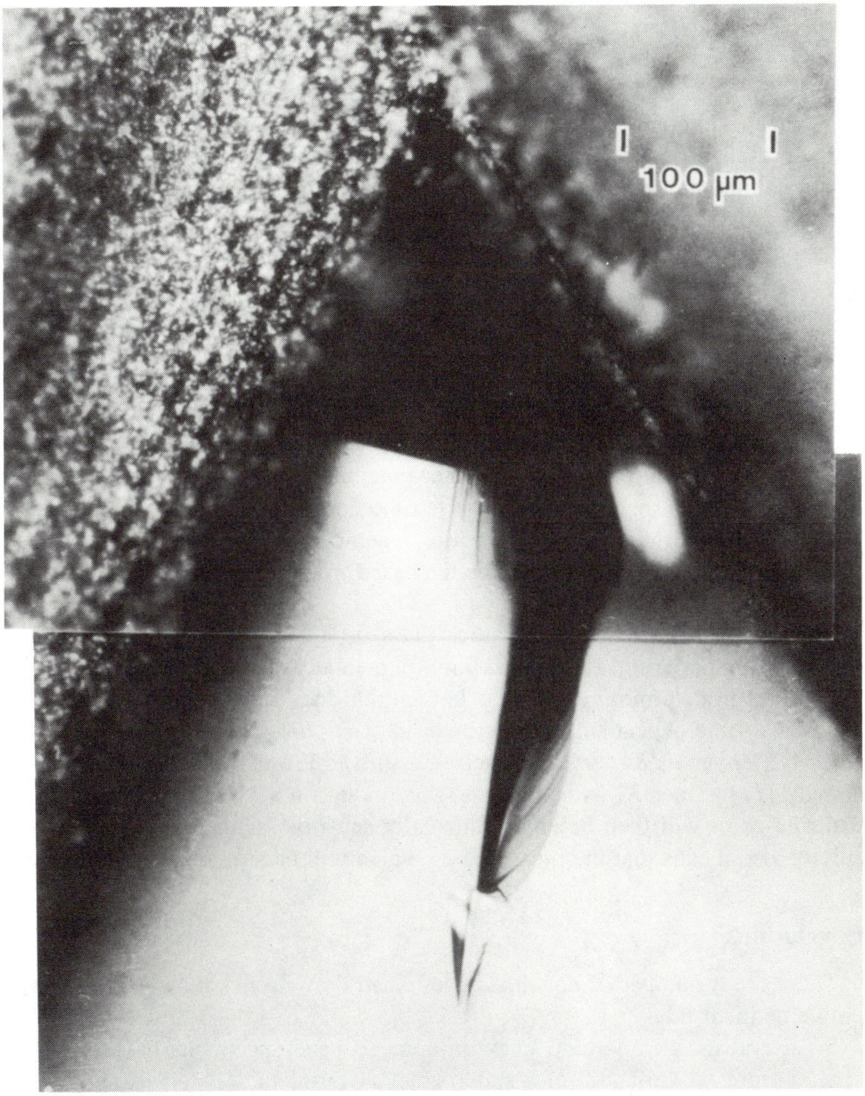

FIG. 3—*Fractograph of a vitreous* SiO_2 *specimen in which a crack had been initiated with a razor blade prior to testing.*

imens and led to a significant overestimate of K_{Ic}. ''Precracking'' by artificially inducing damage into the tip of the chevron by either grinding with SiC abrasive paper or tapping with a razor blade did not ensure that a valid test could be performed. No predictions could be made of whether a test would be valid, despite the fact that fractographic observations of failed SiO_2 specimens showed that a crack had been present from the outset, even when stable growth was not observed (Fig. 3). These observations suggest that the problem is not the difficulty of initiating a crack, but rather of propagating a ''proper'' chevron crack (that is, a crack that is in the proper plane and not pinned by the notch groove). Indeed, in discussing straight-through notched bend specimens, Munz et al [*1*] state the importance of notch width gap and notch preparation for obtaining valid K_{Ic}-values. Although a chevron notch alleviates these factors somewhat, results of the present study suggest that notch width gap and notch preparation can still be critical and are most likely the source of problem. In particular, the chevron-notch width gap in these specimens is 0.5 mm, but, relative to specimen dimensions, this notch width gap could be effectively yeilding a notch strength in the ''over-load'' tests (Type 1 mode). In support of this contention, Barker [*9*] recently recommended that for short-rod and short-bar chevron specimens the notch width gap should be less than 0.03 B. The gap width in our studies was $N = 0.125B$, whereas the gap width for the specimens used by Munz et al [*1*] was 0.033 B (0.135 mm). This difference is significant because their specimens had nominally the same major dimensions as ours but yet they reported no unstable fractures in their tests. Another potential problem is that our specimen dimensions give a specimen that is quite rigid compared to the compliant testing machine. The energy stored in the testing machine may contribute to much more extensive crack popin than would be the case for a stiffer experimental arrangement.

The data taken for soda-lime-silica glass in H_2O and vitreous SiO_2 in 50% relative humidity argon suggest that environmentally enhanced crack growth can play a part in determining K_{Ic}. It is known that these glasses are susceptible to subcritical time-dependent crack growth in H_2O [*10*]; the position of the crack velocity (V) versus K_I-curve for each will shift to lower K_I with increasing H_2O activity [*11*]. Since K_{Ic} is actually just one point on a V-K_I curve, one can see that this point will then be environmentally sensitive. This environmental sensitivity also means that the measured K_{Ic}-value will be stressing rate dependent.

Conclusions

There are a number of conclusions one can draw from this work despite the problems incurred.

First, and most important, chevron-notched bend specimens will not necessarily ensure the initiation of a stable crack. Considerable overloads may have to be applied, at which point catastrophic failure occurs. It is hypothesized that this instability arises when the chevron-notch width gap exceeds some critical fraction of the specimen dimensions. It will be important to settle this point before chevron-notch testing can be considered to give reliable results. In the interim, a provisional value of $N < 0.03\ B$, as suggested by Barker [9] for chevron-notched short cantilever specimens, might be adopted. Another factor contributing to this instability is the energy stored in both the specimen and the testing machine.

Secondly, a smooth decrease in observed specimen compliance must appear before a test should be accepted as valid. Otherwise, considerable overestimation of K_{Ic} can occur. This point suggests that any data obtained under conditions where a loading curve cannot be recorded should only be accepted with some degree of caution.

Finally, tests on glass performed in moist environments, will generally yield lower values of K_{Ic} because moisture-assisted crack growth shifts the crack velocity curve to lower K_I-values. This is true not only for glass but for any material that exhibits time-dependent subcritical crack growth. This environmental dependence also means that measured values of K_{Ic} can be also a function of loading rate.

Acknowledgment

This work was supported in part by the U.S. Department of Energy, Advanced Research and Technology Division, Fossil Energy Materials Program, under interagency agreement DE-A105-800R20679.

References

[1] Munz, D., Bubsey, R. T., and Shannon, J. L., Jr., *Journal of the American Ceramic Society,* Vol. 63, Nos. 5–6, 1980, pp. 300–305.
[2] Munz, D., Bubsey, R. T., and Shannon, J. L., *Journal of Testing and Evaluation,* Vol. 8, No. 3, 1980, pp. 103–107.
[3] Shannon, J. L., Jr., and Munz, D. G., this publication, pp. 270–280.
[4] Krause, R. F., Jr., and Fuller, E. R., Jr., this publication, pp. 309–323.
[5] Munz, D. G., Shannon, J. L., Jr., and Bubsey, R. T., *International Journal of Fracture,* Vol. 16, 1980, pp. R137–R141.
[6] Bluhm, J. I., *Engineering Fracture Mechanics,* Vol. 7, 1975, pp. 593–604; in *Fracture 1977,* ICF4, D. M. R. Taplin, Ed., Waterloo Press, Waterloo, Ont. 1977, Vol. 3, pp. 409–417.
[7] Srawley, J. E. and Gross, B., in *Cracks and Fracture, ASTM STP 601,* American Society for Testing and Materials, Philadelphia, 1976, pp. 559–579.
[8] Wiederhorn, S. M., *Journal of the American Ceramic Society,* Vol. 52, 1969, pp. 99–105.
[9] Barker, L. M., *Engineering Fracture Mechanics,* Vol. 17, 1983, pp. 289–312.
[10] Wiederhorn, S. M. and Bolz, L. H., *Journal of the American Ceramic Society,* Vol. 53, 1970, pp. 543–548.
[11] Wiederhorn, S. M., Freiman, S. W., Fuller, E. R., Jr., and Simmons, C. J., *Journal of Material Science,* Vol. 17, 1982, pp. 3460–3478.

Wu Shang-Xian[1]

Compliance and Stress-Intensity Factor of Chevron-Notched Three-Point Bend Specimen

REFERENCE: Wu Shang-Xian, "**Compliance and Stress-Intensity Factor of Chevron-Notched Three-Point Bend Specimen,**" *Chevron-Notched Specimens: Testing and Stress Analysis, ASTM STP 855*, J. H. Underwood, S. W. Freiman, and F. I. Baratta, Eds., American Society for Testing and Materials, Philadelphia, 1984, pp. 176–192.

ABSTRACT: An analytical dimensionless compliance formula and normalized load versus load-point displacement plot of chevron-notched three-point bend specimen are derived by use of Bluhm's slice model. The stress intensity factor coefficients are calculated for specimen geometries of W/B = 1.5 and 2, θ = 40° to 90°, and are fitted as polynomials. Fracture-toughness values K_{Icv} of two heat-treatment conditions of GCr15 bearing steel are determined by chevron-notched specimens and compared with the ASTM E 399 valid fracture-toughness values. The comparisons show that for GCr15-I specimens with flat R-curves, K_{Icv} and K_{Ic} are in agreement, but for the GCr15-IV specimens with rising R-curves, K_{Icv} are near to K_{max}, instead of K_{Ic}. The effect of machining tolerance of chevron notch on the measurements is analyzed.

KEY WORDS: fracture toughness, compliance, stress-intensity factor, fracture test, chevron-notched three-point bend specimen, bearing steel

The usefulness and advantage of chevron-notched specimens for the determination of plane-strain fracture toughness of extremely brittle materials have been shown by a large number of investigations [*1–17*]. With the chevron-notched specimen, fracture toughness is determined from maximum load P_{max} with no need for a fatigue precracking operation. The chevron-notched specimen types that have been studied are: short rod [*1–6,10*], short bar [*4–10*], four-point bend [*11–13*], and three-point bend [*14–17*]. The three-point bend specimen is one of the most useful fracture mechanics test specimens. Chevron-notched three-point bend specimens are easy to machine and do not need specially designed loading devices.

In this paper, an analytical formula of dimensionless compliance of chevron-notched three-point bend specimen is derived by use of Bluhm's slice model. The normalized load versus load-point displacement plots are calculated. The

[1]Engineer, Iron and Steel Research Institute, Anshan Iron and Steel Complex, Anshan, Liaoning Province, China.

stress-intensity factor coefficient formulas are given for a wide range of specimens and notch geometries. The plane-strain fracture toughness K_{Icv} determined by chevron-notched specimens are compared with the standard K_{Ic}-values determined according to ASTM Standard Test for Plane-Strain Fracture Toughness of Metallic Materials, (E 399-83) for materials with flat and rising plane-strain crack-growth resistance curves (R-curves). The method of machining the chevron notch is described, and the effect of notch machining tolerance on measurement result is discussed.

Basic Relations

The Irwin-Kies relation for chevron-notched three-point bend specimen can be derived by use of the energy approach of linear elastic fracture mechanics [*8,18*] as follows

$$K_{Ic} = \frac{P}{B\sqrt{W}} Y(\alpha_0, \alpha_1, \alpha) \tag{1}$$

$$Y(\alpha_0, \alpha_1, \alpha) = \left\{ \frac{1}{2} \frac{dC_v(\alpha)}{d\alpha} \frac{\alpha_1 - \alpha_0}{\alpha - \alpha_0} \right\}^{1/2} \tag{2}$$

where

$\alpha_0 = a_0/W$,
$\alpha_1 = a_1/W$, and
$\alpha = a/W$,

the relative lengths of chevron notch and crack, as shown in Fig. 1*b*. When the crack grows to length a, the load applied on the specimen is P, and the crack front becomes trapezoidal, as shown in Fig. 1*b*. $C_v(\alpha)$ is the dimensionless compliance of the specimen with the trapezoidal crack front, $C_v(\alpha) = BE'q/P$, q is load-point displacement and $E' = E/(1 - \nu^2)$ for plane strain.

When a chevron-notched specimen is loaded, as shown in Fig. 1*a*, a crack develops at the chevron tip. During the early stage of crack growth, $Y(\alpha_0, \alpha_1, \alpha)$ decreases with increasing α. If the crack-growth resistance curve (R-curve) of material is flat, that is, K_{Ic} remains constant during crack growth, load P must be increased in order to extend the crack, so the crack growth is stable. When the crack length α increases to a critical value α_c, the $Y(\alpha_0, \alpha_1, \alpha)$ reaches a minimum $Y_c(\alpha_0, \alpha_1) = Y(\alpha_0, \alpha_1, \alpha_c)$, and, at the same time, the load P reaches a maximum P_{max}. Thereafter, the crack growth becomes unstable. Therefore, the K_{Ic} can be calculated from P_{max} and $Y_c(\alpha_0, \alpha_1)$

$$K_{Ic} = \frac{P_{max}}{B\sqrt{W}} Y_c(\alpha_0, \alpha_1) \tag{3}$$

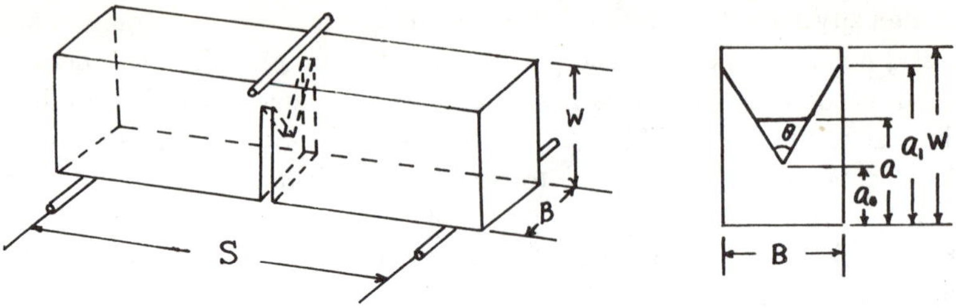

FIG. 1—*Geometry of chevron-notched three-point bend specimens.*

Seeking the minimum of $Y(\alpha_0, \alpha_1, \alpha)$ from Eq 2 leads to the equation determining the dimensionless critical crack length α_c

$$\frac{1}{\alpha - \alpha_0} \frac{dC_v(\alpha)}{d\alpha} = \frac{d^2C_v(\alpha)}{d^2\alpha} \tag{4}$$

Substituting the α_c solved from Eq 4 into Eq 2 gives the stress-intensity factor coefficient

$$Y_c(\alpha_0, \alpha_1) = Y(\alpha_0, \alpha_1, \alpha_c) = \left\{\frac{1}{2} \frac{dC_v(\alpha)}{d\alpha} \frac{\alpha_1 - \alpha_0}{\alpha - \alpha_0}\right\}^{1/2}_{\alpha = \alpha_c} \tag{5}$$

There are three methods for determination of $Y_c(\alpha_0, \alpha_1)$ coefficient. The first one is the direct experimental calibration of $Y_c(\alpha_0, \alpha_1)$ coefficient with K_{Ic}-value obtained by ASTM E 399 standard method, as done in Refs *1,14,15*. The direct calibration is dependent on the behavior of the calibrated material and is restricted to a single specimen geometry. The second method is the experimental calibration of dimensionless compliance $C_v(\alpha)$, as done in Refs *4,5*. The compliance calibration is dependent on the loading device used in the calibration. In addition, the results of $Y_c(\alpha_0, \alpha_1)$ are dependent on the fitting functions of $C_v(\alpha)$.

The third method is a calculation of compliance $C_v(\alpha)$ of a chevron-notched specimen from the dimensionless compliance $C_s(\alpha)$ of a straight-through crack specimen with some approximate methods. The simplest approximate method is the straight-through-crack assumption (STCA) proposed by Munz et al [*11,12*]

$$\frac{dC_v(\alpha)}{d\alpha} = \frac{dC_s(\alpha)}{d\alpha} \tag{6}$$

In Refs *8, 12,* and *16* the STCA were used to calculate the stress-intensity factor coefficients $Y_c(\alpha_0, \alpha_1)$ for chevron-notched short-bar, four-point bend, and three-point bend specimens, respectively.

A more refined approximate method is the slice model proposed by Bluhm [*19,20*]

$$\frac{1}{C_v(\alpha)} = \frac{\alpha - \alpha_0}{\alpha_1 - \alpha_0}\frac{1}{C_s(\alpha)} + \frac{k}{\alpha_1 - \alpha_0}\int_{\alpha}^{\alpha_1}\frac{1}{C_s(\xi)}\,d\xi \tag{7}$$

where k is the shear transfer coefficient. The slice model was applied to the chevron-notched four-point bend [*12*] and three-point bend specimens [*17*].

Compliance and Load-Deflection Plot

In Ref *17* we selected the following compliance formula for straight-through crack three-point bend specimen proposed by Chen et al [*21*]

$$C_s(\alpha) = \gamma + \beta\tan^2\left(\frac{\pi\alpha}{2}\right) \tag{8}$$

with

$$\gamma = \frac{1}{4}\left(\frac{S}{W}\right)^3\left[1 + 3(1 + \nu)\left(\frac{W}{S}\right)^2\right] \tag{9}$$

$$\beta = \frac{2}{\pi}\left(\frac{S}{4W}\right)^2\left[7.31 + 0.21\sqrt{\frac{S}{W} - 2.9}\right]^2 \tag{10}$$

For a specimen of $S/W = 4$ and $\nu = 0.3$

$$C_s(\alpha) = 19.9 + 36.097\tan^2\left(\frac{\pi\alpha}{2}\right) \tag{11}$$

As pointed up in Ref *17*, the compliance formula (Eq 11) and the stress-intensity factor coefficient derived from it according to the Irwin-Kies relation are accurate enough. The main advantage of Eq 8 is that it makes the integral in Eq 7 intergradable, so that the uncertainty related to the fitting function of $C_v(\alpha)$ is avoided.

Substituting Eq 8 into Eq 7 leads to an analytical expression of the dimensionless compliance of the chevron-notched three-point bend specimen

$$\frac{1}{C_v(\alpha)} = \frac{\alpha - \alpha_0}{\alpha_1 - \alpha_0}\frac{1}{\gamma + \beta\tan^2\pi\alpha/2} + \frac{k}{(\alpha_1 - \alpha_0)(\gamma - \beta)}\Bigg\{(\alpha_1 - \alpha) - \frac{2}{\pi}\sqrt{\frac{\beta}{\gamma}}\left[\arctan\left(\sqrt{\frac{\beta}{\gamma}}\tan\frac{\pi\alpha_1}{2}\right) - \arctan\left(\sqrt{\frac{\beta}{\gamma}}\tan\frac{\pi\alpha}{2}\right)\right]\Bigg\} \tag{12}$$

According to Ref *20*

$$k = \begin{cases} 1 + 0.444(\alpha_1)^{3.12} & \text{for } \phi \geq 1 \\ 1 + (\alpha_1)^{3.12}\,(2.236\phi - 4.744\phi^2 + 4.699\phi^3 - 1.77\phi^4) & \text{for } \phi < 1 \end{cases} \tag{13}$$

where

$$\phi = \tfrac{1}{2}(\pi - \theta).$$

As shown in Ref *17,* the compliance values calculated from Eq 12 are in good agreement with the experimentally calibrated values given by Bluhm [*19*].

For chevron-notched three-point bend specimens of $S/W = 4$, $W/B = 1.5$, $\theta = 60°$, $\alpha_0 = 0.1, 0.2, 0.3, 0.4$, the dimensionless compliances calculated from Eq 12 are represented in Fig. 2. The $Y(\alpha_0, \alpha_1, \alpha)$ coefficients calculated from Eq 5 are represented in Fig. 3.

From Eq 1 and the definition of $C_v(\alpha)$, $C_v(\alpha) = qE'B/P$, the normalized load $P/K_{Ic}B\sqrt{W}$ and normalized load-point displacement $qE'/K_{Ic}\sqrt{W}$ can be expressed as

$$\frac{P}{K_{Ic}B\sqrt{W}} = \frac{1}{Y(\alpha_0, \alpha_1, \alpha)} \tag{14}$$

$$\frac{qE'}{K_{Ic}\sqrt{W}} = \frac{C_v(\alpha)}{Y(\alpha_0, \alpha_1, \alpha)} \tag{15}$$

The right sides of Eqs 14 and 15 are known, so for a given α_0, α_1, and α, the normalized load and load-point displacement can be calculated. Therefore, the normalized load versus load-point displacement plot during loading and crack extension can be also calculated. Figure 4 shows the normalized load versus load-point displacement calculated for the specimen geometries in Figs. 2 and 3. It can be seen that the maximum normalized loads are at about the same normalized displacement. This is true for other specimen and notch geometries. For $0.1 \leq \alpha_0 \leq 0.5$, the maximum normalized loads appear at the normalized load-point displacement $qE'/K_{Ic}\sqrt{W} \approx 4 \sim 5$, and do not vary strongly with W/B and θ.

The normalized load versus load-point displacement plots in Fig. 4 are calculated for the specimens with ideal chevron cracks; therefore, the crack extends from the beginning of loading. For the practical chevron-notched specimen, the load-displacement plot has an initial linear portion. When crack develops at the chevron tip, the plot begins to deviate from linearity.

The load versus displacement plots in Fig. 4 are calculated for elastic conditions and for material having a flat R-curve. If the R-curve of the material is rising,

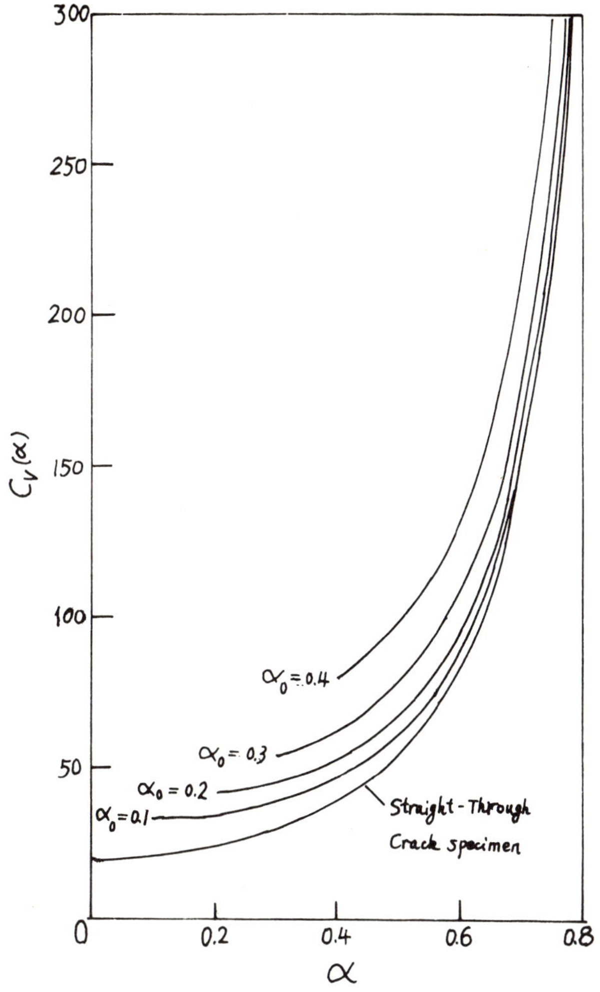

FIG. 2—*Dimensionless compliance of straight-through crack and chevron-notched specimens of* S/W = *4,* W/B = *1.5,* θ = *60°,* α_0 = *0.1, 0.2, 0.3, 0.4.*

the plot will move upward and towards the right. Plastic deformation in specimen will make the plot to move towards the right.

Stress-Intensity Factor Coefficient

Stress-intensity factor coefficients were calculated by use of compliance formula (Eq 12) and Eqs 4 and 5 for specimen geometries of $S/W = 4$, $W/B = 1.5$ and 2, $\theta = 40°, 50°, 60°, 70°, 80°, 90°$. In designing and machining the chevron-

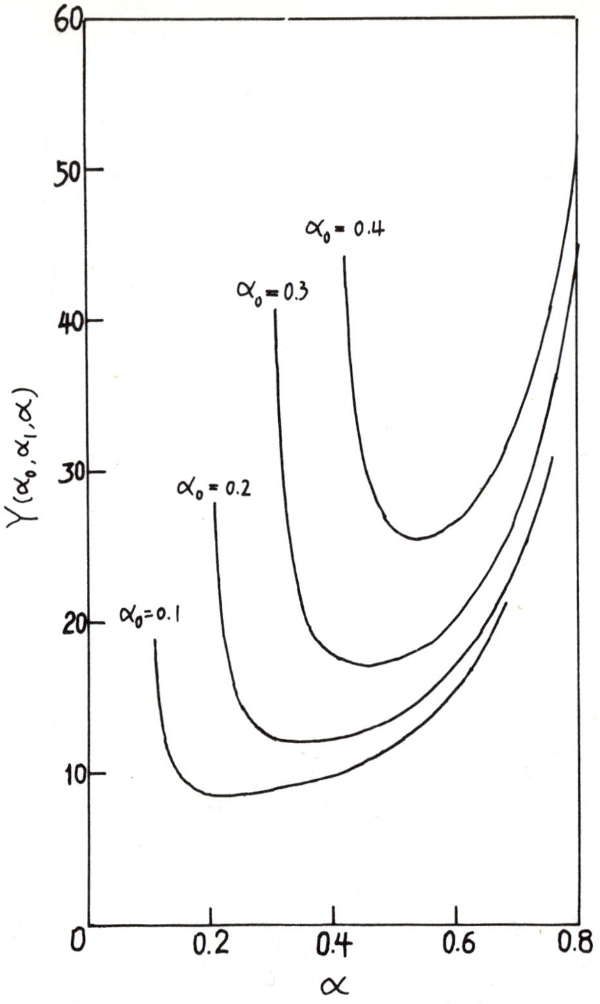

FIG. 3—$Y(\alpha_0, \alpha_1, \alpha)$-*values for chevron-notched three-point bend specimens of* W/S = *4,* W/B = *1.5,* θ = *60°,* α_0 = *0.1, 0.2, 0.3, 0.4.*

notched specimen, we control the chevron angle θ and the depth of chevron tip α_0; therefore, the calculated results are expressed as function of θ and α_0. For given W/B and θ, $Y_c(\theta, \alpha_0)$ were calculated for α_0 from 0 to $1 - B/2W \cot \theta/2$ in 0.01 steps. The calculated results are represented by the curves in Figs. 5 and 6, and are fitted by following polynomials

$$Y_c(\theta, \alpha_0) = d_0 + d_1\alpha_0 + d_2\alpha_0^2 + d_3\alpha_0^3 + d_4\alpha_0^4 + d_5\alpha_0^5 \tag{16}$$

The d's coefficients are given in Table 1.

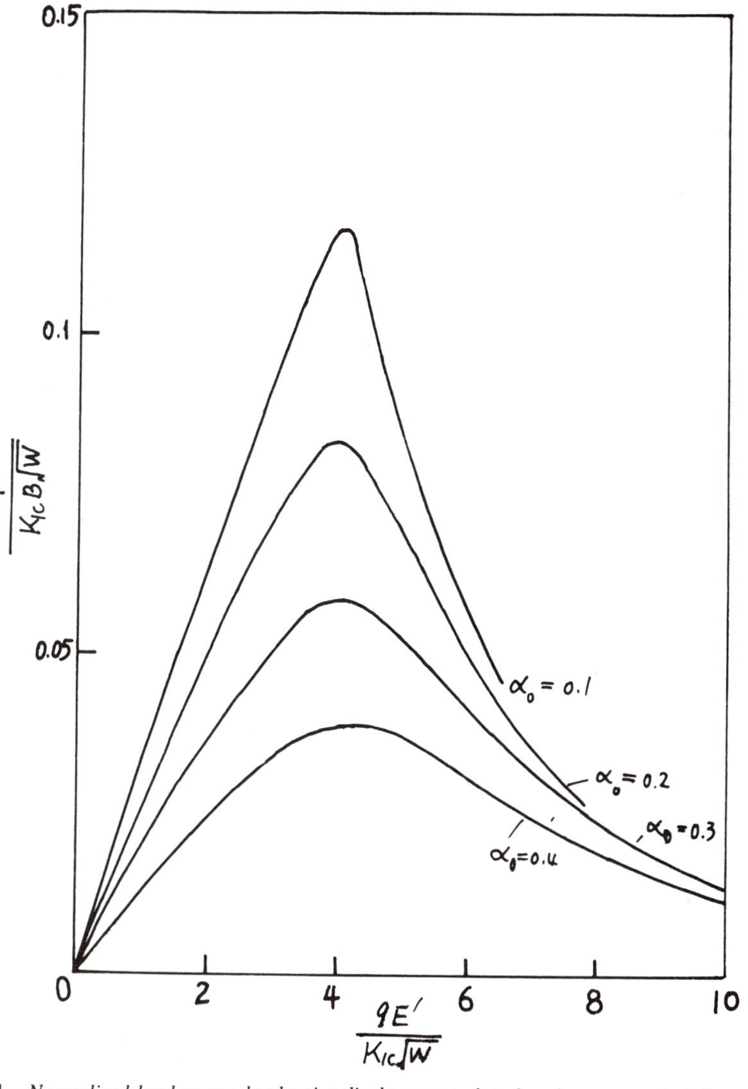

FIG. 4—*Normalized load versus load point displacement plots for chevron-notched three-point bend specimens of* S/W = *4*, W/B = *1.5*, θ = *60°*, α_0 = *0.1, 0.2, 0.3, 0.4.*

In addition, the calculation was also made for the specimen geometry studied by Shih [*15*]. Shih experimentally calibrated stress-intensity factor coefficient for the chevron-notched three-point bend specimen of $B = 0.55$ in., $W = 1$ in., $a_0 = 0.3$ in., and $a_1 = 0.6$ in. With the symbols of the present paper, the $Y_c(\alpha_0, \alpha_1)$-value calculated by Shih is 10.16. Our calculated value of $Y_c(\alpha_0, \alpha_1)$ for Shih's specimen is 10.60, which is in good agreement with the calibrated value.

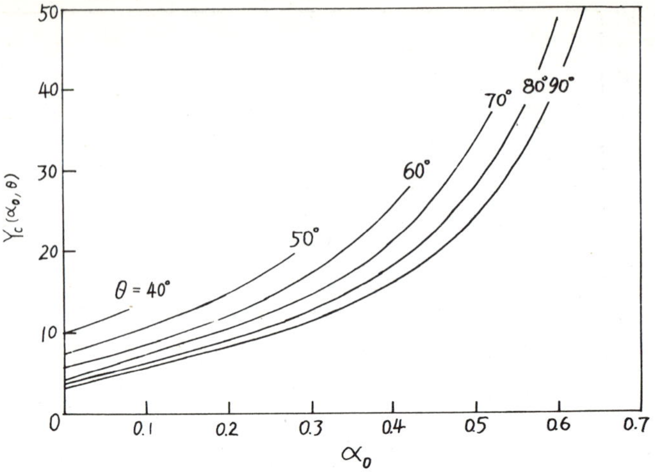

FIG. 5—*Calculated stress-intensity factor coefficients* $Y_c(\theta, \alpha_0)$ *for specimens of* W/B = *1.5.*

Machining Method and Machining Tolerance of the Chevron Notch

The chevron notch was machined by a program-controlled electrodischarge machine (EDM). The specimen was mounted on a specimen stage, which has an oblique plane at an angle of $\theta/2$. The specimen stage design is as shown in Fig. 7. The chevron notch is finished by machining twice, and the feed distance in each machining is $1 = \alpha_1 W \sin \theta/2 = (\alpha_0 + B/2W \cot \theta/2)$, as shown in Fig. 7. Figure 8 shows the fracture appearances of various chevron-notched specimens machined by this method.

The two sides of the chevron notch machined by the preceding method make

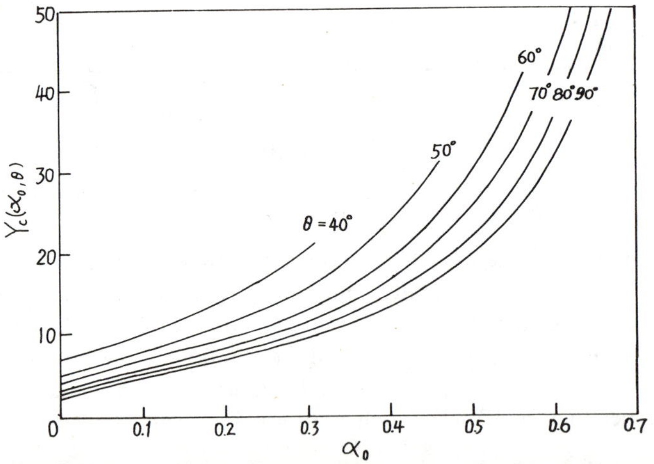

FIG. 6—*Calculated stress-intensity factor coefficients* $Y_c(\theta, \alpha_0)$ *for specimens of* W/B = 2.

TABLE 1—*Values of* d*'s coefficients.*

W/B	θ	d_0	d_1	d_2	d_3	d_4	d_5
1.5	40°	10.05	35.60	31.12	122.1	0	0
	50°	7.314	31.06	10.81	90.38	164.3	0
	60°	5.639	27.44	18.93	−43.42	338.9	0
	70°	4.428	30.61	−62.06	401.2	−874.3	1075
	80°	3.567	35.18	−142.9	783.6	−1703	1630
	90°	2.809	44.51	−269.6	1338	−2736	2242
2	40°	6.957	30.45	9.780	72.24	195.2	0
	50°	5.187	25.68	30.53	−109.5	404.6	0
	60°	3.987	32.12	−94.71	55.74	−1233	1320
	70°	3.087	40.59	−218.6	1122	−2350	2026
	80°	2.282	57.01	−420.8	1972	−3841	2880
	90°	1.488	77.60	−648.3	2848	−5225	3592

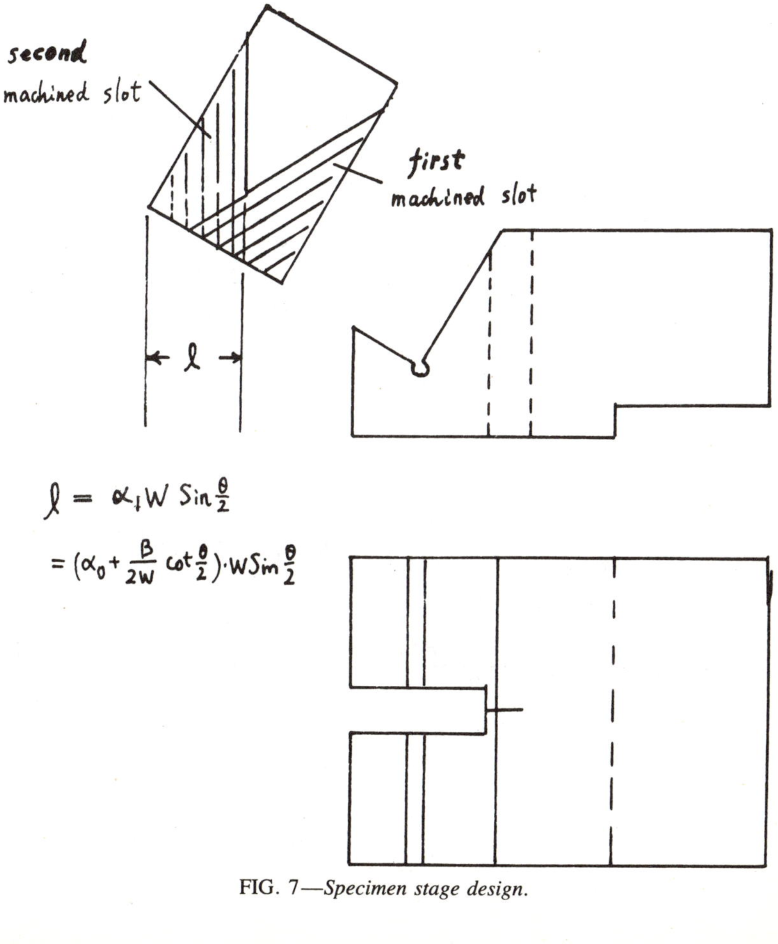

FIG. 7—*Specimen stage design.*

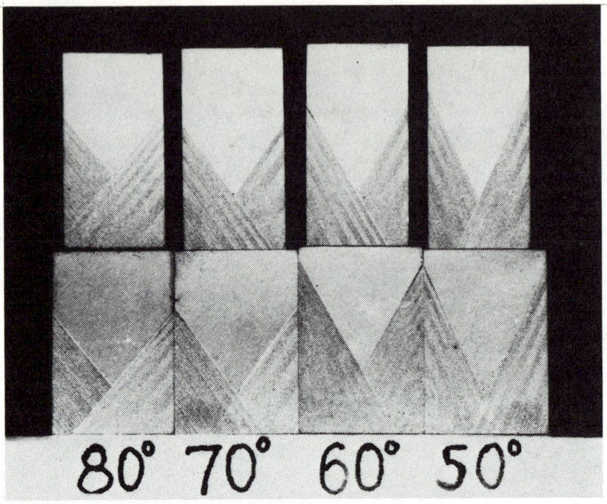

FIG. 8—*Fracture appearances of various chevron-notched three-point bend specimens.*

exactly an angle of $\theta/2$ with the midthickness plane of the specimen. The angle included between the two sides is exactly θ. Therefore, there is little or no variation in angle θ, but a difference between the two feeding lengths may cause a deviation of the chevron notch from the midthickness plane, as shown in Fig. 9. To evaluate the effect of the deviation on the stress-intensity factor coefficient $Y_c(\alpha_0, \alpha_1)$, we derive the dimensionless compliance of the specimen with a deviated chevron notch by use of the slice model. The result is

$$\frac{1}{C_v'(\alpha)} = \frac{1}{C_v(\alpha)} + \Delta(F) \tag{17}$$

where

$$\Delta(F) = \frac{k}{\pi(\alpha - \alpha_0)(\gamma - \beta)}\sqrt{\frac{\beta}{\gamma}}\left[2\arctan\left(\sqrt{\frac{\beta}{\gamma}}\tan\frac{\pi\alpha_1}{2}\right)\right.$$
$$\left. - \arctan\left(\sqrt{\frac{\beta}{\gamma}}\tan\frac{\pi(\alpha_1 + F)}{2}\right) - \arctan\left(\sqrt{\frac{\beta}{\gamma}}\tan\frac{\pi(\alpha_1 - F)}{2}\right)\right] \tag{18}$$

$$F = \frac{f}{W}\cot\frac{\theta}{2} \tag{19}$$

$$\alpha_1 = \frac{a_1}{W} = \frac{B}{2W}\cot\frac{\theta}{2} + \alpha_0 \tag{20}$$

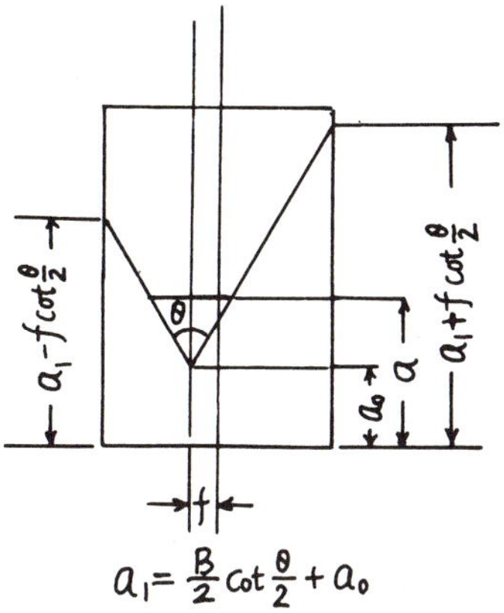

FIG. 9—*Cross section of deviated chevron notch.*

$C_v'(\alpha)$ is the dimensionless compliance of the specimen with chevron-notched deviation of f, as shown in Fig. 9. $C_v(\alpha)$ is the dimensionless compliance of the specimen without deviation as expressed by Eq 12.

Calculating the stress-intensity factor coefficients $Y_c(\theta, \alpha_0, F)$ with the compliance formula (Eq 17) for different F-values and comparing them with the $Y_c(\theta, \alpha_0)$ for $F = 0$, we can evaluate the effect of the chevron-notched deviation. By this method, we have found a correction factor for the chevron-notched deviation

$$p = \begin{cases} 1 & \text{for } F \leqslant 0.035 \\ 1 + 1/2\,(F - 0.035) & \text{for } F > 0.035 \end{cases} \tag{21}$$

For the specimen with a deviation of the chevron notch the stress-intensity factor coefficient formula (Eq 16) should be multiplied by the correction factor p.

Material and Experimental Procedure

The test material, GCr15 bearing steel, contained (in weight percent) 0.94C, 0.51Si, 1.09Mn, 1.40Cr, 0.019P, and 0.046S. After spheroidize annealing, forged square bars of GCr15 steel were machined to specimen blanks of $B = 10$ mm, $W = 20$ mm, and $B = 12$ mm, $W = 18$ mm. The specimen blanks were austenitized at 840°C and subsequently quenched and tempered to two different groups of specimens, GCr15-I (tempered at 220°C) and GCr15-IV (tempered at 160°C). Finally, the specimens were ground. The two groups of specimens have different hardness, fracture toughness, as shown in Table 2.

Some specimens of $B = 10$ mm and $W = 20$ mm were used to measure the standard fracture toughness according to ASTM E 399-83 standard method.

Chevron notches of $\theta = 50°$, 60°, 70°, and 80°, were machined by EDM on

TABLE 2—*Mechanical properties of two groups of specimens.*

Specimen Group	Hardness, HRC	K_{Ic}, MPa $m^{1/2}$ Mean	Standard Deviation	K_{max}, MPa $m^{1/2}$ Mean	Standard Deviation	Number of Valid Tests
GCr15-I	60	12.71	0.38	12.71	0.38	5
GCr15-IV	62	15.32	0.32	16.44	0.38	3

specimens of B = 10 mm, W = 20 mm, and B = 12 mm, W = 18 mm. The slot width of the GCr15-I specimens of B = 12 mm and W = 18 mm and $\theta = 60°$ was 0.16 mm, and the slot width of all other specimens was 0.11 mm. Chevron-notched specimens with different α_0 were loaded on a testing machine with crosshead velocity of 0.3 mm/min. Load versus crack mouth opening displacement (*P-V*) plots were recorded. The lengths of chevron tip and two chevron roots at specimen surfaces were measured on the broken specimens. The α_0 and F were calculated from the measurements.

Results and Discussions

Experimental Results of Standard K_{Ic}

P-V records of GCr15-I standard specimens were as the Fig. 5 Type III of ASTM E 399-83, that is, the *P-V* records remained straight lines before catastrophic fracture. As analyzed in Ref *17,* this type of record shows that the plane-strain R-curve of material is flat and that there are only very small amounts of stable crack growth before rapid fracture. Thus, for the GCr15-I specimens, the K_{max} calculated from maximum load is equal to K_{Ic}.

The *P-V* records of GCr15-IV were as Fig. 5 Type I in ASTM E 399-83. There were large deviations from the initial linear portions on the *P-V* records before load reached a maximum. Among the six tested GCr15-IV standard specimens, three satisfied the requirement of $P_{max}/P_Q \leq 1.1$, and these three specimens were used to obtain K_{Ic} and K_{max}. The mean values and standard deviations of K_{Ic} and K_{max} are given in Table 2. For the other three specimens, $P_{max}/P_Q > 1.1$.

As analyzed in Ref *17,* for the high-hardness and low-toughness materials, such as GCr15 bearing steel, the deviation of *P-V* record from the initial linearity results mainly from the stable crack growth. To confirm this view, we determined the plane-strain crack-growth resistance curve of GCr15-IV specimens by a multiple-specimen method. Some standard straight-through precracked GCr15-IV specimens were loaded to predetermined opening displacements and immediately unloaded. The unloaded specimens were fatigue cycled to extend the crack, and, finally, the specimens were broken. The crack lengths a and crack extensions Δa were measured on the broken specimens. K_R-values were calculated from the unloading unloaded points and a K_R-Δa (R-curve) was developed, as shown in Fig. 10. It can be seen that the R-curve of GCr15-IV specimens is rising.

The foregoing results show the specificity of fracture toughness determination of brittle metallic materials: the plane-strain requirement $B \geqslant 2.5(K_Q/\sigma_{ys})^2$ is easily satisfied and the main difficulty is to satisfy the requirement of $P_{max}/P_Q \leqslant 1.1$. Even if $B >> 2.5(K_Q/\sigma_{ys})^2$, the requirements of $P_{max}/P_Q \leqslant 1.1$ is not necessarily satisfied.

Experimental Results of Chevron-Notched Specimens

The *P-V* records show that all chevron-notched specimens experienced stable crack extensions before rapid fractures. For each chevron-notched specimen, a fracture toughness value K_{Icv} was calculated according to Eqs 3 and 16. The experimental results of two groups of specimens are shown in Figs. 11 and 12. In these figures, the K_{Ic} and K_{max} values also are shown.

The experimental results of GCr15-I in Fig. 11 show that when $\alpha_0 \geqslant 0.3$, the K_{Icv} and K_{Ic} are in agreement. For $\alpha_0 < 0.3$, the results obtained by using Eq 16 are an underestimate, which is consistent with the observation in Ref *17*. Therefore, for practical determination of fracture toughness by chevron-notched three-point bend specimen, the depth of chevron notch of $\alpha_0 \geqslant 0.3$ is recommended.

In addition, it can be seen from the results of Figs. 11 and 12, the fracture

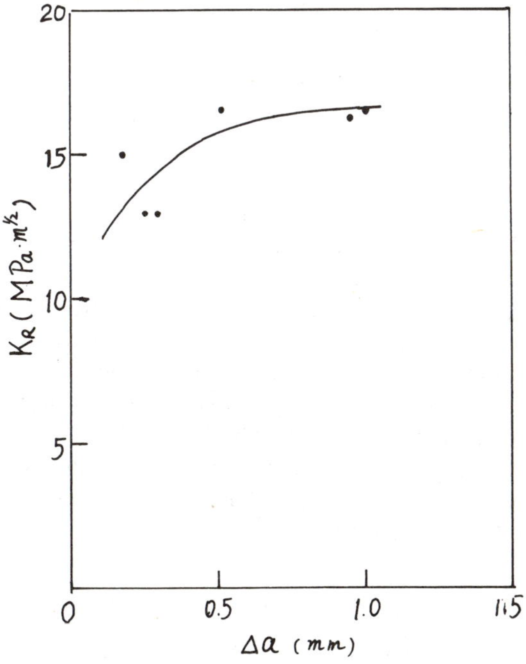

FIG. 10—K_R-Δa (*R-curve*) *of GCr15-IV specimens.*

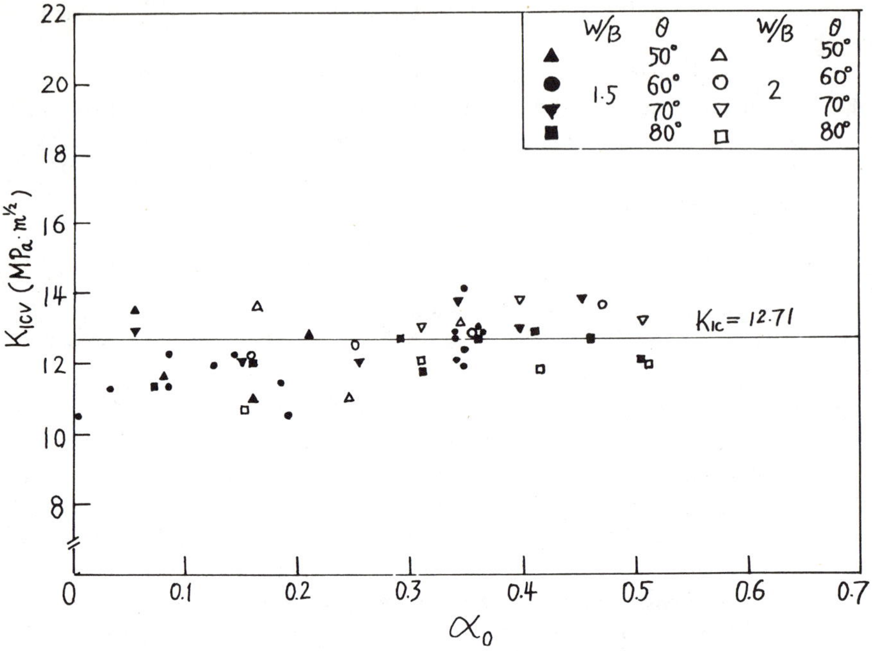

FIG. 11—*Experimental results of chevron-notched specimens of GCr15-I.*

toughness K_{Icv} determined by chevron-notched specimen are somewhat scattered. Therefore, there is further standardization required before the method of chevron-notched three-point bend specimen can become a practical and accurate method to determine fracture toughness.

For GCr15-IV specimens, K_{Icv}-values are near to K_{max}-values, instead of K_{Ic}, and generally, K_{Icv}-values are slightly higher than K_{max}. This result is expected. The chevron-notched specimen experiences a larger amount of stable extension before the crack length reaches the critical length, so the crack growth has reached the steady-state condition at the maximum load, as pointed up by Barker [*18*]. Therefore, the fracture toughness K_{Icv} determined by chevron-notched specimen from the maximum load is the steady-state fracture toughness of material. For the material with flat R-curve, the steady-state fracture toughness is in agreement with the fracture toughness K_{Ic} determined by ASTM E 399-83 standard method. But for the material with rising R-curve, the steady-state value is not in agreement with K_{Ic}, and may be near to or higher than K_{max}. This phenomenon has been noted and analyzed by Munz [*10*].

Conclusions

1. The dimensionless compliance formula and normalized load versus load-point displacement plot of chevron-notched three-point bend specimen are derived by use of Bluhm's slice model.

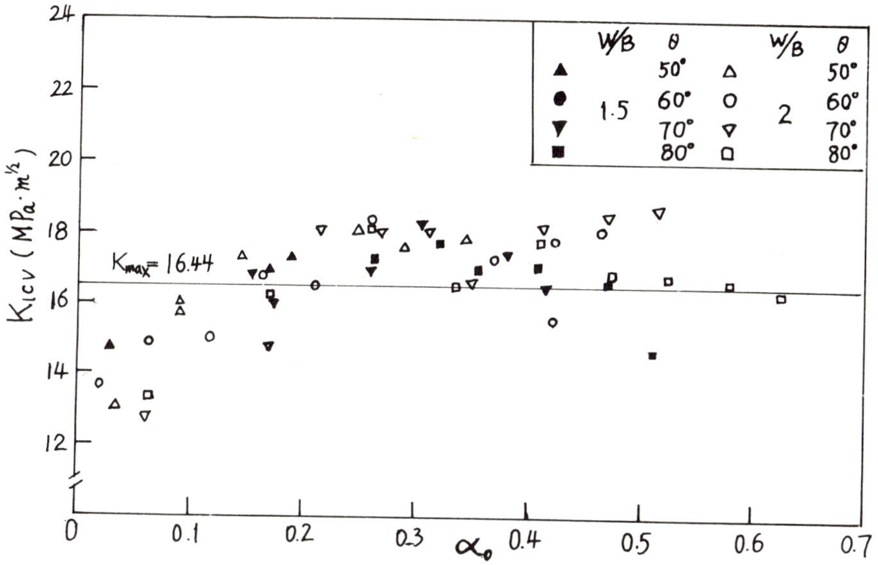

FIG. 12—*Experimental results of chevron-notched specimens of GCr15-IV.*

2. The stress-intensity factor coefficient $Y_c(\theta, \alpha_0)$ are calculated for chevron-notched three-point bend specimens of $W/B = 1.5$ and 2, $\theta = 40°$ to 90°, and are fitted as polynomials.

3. The fracture toughness K_{Icv}-values of two heat-treatment conditions of GCr15 bearing steel are determined by chevron-notched three-point bend specimens of various specimen and notch geometries and compared with the fracture toughness K_{Ic} and K_{max}-values determined according to the ASTM E 399-83 standard method. For GCr15-I specimens, K_{Icv} and K_{Ic} are in agreement. For GCr15-IV specimens, K_{Icv} is slightly higher than K_{max}.

4. The fracture toughness determined by chevron-notched specimen from maximum load is the steady-state fracture toughness of material. For the materials with flat R-curve, the steady-state value is in agreement with K_{Ic}. For the materials with rising R-curve, the steady-state fracture toughness is equal to or higher than K_{max}.

5. The stress-intensity factor coefficient formula (Eq 16) is more accurate for $\alpha_0 \geq 0.3$. Therefore, chevron-notched three-point bend specimen with notch depth of $\alpha_0 \geq 0.3$ is recommended for the determination of fracture toughness.

6. The experimental results of fracture toughness K_{Icv} determined by chevron-notched three-point bend specimens are somewhat scattered. Further standardization has to be done for this type of specimen.

References

[1] Barker, L. M., *Engineering Fracture Mechanics,* Vol. 9, 1977, pp. 361–369.

[2] Barker, L. M. in *Proceedings,* Second International Conference on Mechanical Behavior of Materials, American Society for Metals, Metal Park, Ohio, 1976, pp. 1547–1551.

[3] Barker, L. M. and Baratta, F. I., *Journal of Testing and Evaluation,* Vol. 8, 1980, pp. 97–102.
[4] Bubsey, R. T., Munz, D., Pierce, W. S., and Shannon, J. L., Jr., *International Journal of Fracture,* Vol. 18, 1982, pp. 125–133.
[5] Barker, L. M. in *Fracture Mechanics: Thirteenth Conference, ASTM STP 743,* Richard Roberts, Ed., American Society for Testing and Materials, Philadelphia, 1981, pp. 456–475.
[6] Shannon, J. L., Jr., Bubsey, R. T., Pierce, W. S., and Munz, D., *International Journal of Fracture,* Vol. 19, 1982, pp. R55–R58.
[7] Barker, L. M. in *Fracture Mechanics Applied to Brittle Materials, ASTM STP 678,* American Society for Testing and Materials, Philadelphia, 1979, pp. 73–82.
[8] Munz, D., Bubsey, R. T., and Srawley, J. E., *International Journal of Fracture,* Vol. 16, 1980, pp. 359–374.
[9] Munz, D., Bubsey, R. T., and Shannon, J. L., Jr., *Journal of Testing and Evaluation,* Vol. 8, 1980, pp. 103–107.
[10] Munz, D., *Engineering Fracture Mechanics,* Vol. 15, 1981, pp. 231–236.
[11] Munz, D., Bubsey, R. T., and Shannon, J. L., Jr., *Journal of the American Ceramic Society,* Vol. 6, 1980, pp. 300–305.
[12] Munz, D., Shannon, J. L., Jr., and Bubsey, R. T., *International Journal of Fracture,* Vol. 16, 1980, pp. R137–R141.
[13] Munz, D., Himsolt, G., and Eschweiler, J. in *Fracture Mechanics Methods for Ceramics, Rock, and Concrete, ASTM STP 745,* S. W. Freiman and E. R. Fuller, Eds., American Society for Testing and Materials, Philadelphia, 1981, pp. 69–84.
[14] Shih, T. T., *Journal of Testing and Evaluation,* Vol. 9, 1981, pp. 50–55.
[15] Shih, T. T., *Engineering Fracture Mechanics,* Vol. 14, 1981, pp. 821–832.
[16] Wu Shang-Xian, *International Journal of Fracture,* Vol. 19, 1982, pp. R27–R30.
[17] Wu Shang-Xian, "Fracture Toughness Determination of Bearing Steel Using Chevron-Notch Three Point Bend Specimen," to be published in *Engineering Fracture Mechanics.*
[18] Barker, L. M., *International Journal of Fracture,* Vol. 15, 1979, pp. 515–536.
[19] Bluhm, J. I., *Engineering Fracture Mechanics,* Vol. 7, 1975, pp. 593–604.
[20] Bluhm, J. I. in *Fracture 1977,* Vol. 3, pp. 409–417, D. M. R. Taplin, Ed., University of Waterloo Press, Waterloo, Ont., Canada, 1977.
[21] Chen Chih, Yao Heng, and Chin Chih-Ying, *Kexue Tongbao,* Vol. 19, 1974, pp. 41–45 (in Chinese).

Wang Chizhi,[1] *Yuan Maochan,*[1] *and Chen Tzeguang*[1]

An Investigation on the Method for Determination of Fracture Toughness K_{Ic} of Metallic Materials with Chevron-Notched Short-Rod and Short-Bar Specimens

---

REFERENCE: Wang Chizhi, Yuan Maochan, and Chen Tzeguang, "**An Investigation on the Method for Determination of Fracture Toughness K_{Ic} of Metallic Materials with Chevron-Notched Short-Rod and Short-Bar Specimens,**" *Chevron-Notched Specimens: Testing and Stress Analysis, ASTM STP 855,* J. H. Underwood, S. W. Freiman, and F. I. Baratta, Eds., American Society for Testing and Materials, Philadelphia, 1984, pp. 193–204.

ABSTRACT: By means of an expression proposed by the authors, dimensionless critical crack lengths and stress-intensity factor coefficients for chevron-notched short-rod specimens were calculated based on the straight-through crack assumption. Calculation was also done for chevron-notched short-bar specimens in an extended range including short bar with rectangular cross section. Three configurations of chevron-notched specimens were used to test fracture toughness K_{Ic} for eight metallic materials. Some results were compared with values from standard compact tension and three-point bend specimens. Good agreement was obtained when both stable crack extension and a very small plasticity factor were achieved. The role of load-displacement plot is emphasized. A discussion about some specimen parameters such as dimensionless initial crack length and width to half height (or width to radius) ratio is given.

KEY WORDS: stress-intensity factor, chevron notch, short rod, short bar, fracture toughness, steels, aluminum alloys

A new kind of fracture toughness specimen with chevron-notched, first proposed by Barker in short rod [*1,2*] and then by other authors in short bar [*3*], three-point bend specimens [*4*], and four-point bend specimens [*5*], greatly simplifies the testing procedure of fracture toughness K_{Ic} (mainly for brittle materials). No specimen fatigue precracking is needed since a sharp crack can be initiated from the apex of V-shaped ligament in loading process. Stress-intensity factor (SIF) coefficients can be calculated beforehand without the necessity of crack length measurement, because at the critical point the crack length is a constant for a specified geometry.

[1]Assistant, professor, and instructor, respectively, First Department of Mechanical Engineering, Chongqing University, Chongqing, Sichuan, China.

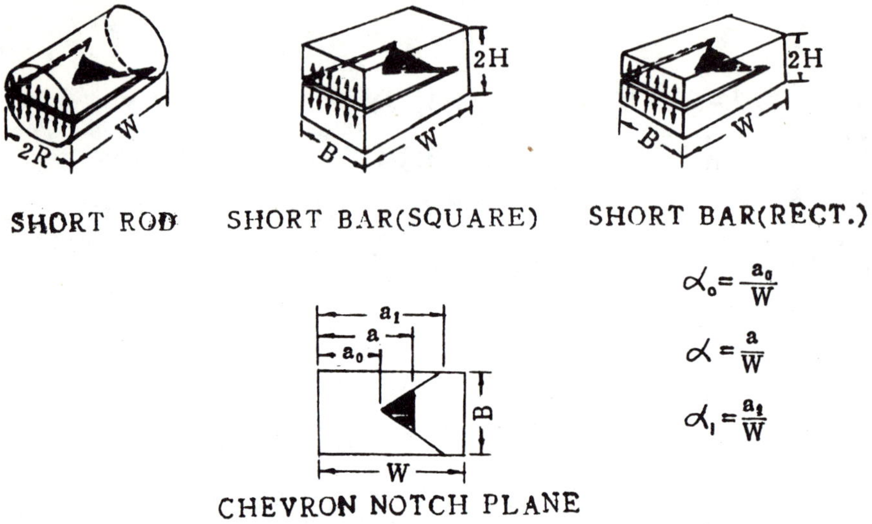

FIG. 1—*Three configurations of chevron-notched specimen.*

In this paper, the authors give some analytical results concerning calculation of dimensionless critical crack lengths and SIF coefficients for chevron-notched short-rod and short-bar specimens. The authors successfully applied straight-through crack assumption (STCA) [*3*] to short-rod specimens. Short rod as well as short bar with both square and rectangular cross section (Fig. 1) were used to test K_{Ic} for eight metallic materials (steels and aluminum alloys). Comparison is made with standard tests using compact tension (CT) and three-point bend (3PB) specimens. Five materials were tested successfully with chevron-notched specimens, their toughness values cover a wide range from less than 30 MPa$\sqrt{m}$ up to over 100 MPa$\sqrt{m}$. Three materials were not successful because of catastrophic failure (drastic crack jumps) or excessive plasticity of specimen.

Analytical Results

ASTM Test for Plain-Strain Fracture Toughness of Metallic Materials (E 399-81) prescribes that a secant through the origin with a 5% decrease in slope relative to the initial tangent should be drawn in the load-crack mouth opening displacement test record. The intercept load P_5 corresponds to 2% crack growth from the initial length. Usually this load, or possibly the maximum load which precedes P_5, is used to calculate fracture toughness K_{Ic}.

With chevron-notched short-rod and short-bar specimens, K_{Ic} is calculated from maximum load P_{max}

$$K_{Ic} = \frac{P_{max}}{B\sqrt{W}} \cdot Y_m{}^* \qquad (1)$$

where

B = thickness of specimen,
W = width of specimen, and
Y_m^* = minimum SIF coefficient.

The critical point at which K_{Ic} is measured with chevron-notched specimen corresponds to a much larger crack extension than that in ASTM Method E 399. Fortunately, for high-strength metallic materials, the K_R resistance curves are saturated after a small crack extension, for example, saturation is reached for four steels in Ref *6* when crack extension is only 0.2 to 0.5 mm. In such cases, two different measuring critical points are essentially identical.

STCA method proposed by Munz et al [*3*] can be used to determine critical crack length and SIF coefficient for chevron-notched specimen only if the SIF expression for corresponding straight-through cracked specimen is known. From STCA there is

$$Y^* = Y\left[\frac{\alpha_1 - \alpha_0}{\alpha - \alpha_0}\right]^{1/2} \tag{2}$$

where Y^* and Y are SIF coefficients for chevron-notched and straight-through cracked specimens, respectively, and α_0, α, α_1 are dimensionless initial, intermediate, and surface crack (notch) lengths, respectively, see Fig. 1.

Substitute Eq 2 into critical condition $dY^*/d\alpha = 0$, we obtained

$$\frac{dY}{d\alpha} - \frac{1}{2} \cdot \frac{1}{\alpha - \alpha_0} \cdot Y = 0 \tag{3}$$

The dimensionless critical crack length α_c is the root of Eq 3. Substituting α_c for α in Eq 2, Y_m^* is obtained.

Y for straight-through cracked bar was given in Ref *3*, whereas Y for straight-through cracked rod, that is, Y_{rod} was lacking. But Y_{rod} has been obtained through slice synthesis in Ref *8* as

$$Y_{rod} = \frac{\alpha}{(1 - \alpha)^{3/2}} \cdot \left\{\ln\left[\exp\left(\frac{2.702}{\alpha} + 1.628 + \exp(F_2)\right)\right]\right\} \tag{4a}$$

and

$$F_2 = \frac{M}{\sum_{n=1}^{M}\left\{\sqrt{12\frac{W}{R}\cdot\frac{M}{\sqrt{M^2-(n-1)^2}}\cdot(1-\alpha)^3\cdot\left(1+\frac{0.679}{\alpha}\cdot\frac{R}{W}\cdot\frac{\sqrt{M^2-(n-1)^2}}{M}\right)}\right\}^{-1}} \tag{4b}$$

where

R = radius of rod, and
M = being a parameter.

Calculated values of Y_m^* for chevron-notched short-rod specimen through Eqs 2 to 4 converge rapidly when M increases. It is sufficient to take $M = 8$.

In order to ease calculation of Y_m^*, it is desirable to have polynomial expressions, which have been obtained through fitting Y_m^*-values from computer output. The short-bar specimen with a rectangular cross section used in this investigation is out of the bounding range of the expression for Y_m^* in Ref *3*, so a similar expression for short-bar specimen with an extended bounding range was obtained as

$$Y_m^* = \left\{11.304 + 2.121\frac{W}{H} + 0.664\left(\frac{W}{H}\right)^2 + \left[-80.428 + 17.533\frac{W}{H} - 0.336\left(\frac{W}{H}\right)^2\right]\alpha_0 + \left[255.713 - 63.641\frac{W}{H} + 4.695\left(\frac{W}{H}\right)^2\right]\alpha_0^2\right\} \cdot \left[\frac{\alpha_1 - \alpha_0}{1 - \alpha_0}\right]^{1/2} \tag{5}$$

for $3 \leqslant W/H \leqslant 4.6$ and $0.2 \leqslant \alpha_0 \leqslant 0.5$. In Eq 5, H denotes half height, see Fig. 1. The difference between Eq 5 and computer output in the whole bounding range is within -1 to 1.3%.

An expression for short-rod specimen was obtained as

$$Y_m^* = \left\{7.573 + 4.748\frac{W}{R} + 0.625\left(\frac{W}{R}\right)^2 + \left[-66.728 + 13.236\frac{W}{R} + 0.55\left(\frac{W}{R}\right)^2\right]\alpha_0 + \left[216.365 - 53.459\frac{W}{R} + 3.542\left(\frac{W}{R}\right)^2\right]\alpha_0^2\right\} \cdot \left[\frac{\alpha_1 - \alpha_0}{1 - \alpha_0}\right]^{1/2} \tag{6}$$

for $3 \leqslant W/R \leqslant 4$ and $0.2 \leqslant \alpha_0 \leqslant 0.5$. The difference between Eq 6 and computer output in the whole bounding range is within $\pm 0.7\%$. By the time this work was finished there had been no polynomial expression of Y_m^* for short-rod specimens. Later on we saw a similar expression from compliance calibration

TABLE 1—*Compositions of tested materials.*

Steel	Composition, %									
	C	Si	Mn	P	S	Ni	Cr	Cu	Mo	V
30Cr-Ni-2Mo-V-A	0.28	0.19	0.48	0.13	0.006	2.27	0.77	≤0.15	0.25	0.22
60Si-2Mn-A	0.61	1.88	0.80	0.018	0.012	0.10	0.27	0.10		
30Cr-Mn-Si-A	0.3	1.05	0.95				0.95			
20G	0.17	0.21	0.57	0.016	0.021					

Aluminum Alloy	Composition, %					
	Cu	Mg	Mn	Zn	Cr	Si
LC4CZ	1.4 to 2.0	1.8 to 2.8	0.2 to 0.6	5.0 to 7.0	0.10 to 0.25	
LY11CZ	4.53	0.77	0.49			
LD10	3.9 to 4.8	0.4 to 0.8	0.4 to 1.0			0.6 to 1.2
LY12CZ	3.8 to 4.9	1.2 to 1.8	0.3 to 0.9			

[7]. A whole range comparison is made between them, and, generally speaking, ours is slightly larger, the difference being between −0.7 to 5.8%.

Experimental Materials and Procedure

The materials investigated were four steels and four aluminum alloys. The material types (Chinese standard) and compositions are given in Table 1, their heat treatments and yield strengths are listed in Table 2.

Three configurations of chevron-notched specimens were used, they are short-rod specimens ($W/R = 3.4$, $\alpha_0 = 0.49$, $\alpha_1 = 1$), short-bar specimens with

TABLE 2—*Heat treatments and yield strengths of tested materials.*

Material Type	Heat Treatment	Yield at 0.2% Offset, MPa
	STEELS	
30Cr-Ni-2Mo-V-A	held at 940°C for 1 h, quenched, tempered again held at 930°C for 2 h, quenched, annealed	1382
60Si-2Mn-A	held at 870°C, quenched, tempered	1648
30Cr-Mn-Si-A	held at 930°C for 1 h, quenched, tempered	1177
20G	as rolled	245
	ALUMINUM ALLOYS	
LC4CZ	held at 470°C for 1 h, quenched, held at 122°C for 24 h	412
LY11CZ	held at 500°C for 1 h, quenched	324
LD10	held at 503°C for 1 h, quenched, held at 155°C for 12 h	353
LY12CZ	held at 498°C for 1 h, quenched	255

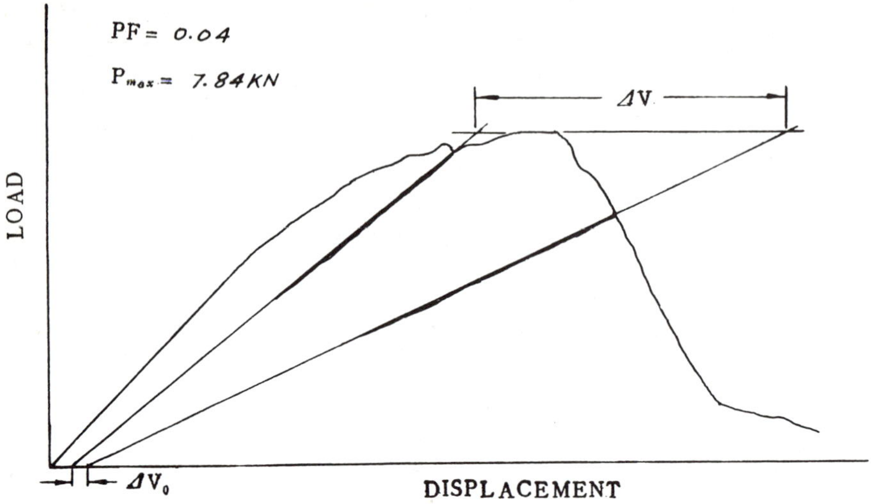

FIG. 2—*Typical load versus load-point displacement curves of a chevron-notched short-rod specimen* (B = *20 mm) for steel 30Cr-Ni-2Mo-V-A.*

square cross section ($W/R = 3.4, \alpha_0 = 0.49, \alpha_1 = 1$), and short-bar specimens with rectangular cross section [B: $2H = 4:3$, $W/H = 4.53$, $\alpha_0 = 0.49$, $\alpha_1 = 1$). Their values of Y_m* are 41.55, 36.56, and 45.98, respectively. Parameters of three configurations were purposefully chosen to ease preparation of chevron-notched on a simple electric spark wire cutting machine. Notches on opposite sides of a specimen were cut separately but their coplanarity was assured. The width of the chevron-notch is about 0.14 mm. During testing, load-displacement plots were recorded with two partial unloadings near maximum load. A plasticity factor (PF) is calculated from two intercepts of the extended unloading lines on abscissa and the horizontal line corresponding to maximum load, see Fig. 2, PF $= \Delta V_0/\Delta V$. Calibrations for displacement and load are not necessary. Our analysis is slightly different from Barker's [2]. PF is not used as a correction to tested value but rather as a criterion for valid tests.

Tested Results Compared with Values from CT and 3PB Specimens of E 399

For two high-strength steels, 30Cr-Ni-2Mo-V-A and 60Si-2Mn-A, five pieces of CT specimens for each material were prepared with a thickness of 30 mm and crack orientation R-L. Valid values of K_{Ic} were obtained according to ASTM E 399-81 for every specimen. Then two fractured halves of every CT specimen were used to prepare a short-rod and a short-bar specimen, respectively, with thickness B changed to 28 mm and the same crack orientation guaranteed. Tests were performed following the procedure mentioned previously. Comparison of results from short-rod and short-bar specimens with K_{Ic} values from CT specimens are shown in Fig. 3 and Fig. 4, where every pair composed of a black dot

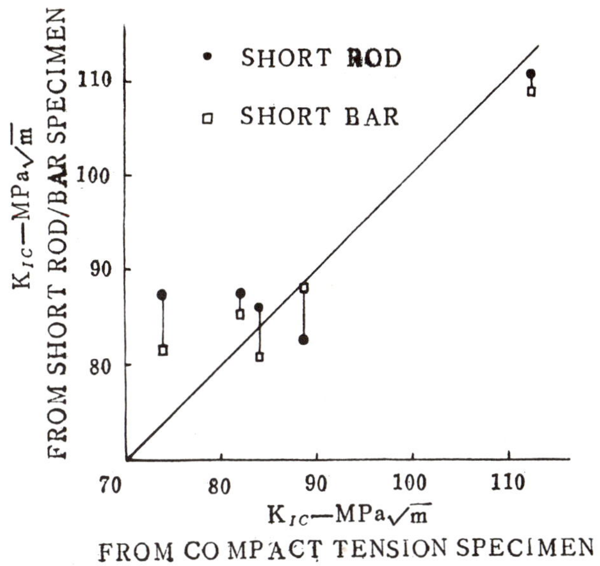

FIG. 3—*Comparison of test results of* K_{Ic} *from short-rod and short-bar specimens with* K_{Ic}*-values from compact tension specimens for steel 30Cr-Ni-2Mo-V-A.*

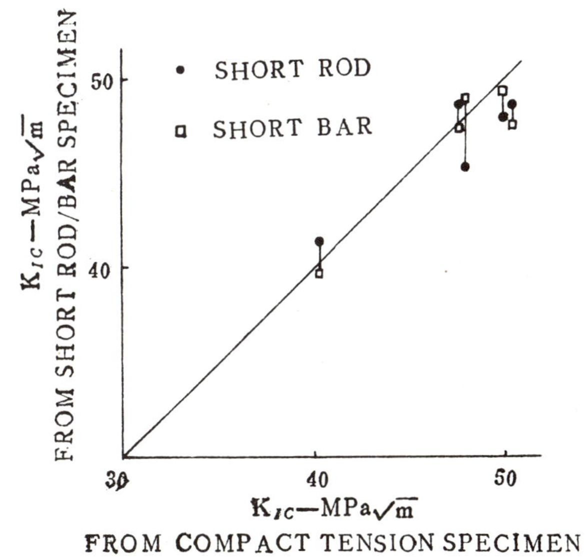

FIG. 4—*Comparison of test results of* K_{Ic} *from short-rod and short-bar specimens with* K_{Ic}*-values from compact tension specimens for steel 60Si-2Mn-A.*

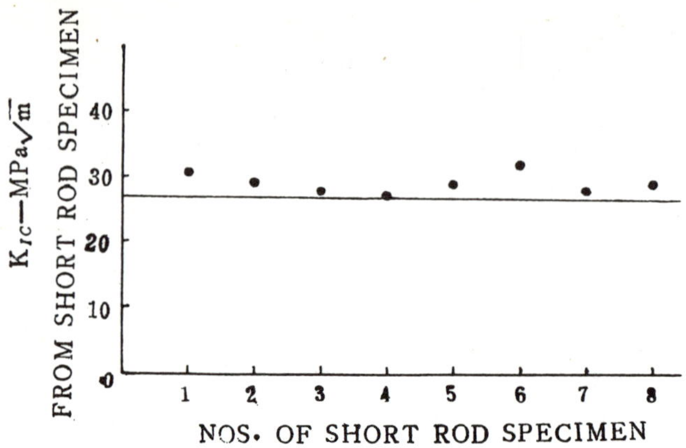

FIG. 5—*Comparison of test results of* K_{Ic} *from short-rod specimens with* K_{Ic}*-values from a three-point bend specimen for aluminum alloy LC4CZ.*

(denoting short rod) and a small square (denoting short bar) is linked by a vertical line to imply their being taken from the identical CT specimen. If K_{Ic} from short-rod (or short-bar) specimen agrees completely with that from CT specimen, the dot (or square) should fall on the inclined line (45° with abscissa), the vertical distance of a dot (or square) to the inclined line indicates the difference between short-rod (or short-bar) and CT specimens. Among all the differences, the largest two are 18 and 10% for short rod and short bar, respectively, the rest 18 differences are between −5.5 to 7.2%.

Comparison is also made between short-rod (bar) specimens and 3PB specimen, see Figs. 5 and 6, where horizontal lines corresponds to valid K_{Ic}-values obtained for aluminum alloy LC4CZ with 3PB specimens according to E 399,

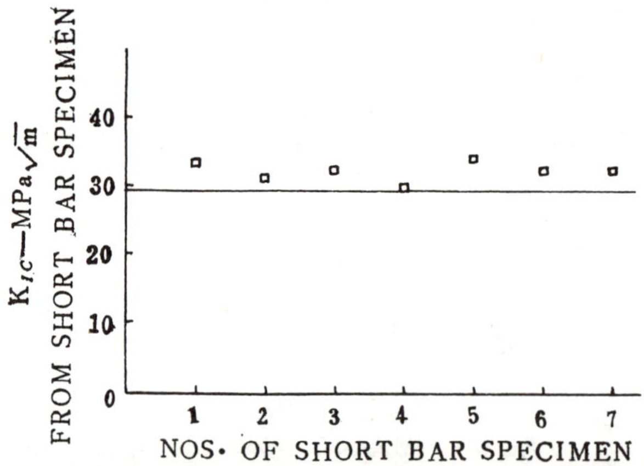

FIG. 6—*Comparison of test results of* K_{Ic} *from short-bar specimens with* K_{Ic}*-values from a three-point bend specimen for aluminum alloy LC4CZ.*

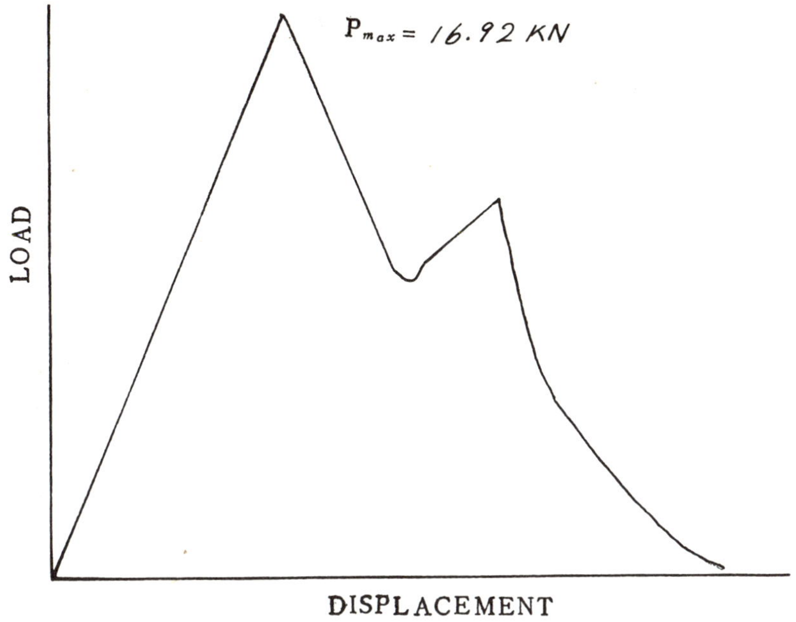

FIG. 7—*Typical load versus load point displacement curve of a chevron-notched short-bar specimen* (B = *28 mm) for steel 30Cr-Mn-Si-A.*

dots and squares represent short-rod and short-bar specimens which were prepared from the fractured halves of 3PB specimens. All differences are less than 10%, the results from short-rod and short-bar specimens are higher than K_{Ic}-values from 3PB specimens.

Discussion

All the previously mentioned successful tests with chevron-notched specimens have three things in common: (1) specimen thickness $B \geqslant 2.5(K_{Ic}/\sigma_{ys})^2$, (2) stable crack extension is realized, and (3) the plasticity factor is very small (PF $\leqslant$ 0.04). Explanations are as follows.

Since K_{Ic} is the plane-strain fracture toughness, the central plane-strain region should constitute a dominant part in the crack plane of the specimen. Although chevron-notched specimens have beneficial effect to the satisfaction of plane-strain conditions, a minimum requirement for specimen thickness should be set. Condition (1) is reasonable for it is easy for brittle material to satisfy; still, chevron-notched short-rod or short-bar specimens are several times smaller in volume than CT specimen of same thickness.

Many authors [*1,3,8*] pointed up analytically that stable crack extension can be realized in chevron-notched specimens, so that tiresome fatigue precracking of specimens can be eliminated. But in reality, catastrophic failure or sudden crack jumps may occur in short-rod or short-bar specimens, such phenomena

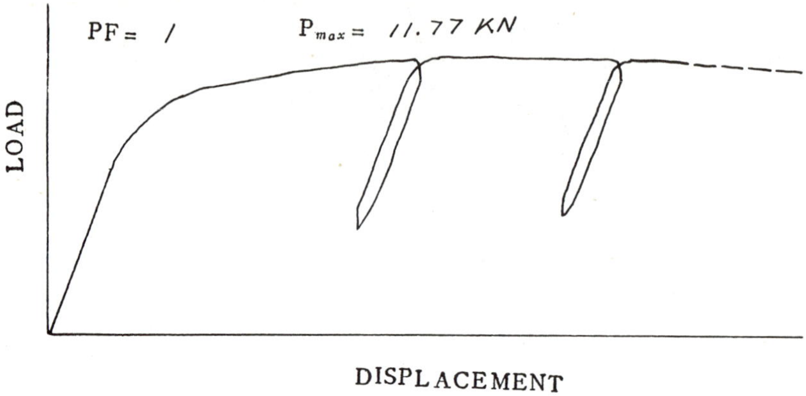

FIG. 8—*Typical load versus load-point displacement curve of a chevron-notched short-rod specimen* (B = *28 mm) for steel 20G.*

took place in steel 30Cr-Mn-Si-A we tested, see Fig. 7, in which results from short-rod and short-bar specimen deviate in a random way from the valid value K_{Ic} = 144.8 MPa$\sqrt{m}$ obtained with 3PB specimen. The fact that no crack extension takes place before maximum load or that crack jumps rapidly violates the basic principle of stable crack extension for the methodology with chevron-notched specimens, test results in these circumstances should be discarded. Load-displacement plots play an important role in monitoring stable crack extension.

Load-displacement plots are also useful to determine PF, which is an approximate indication of the degree to which a specimen behaves elastically, see Fig. 2. When PF → 0 that is $\Delta V_0 \rightarrow 0$, it means that the change in slopes of unloading lines is solely due to crack extension not the plastic zone at the crack tip. A very small PF can be expected for brittle materials, approximately PF $\leq$ 0.04 in this investigation for successful tests. Another extreme case (PF → 1) was observed in steel 20G, its load-displacement plots are shown in Fig. 8. The two unloading lines are nearly parallel, indicating that the ligament of specimen yields before final fracture. Tests with aluminum alloy LY12CZ also showed large PF (PF = 0.45). Test results are much higher (for LY12CZ) or lower (for 20G) than toughness values converted from J_{Ic}. Tests of aluminum alloys LYIICZ and LDI0 satisfy these three conditions. Test values (average) are 30.4 and 22.4 MPa$\sqrt{m}$, respectively.

A well-established method of fracture toughness testing should not subject to obvious effects from specimen geometry if its prescribed requirements are satisfied. We performed experiments in this aspect with aluminum alloy LYIICZ. Specimens of different configurations and sizes had almost the same toughness value as shown in Fig. 9. It can be seen that the rising trend is minimal in tested values as specimen thickness is increased, while the effect of specimen configuration is not obvious. Such a conclusion is supported by the other two groups of specimens in different sizes and configurations tested for steel 30Cr-Ni-Mo-V-A and 60Si-2Mn-A [*8*].

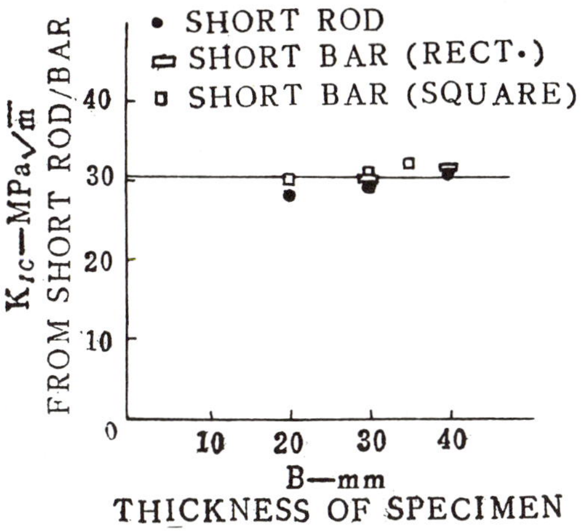

FIG. 9—*Tested results of* K_{Ic} *from chevron-notched specimens of different sizes and configurations for aluminum alloy LYIICZ.*

Among the three specimen configurations, the short bar with a rectangular cross section has longer crack extension ($\Delta\alpha = \alpha_c - \alpha_0$), flatter load-displacement plots as well as more stable specimen behavior than the other two (short rod ranks second before short bar with a square cross section); the comparison is made with thickness held the same. Since calculations showed that smaller initial crack length α_0 and larger W/H (or W/R) correspond to larger $\Delta\alpha$, as illustrated in Table 3, more stable behavior can be expected for these kinds of specimens; hence, they are more desirable.

Conclusions

Based on the analytical and test results as well as the subsequent discussion, the following conclusions were made.

The approximate STCA method has been applied to short-rod specimen with success. Equation 6 agrees well with expression derived from the compliance calibration [*6*]. Equation 5 is an extension in Ref *3* to include short-bar specimens with rectangular cross section.

TABLE 3—*Calculated crack extension* $\Delta\alpha$.

	W/H			
α_0	3	3.5	4	4.6
0.2	0.300	0.333	0.350	0.345
0.3	0.247	0.283	0.314	0.345
0.4	0.189	0.219	0.249	0.278
0.5	0.137	0.158	0.180	0.206

Three conditions are proposed to justify the tests with chevron-notched specimens. Emphasis is placed on the record and analysis of load-displacement plots. Only when stable crack extension and very small plasticity factor (approximately less than 0.04) are achieved, can results be considered reliable. In this case, fracture toughness K_{Ic} of metallic materials determined with short-rod or short-bar specimens is in good agreement with values from CT or 3PB specimens, the differences are normally within 10%. There is a slightly rising trend in test results with increases in specimen thickness observed in one aluminum alloy, but no effect of three specimen configurations exists. Short-bar specimens with a rectangular cross section behave more stably than the other two in this investigation.

Catastrophic failure and sudden crack jumps of short-rod and short-bar specimens of one steel were observed. To avoid the possibility of this fracture mode and to test very brittle materials with ease, it is suggested to choose the specimen parameters of small α_0 and large W/H (or W/R).

References

[1] Barker, L. M., *Engineering Fracture Mechanics,* Vol. 9, No. 2, 1977, pp. 361–369.
[2] Barker, L. M., *International Journal of Fracture,* Vol. 15, No. 6, 1979, pp. 515–536.
[3] Munz, D., Bubsey, R. T., and Srawley, J. E., *International Journal of Fracture,* Vol. 16, No. 4, 1980, pp. 359–374.
[4] Shih, T. T., *Journal of Testing and Evaluation,* Vol. 9, No. 1, 1981, pp. 50–55.
[5] Munz, D., *International Journal of Fracture,* Vol. 16, No. 3, 1980, pp. R137–R141.
[6] Zhu Wuyang, et al, *Testing of Fracture Toughness,* Science Publishing House, Beijing, 1979, p. 32.
[7] Munz, D., *Engineering Fracture Mechanics,* Vol. 15, No. 1-2, 1981, pp. 231–236.
[8] Wang Chizhi, "An Investigation on the Method for Determination of Fracture Toughness K_{Ic} of Materials with Chevron-Notched Short-Rod and Short-Bar Specimens," thesis, Chongqing University, May 1982 (advisors: Yuan Maochan and Chen Tzeguang).

James L. Stokes[1] *and George A. Hayes*[1]

Investigation of Acoustic Emission During Fracture Toughness Testing of Chevron-Notched Specimens

REFERENCE: Stokes, J. L. and Hayes, G. A., "**Investigation of Acoustic Emission During Fracture Toughness Testing of Chevron-Notched Specimens,**" *Chevron-Notched Specimens: Testing and Stress Analysis, ASTM STP 855,* J. H. Underwood, S. W. Freiman, and F. I. Baratta, Eds., American Society for Testing and Materials, Philadelphia, 1984, pp. 205–233.

ABSTRACT: Acoustic emission (AE) monitoring in conjunction with fracture toughness testing of chevron-notched specimens was employed to investigate crack growth in four steels prepared by electroslag-remelt casting. 15-5PH, AISI 4140, D6AC, and AISI 440C were investigated. AE data were recorded and correlated with variations in fracture toughness test behavior and fracture surface topography. AE parameters such as total counts, count rate, amplitude distribution, etc., were used to characterize flow and fracture processes. AE measurements were observed to have good correlation with crack growth parameters and the energetics of discrete cracking events.

KEY WORDS: acoustics, emission, fracture (materials), fracture tests, toughness, cracking (fracturing), steels, steel castings

In recent years there has been considerable interest in the development of a standardized fracture toughness test utilizing chevron-notched specimens [*1–4*]. These investigations have determined that crack growth behavior in chevron-notched specimens depend in part on the microstructural characteristics of a material, as well as inherent material properties, that is, yield strength, fracture toughness, work hardening rate, etc. For example, depending on the material, heat treatment, or test environment, crack growth can occur continuously with applied load, or in discontinuous mode (crack jumps), with varying amounts of plastic yielding occurring at the crack tip. The fracture mode can range from fibrous failure to cleavage cracking.

Recent advances in acoustic emission (AE) technology have shown that AE activity in a metal, regardless of the method of mechanical testing, depends upon metallurgical variables such as composition and heat treatment [*5*]. The purpose of this investigation was to determine the feasibility of using AE for monitoring

[1]Metallurgist and branch head, respectively, Materials Engineering Branch, Naval Weapons Center, China Lake, Calif. 93555.

TABLE 1—*Chemical analyses of materials.*

	Element, weight %											
Alloy	C	Mn	P	S	Si	Cu	Cr	Ni	Mo	Cb	V	Fe
15-5PH	0.04	0.66	0.003	0.004	0.33	3.56	14.95	4.20	...	0.35	...	...[a]
AISI 4140	0.42	0.88	0.004	0.007	0.26	...	1.00	...	0.21	...	...	...[a]
D6AC	0.46	0.74	0.002	0.002	0.29	0.18	1.01	0.59	1.05	...	0.09	...[a]
AISI 440C	1.02	0.50	0.003	0.002	0.22	...	16.50	0.28	0.45	...	...	...[a]

[a]Remainder.

the initiation and propagation of cracks in chevron-notched specimens in order to characterize cracking mechanisms.

Experimental Procedures

Materials

Four different cast steel compositions were tested in this investigation: 15-5PH, AISI 4140, D6AC, and AISI 440C. The composition of each steel is given in Table 1. Castings were prepared by electroslag remelting (ESR) the alloys into ingots approximately 200 mm (8 in.) long by 76 mm (3 in.) in diameter. Each steel was heat treated to a common commercially used condition, then characterized by measuring hardness, yield and ultimate strengths, and percent elongation. The heat treatment given each steel and the resulting mechanical properties are presented in Table 2. (No tensile data are shown for AISI 440C. The specimens were so brittle that they shattered in the grips.) The microstructure of each alloy is shown in Fig. 1. Figures 1*a*, *b*, and *c* show uniform matrixes of tempered martensite typical of the respective alloy and heat treatment. Figure 1*d* (AISI 440C) shows a tempered martensite matrix containing large primary carbides at the grain boundaries, also typical of this alloy in the cast condition.

TABLE 2—*Mechanical properties of fracture toughness specimens.*

Alloy	Heat Treatment	0.2% Offset Yield Strength, MN/m² (ksi)	Ultimate Tensile Strength, MN/m² (ksi)	Elongation %, 25.4 mm (1 in.)	Hardness, HRC
15-5PH	condition H900	1172 (170)	1276 (185)	12	43
AISI 4140	857°C (1575°F)-OQ 385°C (725°F) temper	1441 (209)	1544 (224)	10	47
D6AC	899°C (1650°F)-OQ 427°C (800°F) temper	1469 (213)	1655 (240)	8	49
AISI 440C	1038°C (1900°F) 191°C (375°F) temper	broke in grips (too brittle)		...	57

Fracture Toughness Testing

Chevron-notched fracture toughness specimens (25.4 mm (1 in.) in diameter) were machined from the cast ESR ingots. Test specimen configuration and dimensions are described elsewhere [*1*]. The specimens were heat treated before the thin longitudinal chevron slots were cut. The chevron-notched rods were tested at ambient temperature in air using a TerraTek screw-driven mechanical test machine [2]. Fracture toughness values were determined using the method developed by Barker [*3*]. All fracture surfaces were characterized by means of the scanning electron microscope (SEM). An SEM fractograph of each material is shown in Fig. 2. These are interesting in that they all indicate brittle fracture modes. This is especially apparent in Fig. 2*a* (15-5PH) which shows cleavage exclusively. Although somewhat less well defined, it appears that the AISI 4140 (Fig. 2*b*) and the D6AC (Fig. 2*c*) failed primarily by cleavage also. The fracture illustrated in Fig. 2*d* (AISI 440C) is unique in that it shows cleavage of the second-phase carbides and evidence of intergranular fracture of the matrix.

Acoustic Emission Monitoring

Fracture toughness tests of each steel were acoustically monitored using commercially available Dunegan/Endevco (DE) 3000 series instrumentation. The acoustic pickup was a 100 kHz resonant piezoelectric transducer (Model DE S9204). The signals from the transducer were processed in a signal conditioning unit incorporating a DE 302A amplifier with gain of 50 dB; a DE 1801 preamplifier with gain of 40 dB; band pass filters between 100 kHz and 2 MHz to remove extraneous noise; a DE 920A distribution analyzer and DE 921 amplitude detector to sum the number of acoustic emissions greater than a preselected threshold of 30 dB. AE data were recorded with a Hewlett-Packard (HP) 9825B computer for subsequent playback and analysis.

Because of the geometry of the TerraTek loading fixture, it was necessary to use a spacer between the test specimen and the transducer. The spacer was a 25.4 mm (1 in.) diameter by 12.7 mm (0.5 in.) cylinder of PH13-8Mo, Condition H1125, corrosion resistant steel. The test specimen, spacer, and transducer were coupled with DE water-soluble acoustic couplant.

Results and Discussion

Load Versus Displacement

Figure 3 shows parametric plots of load versus displacement for each alloy. In each case, the specimen was slowly loaded until a crack was intitiated. A continously increasing load was required to advance the crack until it reached a critical length, where the load went through a maximum. Two or more relaxation and reloading cycles were made when the load was near the maximum value to allow calculation of the degree to which linear elastic fracture mechanics conditions had been violated [*3*].

b
a

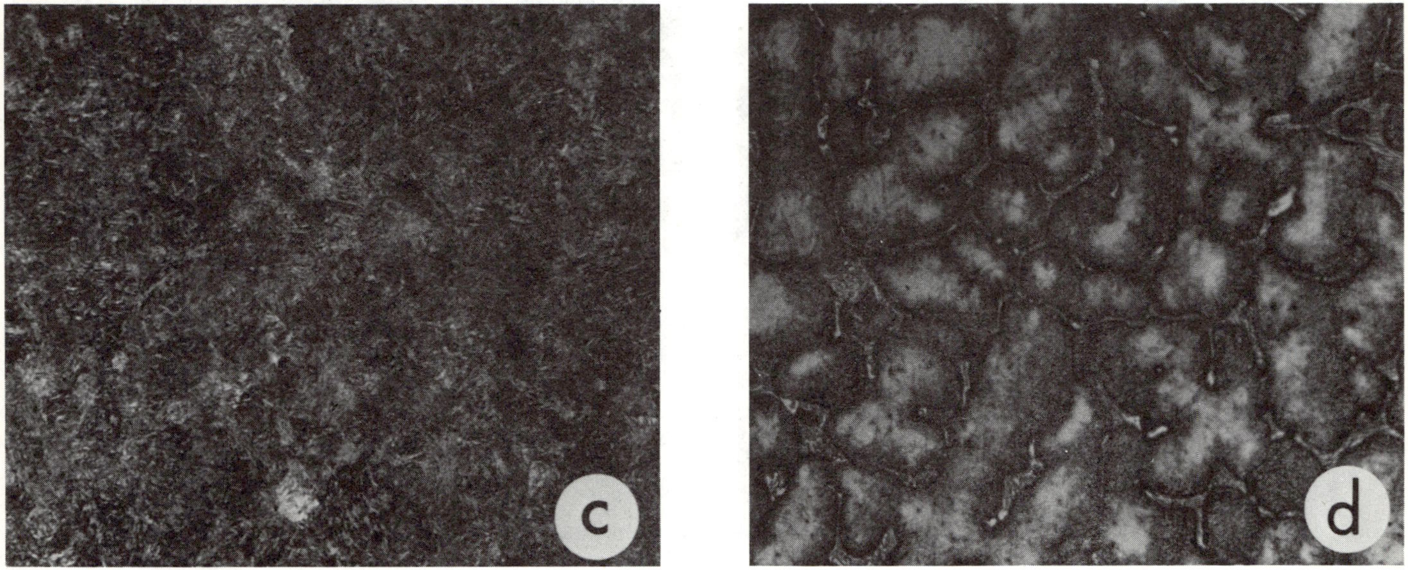

FIG. 1—*Miscrostructures of chevron-notched fracture toughness specimens with original magnification at ×200.* (a) *15-5PH,* (b) *AISI 4140,* (c) *D6AC, and* (d) *AISI 440C.*

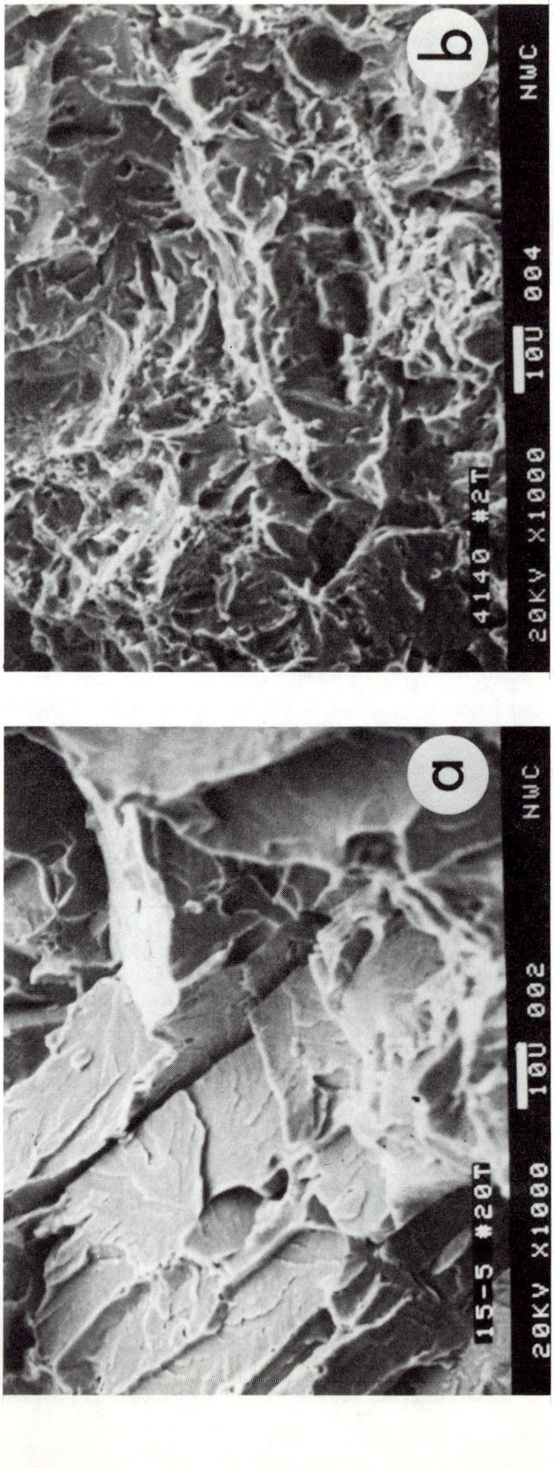
b
4140 #2T
20KV X1000 10U 004 NWC
a
15-5 #20T
20KV X1000 10U 002 NWC

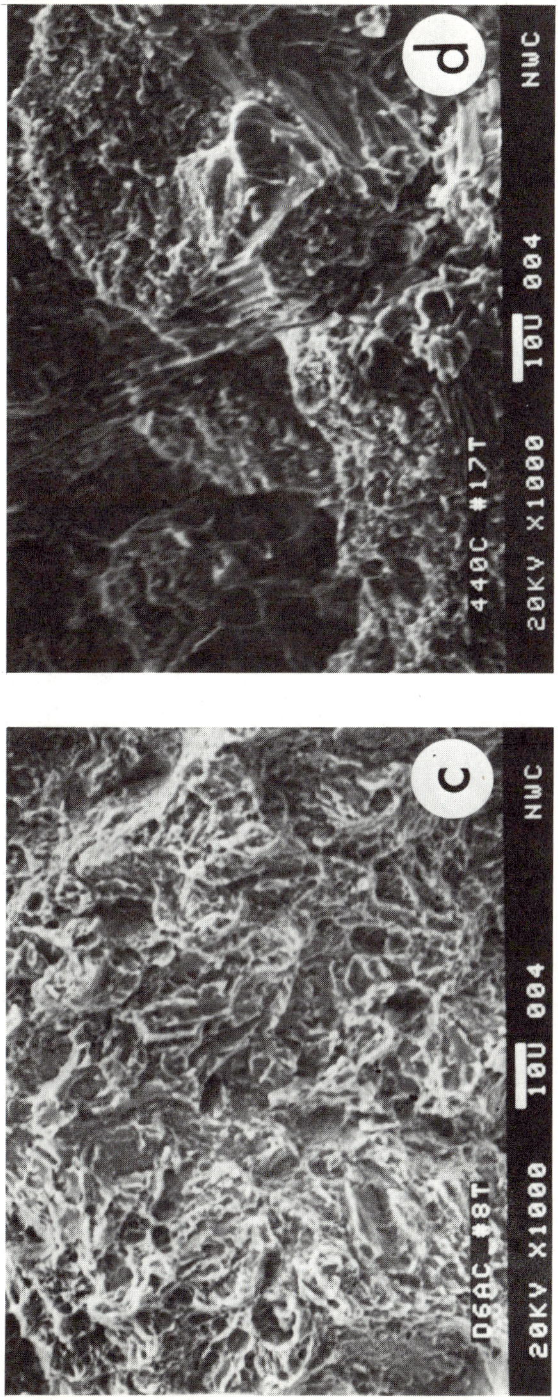

FIG. 2—*SEM fractographs of chevron-notched fracture toughness specimens with original magnification at* $\times 1000$. (a) *15-5PH*, (b) *AISI 4140*, (c) *D6AC*, *and* (d) AISI 440C.

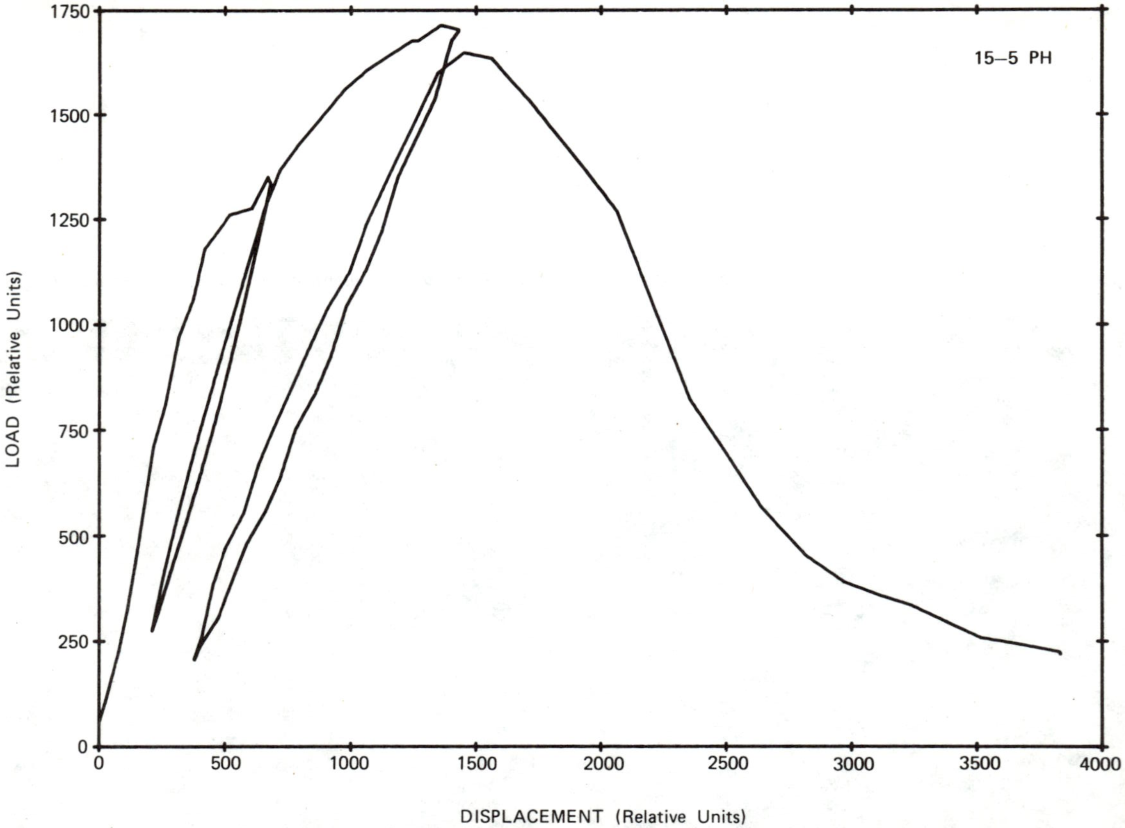

FIG. 3—*Load versus displacement for chevron-notched fracture toughness specimens.* (a) *15-5PH,* (b) *AISI 4140,* (c) *D6AC, and* (d) *AISI 440C.*

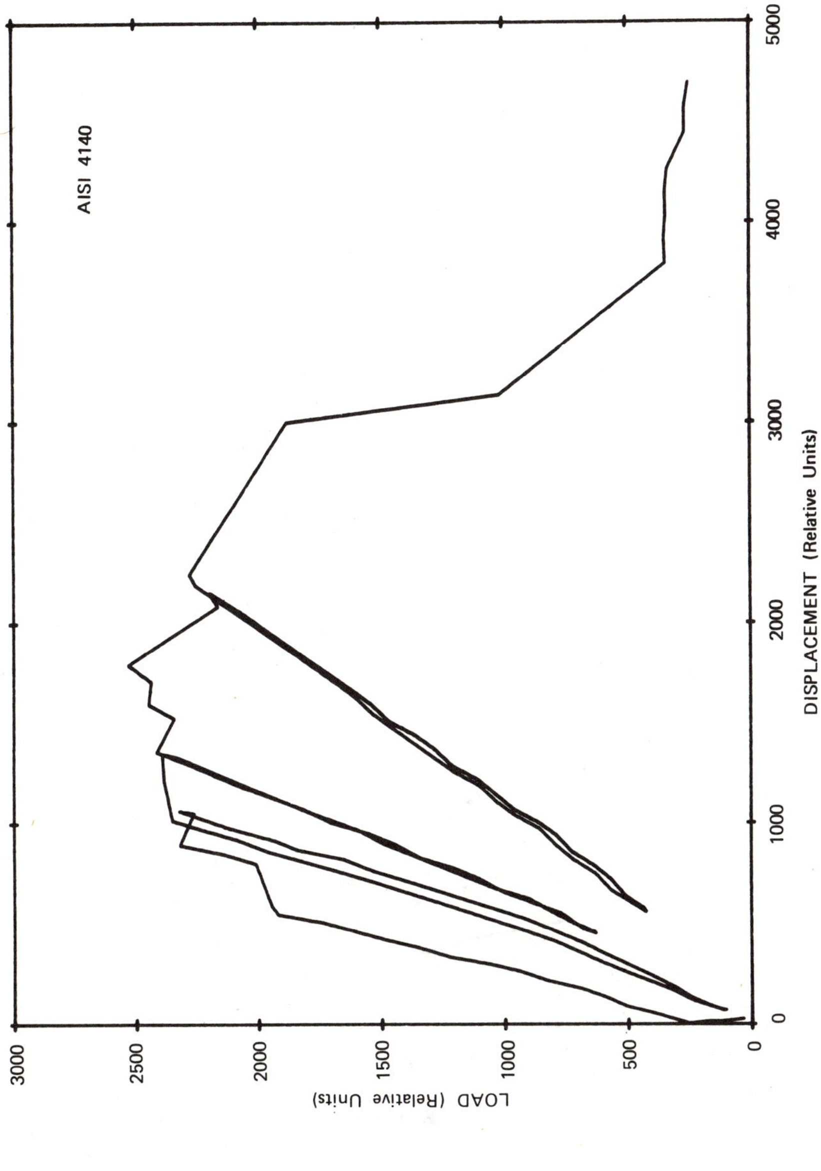

FIG. 3—*Continued.*

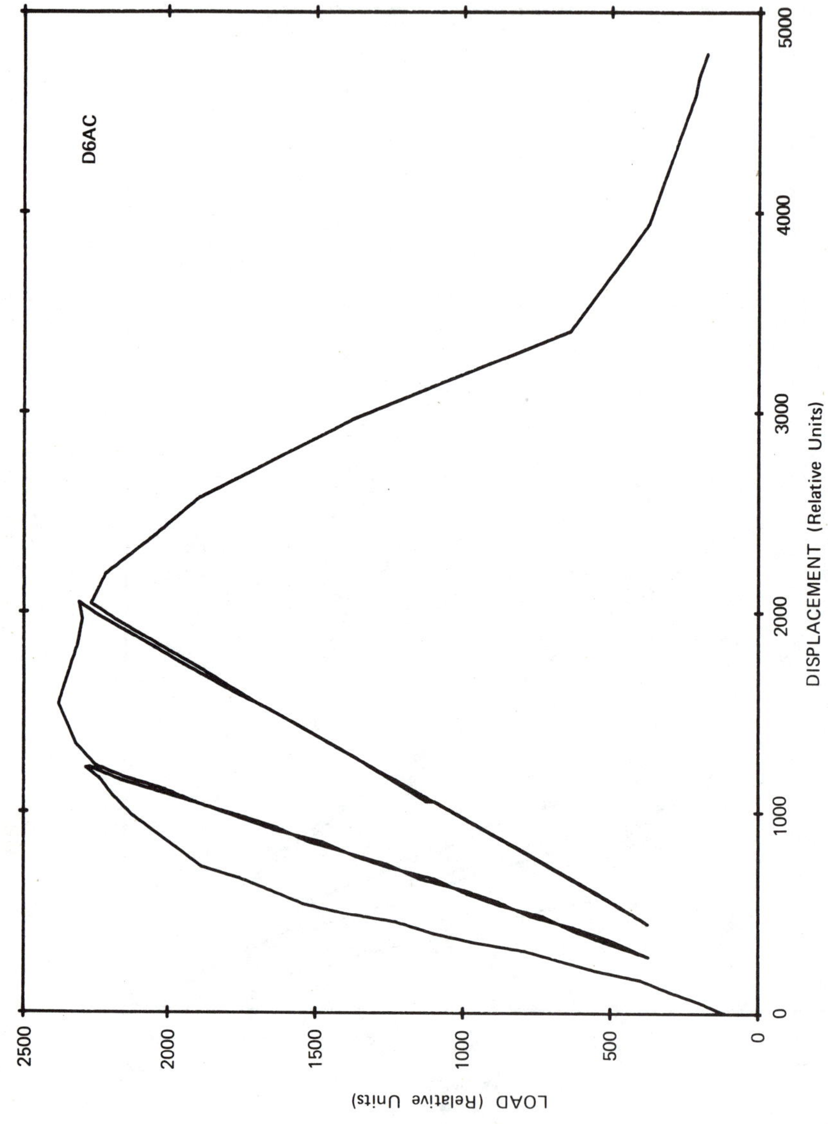

FIG. 3—*Continued.*

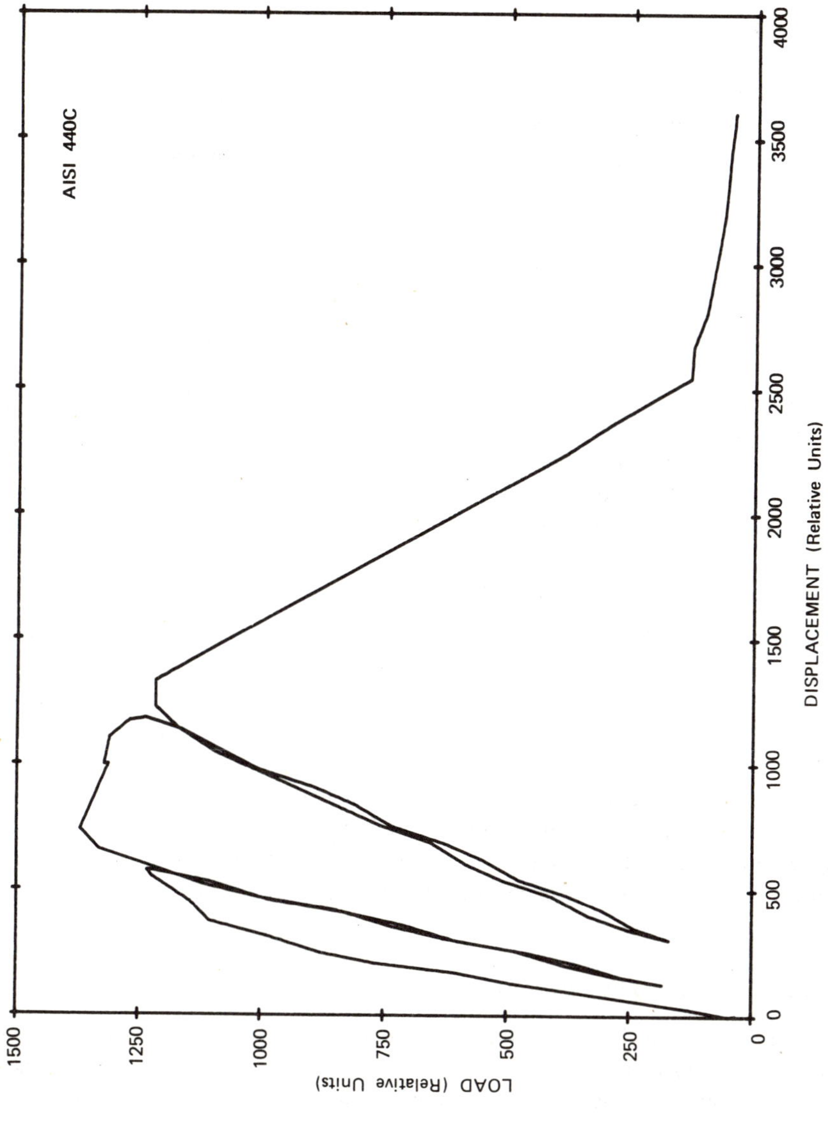

FIG. 3—*Continued.*

For the 15-5PH, D6AC, and AISI 440C, the second unloading represented the end of the test. In the case of the AISI 4140, a third load/unload sequence was made in order to calculate the fracture toughness associated with each substantial crack jump. In the crack jump case, the fracture toughness was taken at the average of several values of fracture toughness calculated after each major crack jump.

The fracture toughness values computed for each alloy are shown in Table 3. The data show that fracture toughness was inversely proportional to strength and hardness. Thus, the AISI 440C at a hardness of HRC 57 exhibited a fracture toughness of 40.2 MPa$\sqrt{m}$ (37 ksi$\sqrt{in.}$), and the 15-5PH at a hardness of HRC 43 exhibited a fracture toughness of 79.2 MPa$\sqrt{m}$ (72 ksi$\sqrt{in.}$). As shown in Table 3, the remaining two alloys were intermediate in hardness and toughness.

Load Versus Acoustic Emission Counts

Figure 4 shows plots of relative load versus cumulative acoustic emission (AE) counts. It can be seen from the plots that although the same criterion for unloading was used in each case, the total AE counts generated up to the first unloading point varied significantly from one alloy to the next. Thus, the toughest alloy, 15-5PH, generated approximately 1.5×10^5 counts up to the first unloading; whereas, the most brittle alloy, AISI 440C, generated approximately 6.2×10^5 counts up to the first unloading point.

Acoustic Emission Count Rate Versus Time (load versus time superimposed)

Figure 5 shows AE count rate versus time. Superimposed on each plot is a plot of load versus time. These plots show that essentially all of the AE was generated during loading. (The AE observed during the final unloading should be ignored. The gross "noise" was generated by the specimen separating rapidly into two halves.)

A comparison of count rates during the first load/unload cycle between the four alloys confirms the trend observed in Fig. 4. That is, the toughest alloy had a relatively low count rate, 2.8×10^5 counts/s during the first cycle, and the most brittle alloy had a much higher count rate, 9.6×10^5 counts/s during the first cycle.

Log-Sum Amplitude Distributions

Figure 6 contains log-sum amplitude distributions showing AE cumulative counts on the vertical axis and amplitude in dB on the horizontal axis. In each case, the threshold was set to exclude signals below 30 dB. Cumulative amplitude distributions were recorded at 6 s intervals throughout each test.

At least four distributions are shown for each test. The first distribution in each figure shows cumulative AE events versus amplitude up to the first unloading point. Thus, the first distribution represents primarily plastic deformation

TABLE 3—*Summary of test results.*

			b-value	
Material	Fracture Toughness, MPa$\sqrt{m}$ (ksi$\sqrt{in.}$)	AE Counts at First Loading	Initial Distribution	Final Distribution
15-5PH	79.0 (72)	1.5×10^5	1.05	0.29
AISI 4140	69.3 (63)	2.3	0.83	0.87/0.36
D6AC	66.8 (61)	4.8	1.18	0.64/0.18/0.31
AISI 440C	40.2 (37)	6.2	0.83	0.53

and crack initiation. The second distribution in each figure is the one immediately following the first. That is, the one recorded 6 s later in the test. For Fig. 6*b* only, a third distribution is shown taken 6 s after the second distribution. The next to last distribution in each figure was recorded at the second unloading point. And finally, the last distribution in each figure is the cumulative amplitude distribution for the entire test.

The significance of these comparisons is that for all four alloys a distinct change in amplitude distribution was observed after the first unloading point. This point would seem to mark the transition from crack initiation to slow crack growth. From this point on, the amplitude distributions remained essentially the same.

According to Pollack [*6*], the shape of the log-sum amplitude distribution can in many cases be related to the fracture toughness of the material. Pollack shows that for most engineering materials the slope, *b*, will be between 0.7 and 1.5 with occasional values as low as 0.4 or as high as 4.0. He states that the lowest *b*-values are found for discontinuous crack growth processes in brittle materials, while plastic deformation prior to crack growth gives relatively high *b*-values.

In each of the Figs. 6*a* through *d*, *b*-values are shown for the first and the last distributions. It will be noted that the final distribution in Fig. 6*b* has two *b*-values, and the final distribution in Fig. 6*c* has three *b*-values. It is also interesting to note that in Figs. 6*b* and *d* representing AISI 4140 and AISI 440C, respectively, the *b*-values for the first distribution were identical.

The *b*-value data are summarized in Table 3. In general, these values are in accordance with Pollack's data showing high values for plastic deformation and low values for slow crack growth. In this case, however, it appears that the higher *b*-values represent a combination of plastic deformation *and* microcracking. It is apparent also that for this series of tests the *b*-values were not proportional to fracture toughness.

Noise

To determine the contribution of noise to the total AE signal, a dummy specimen (without a sharp notch) was instrumented and loaded well into the elastic range several times. The AE response recorded during the final load cycle

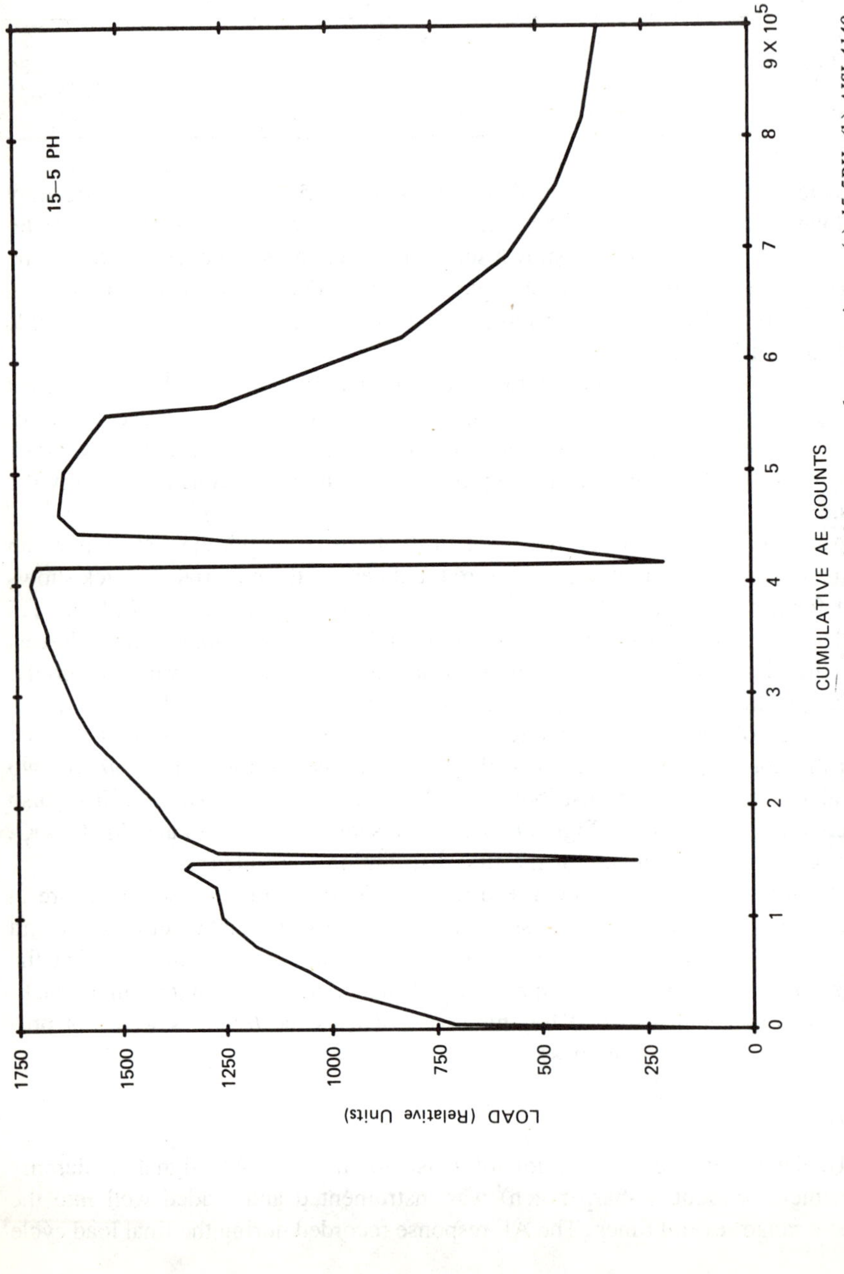

FIG. 4—*Load versus cumulative AC counts for chevron-notched fracture toughness specimens.* (a) *15-5PH,* (b) *AISI 4140,* (c) *D6AC,* and (d) *AISI 440C.*

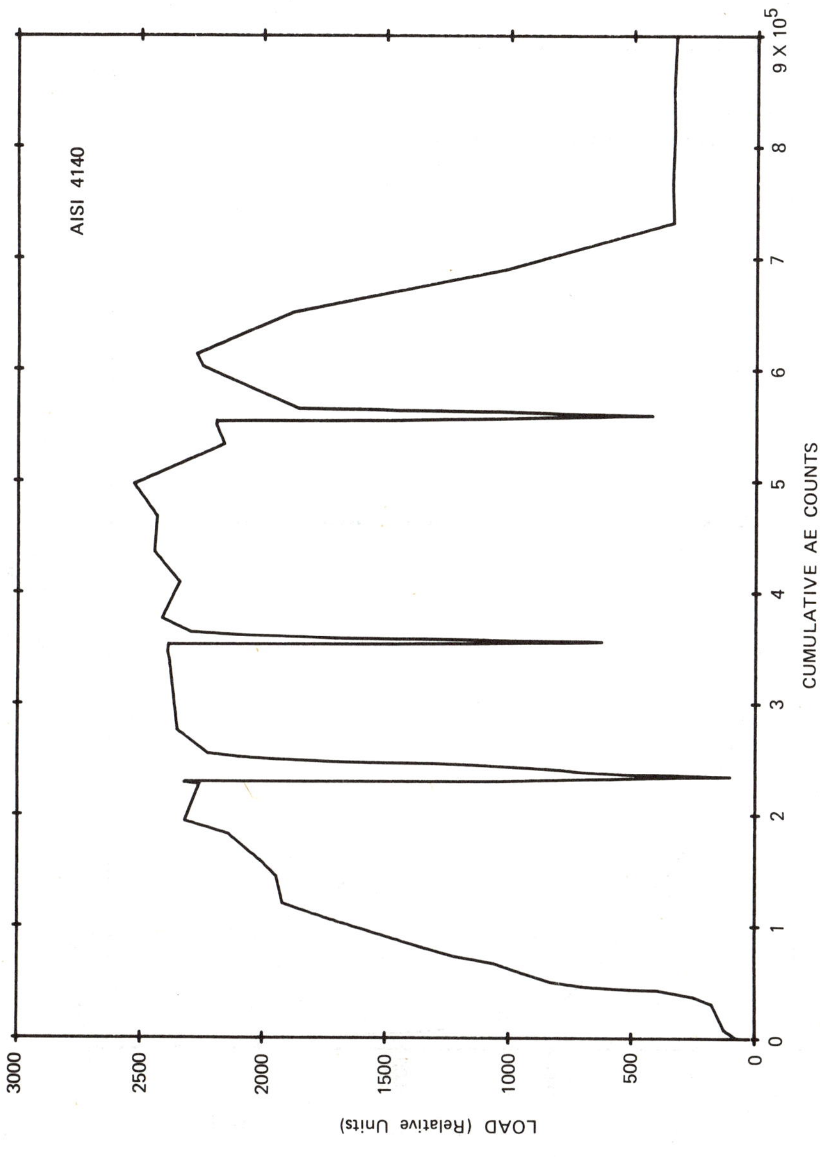

FIG. 4—*Continued.*

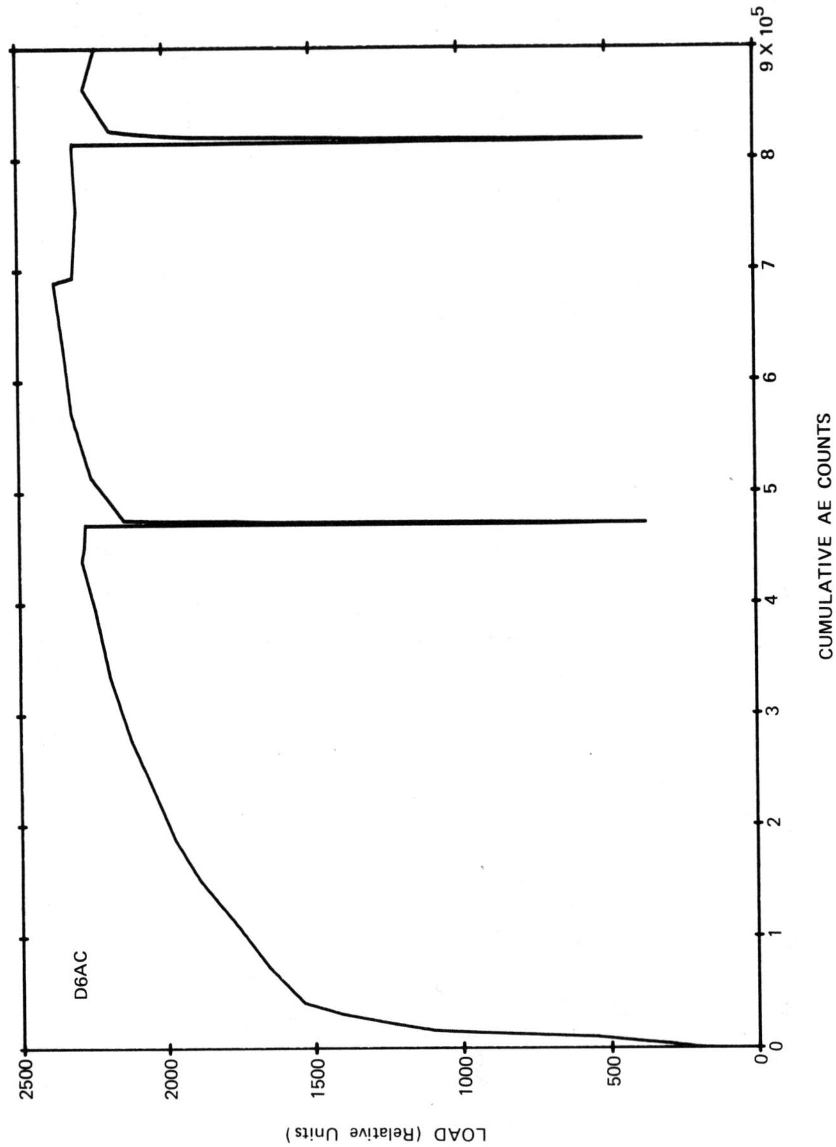

FIG. 4—*Continued.*

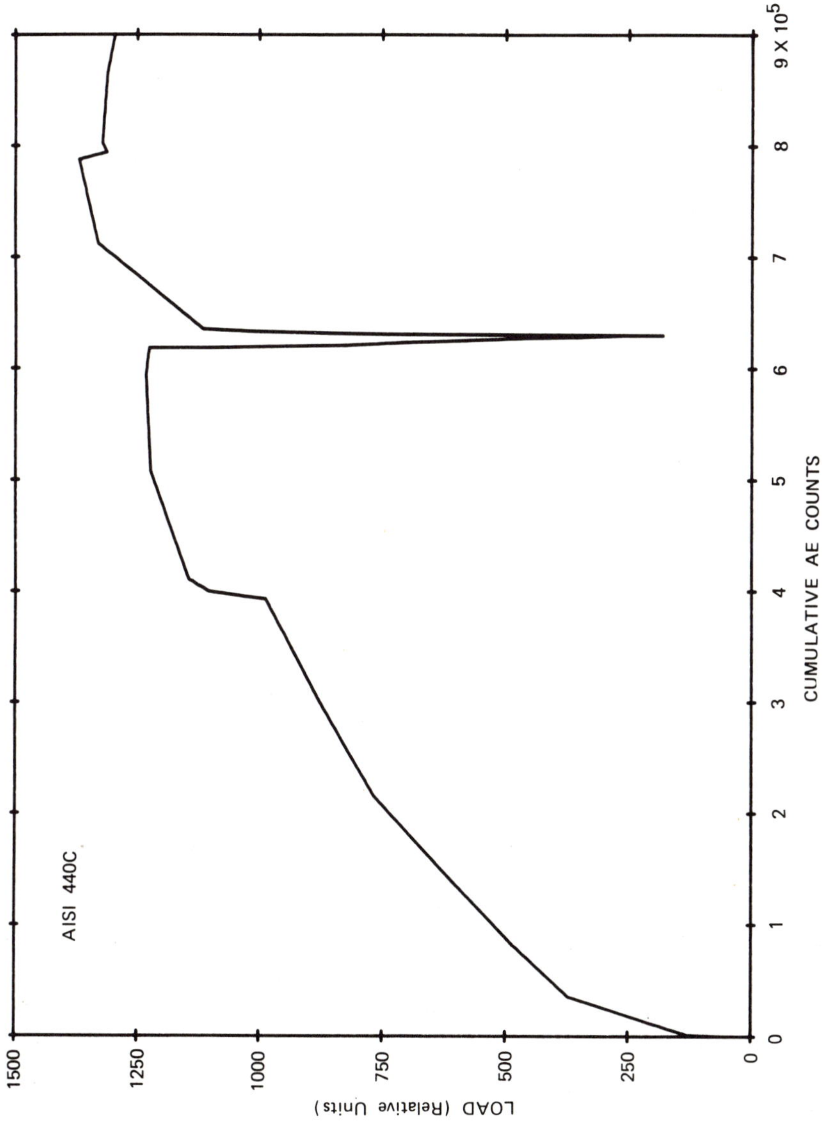

FIG. 4—*Continued.*

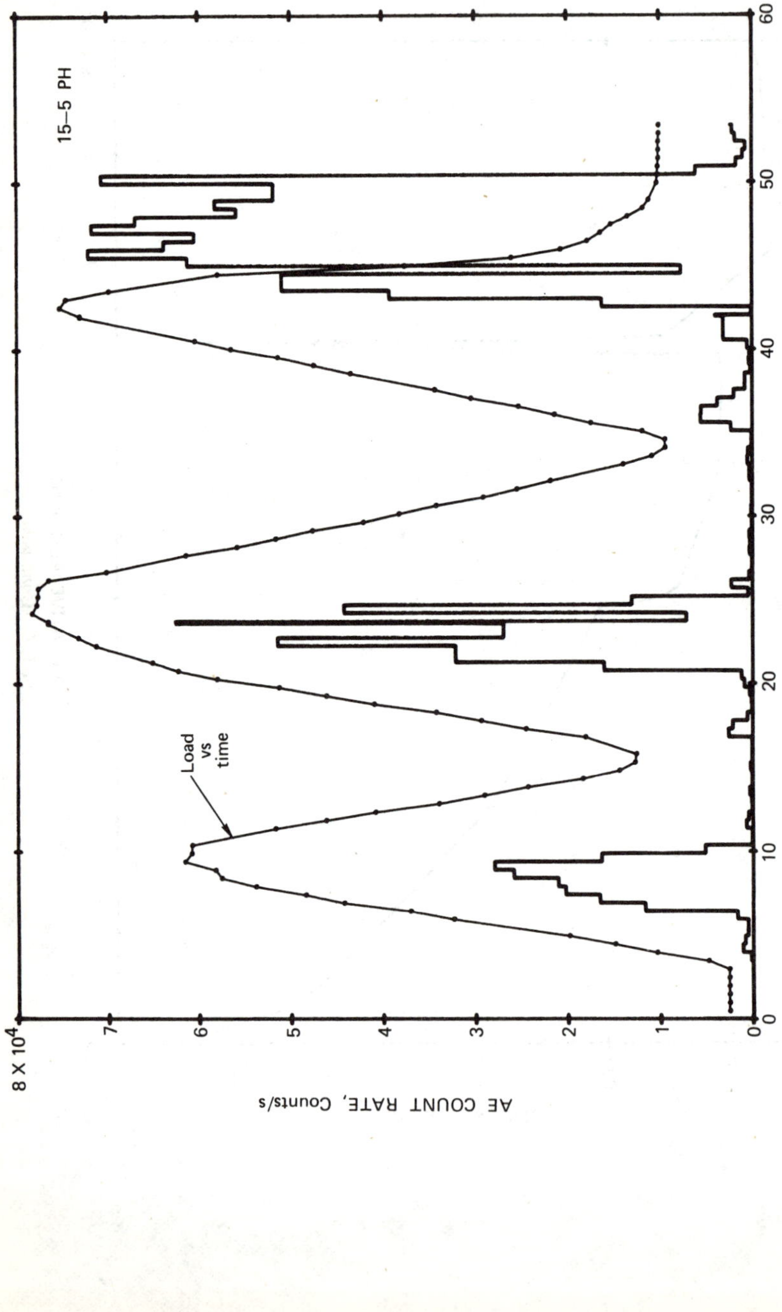

FIG. 5—*AE count rate versus time for chevron-notched fracture toughness specimens, with plot of load versus time superimposed* (dotted line). (a) *15-5PH*, (b) *AISI 4140*, (c) *D6AC*, *and* (d) *AISI 440C*.

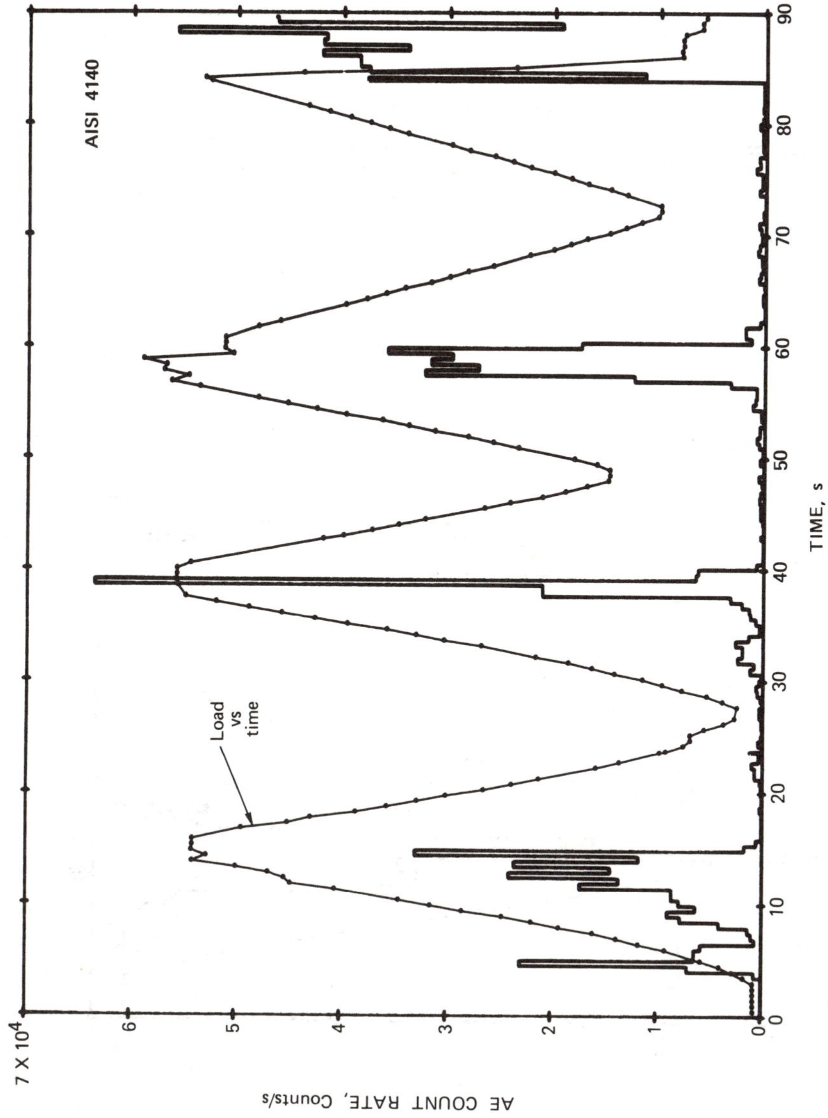

FIG. 5—*Continued.*

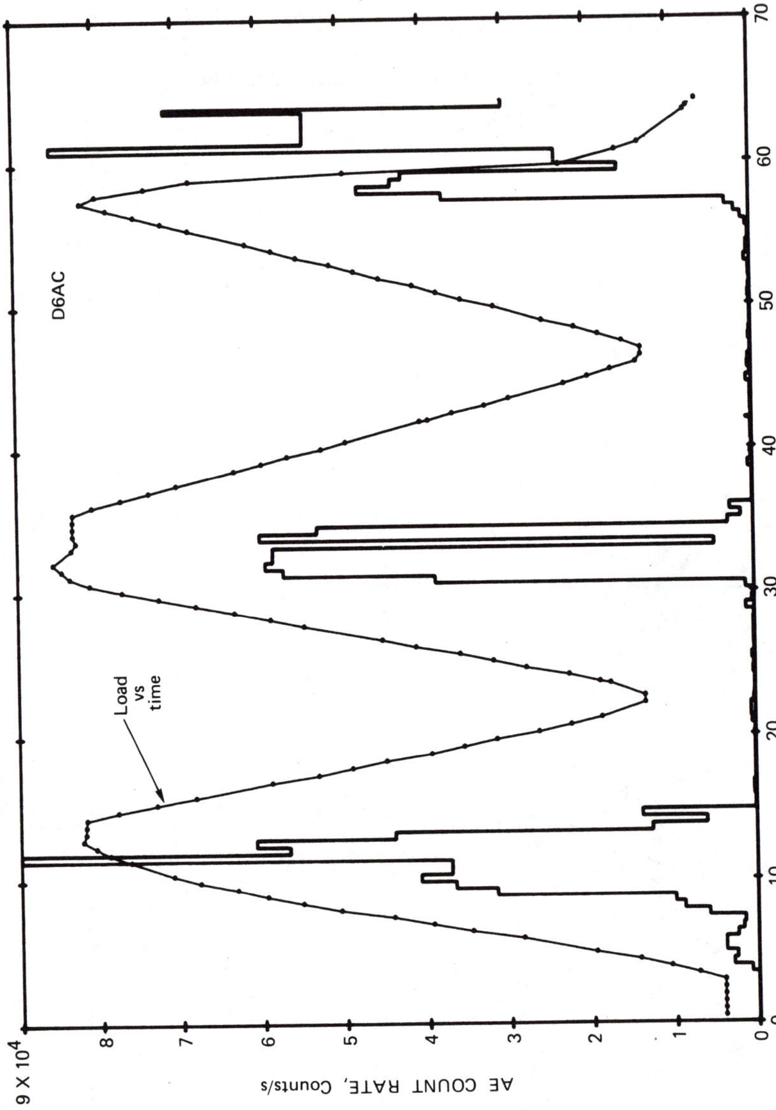

FIG. 5—*Continued.*

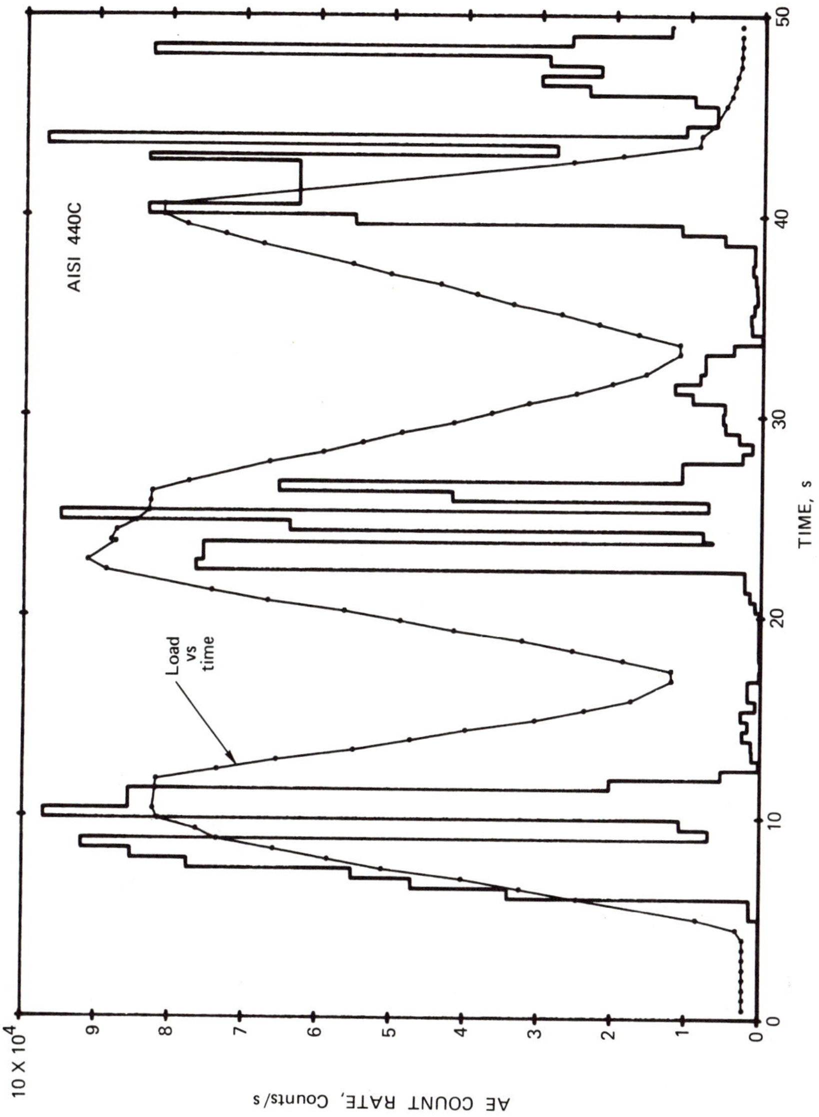

FIG. 5—*Continued.*

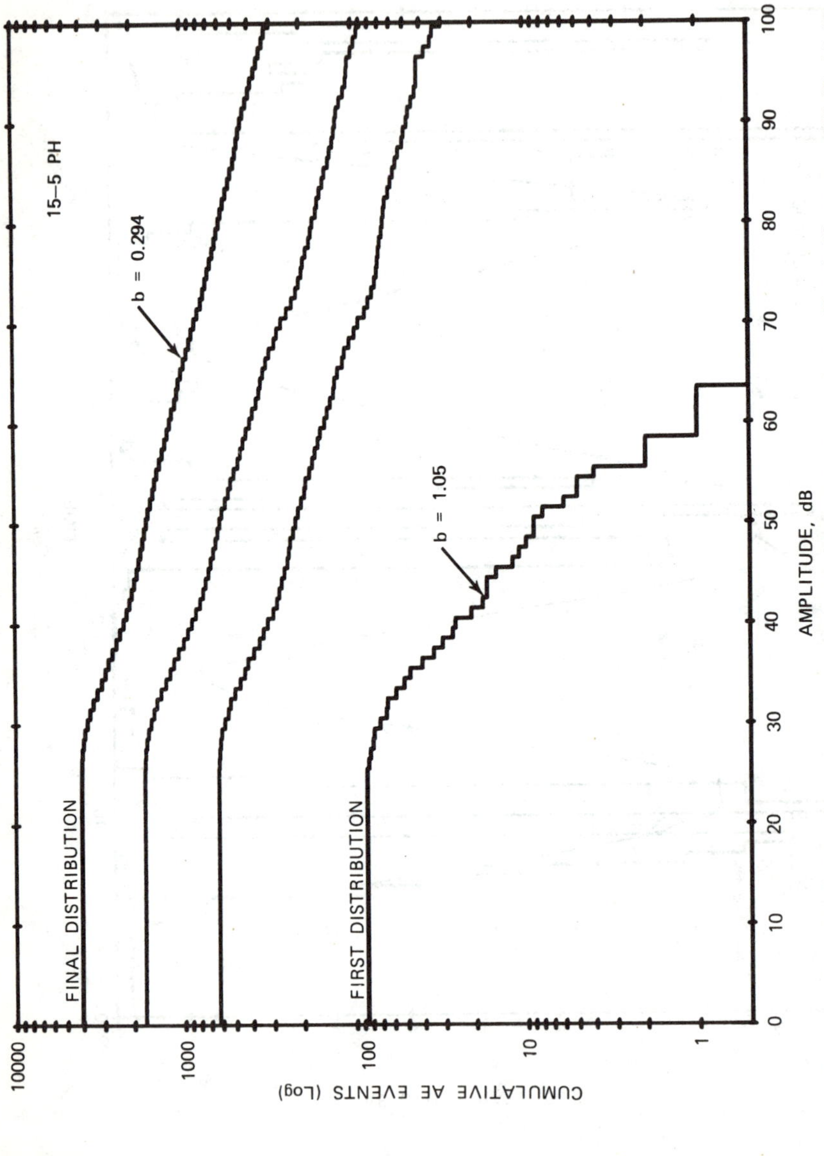

FIG. 6—AE amplitude distributions for chevron-notched fracture toughness specimens. Threshold was set at 30 dB. (a) *15-5PH*, (b) *AISI 4140*, (c) *D6AC*, *and* (d) *AISI 440C*.

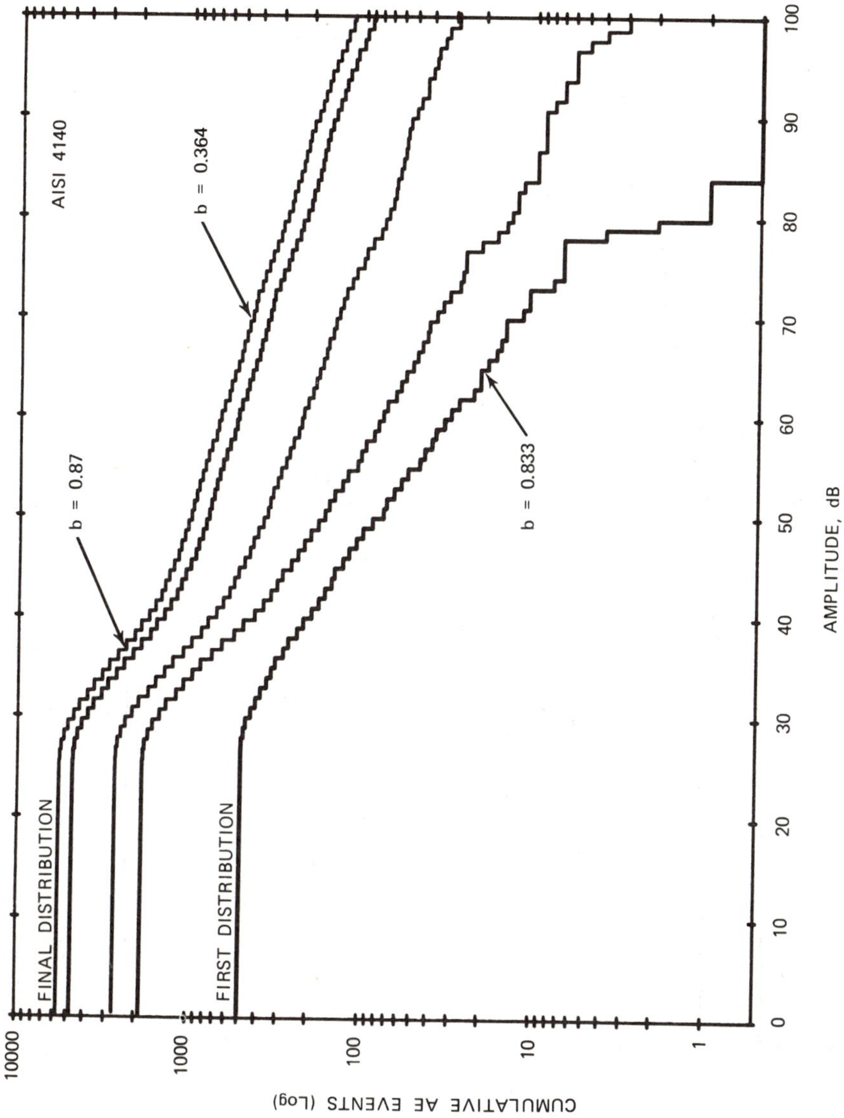

FIG. 6—*Continued.*

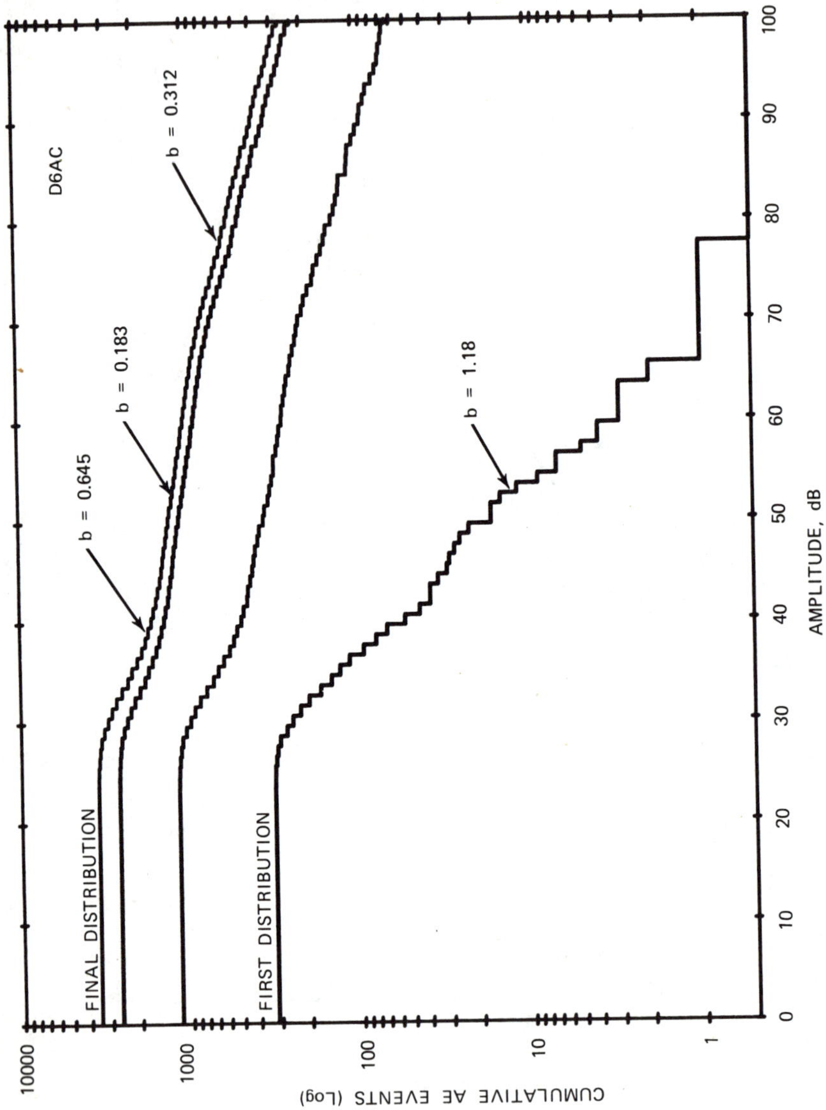

FIG. 6—*Continued.*

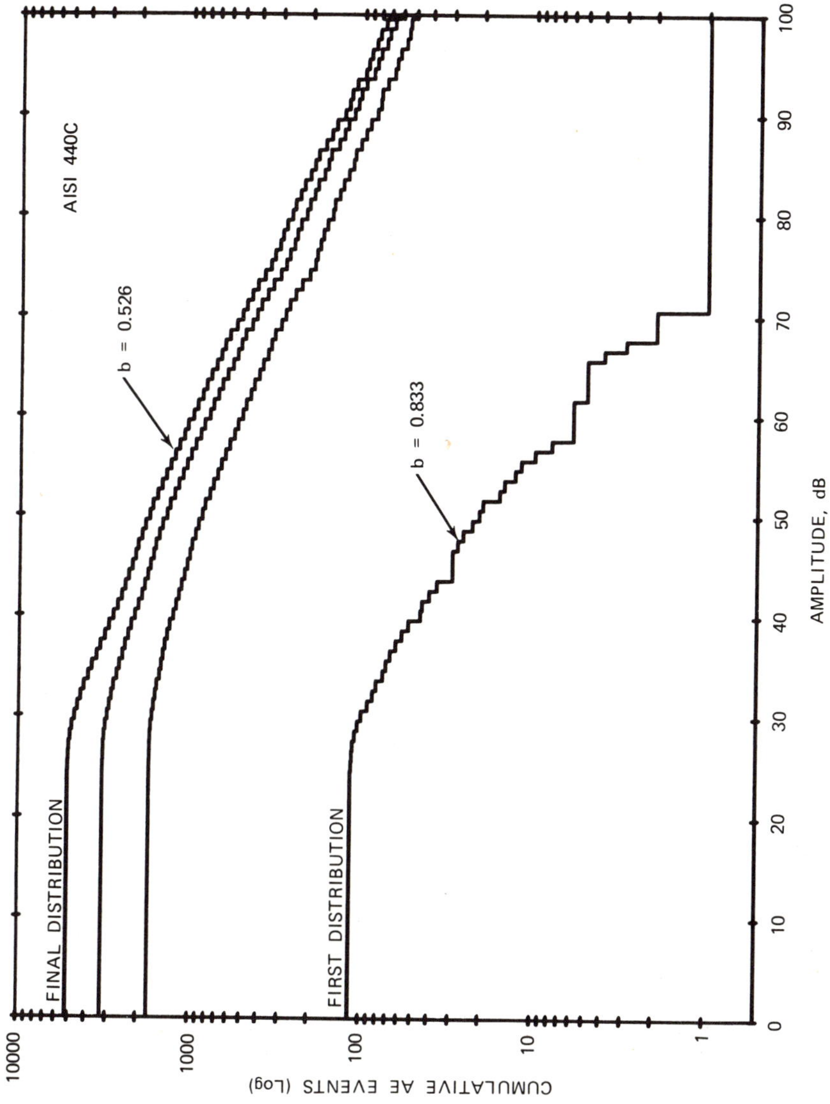

FIG. 6—*Continued.*

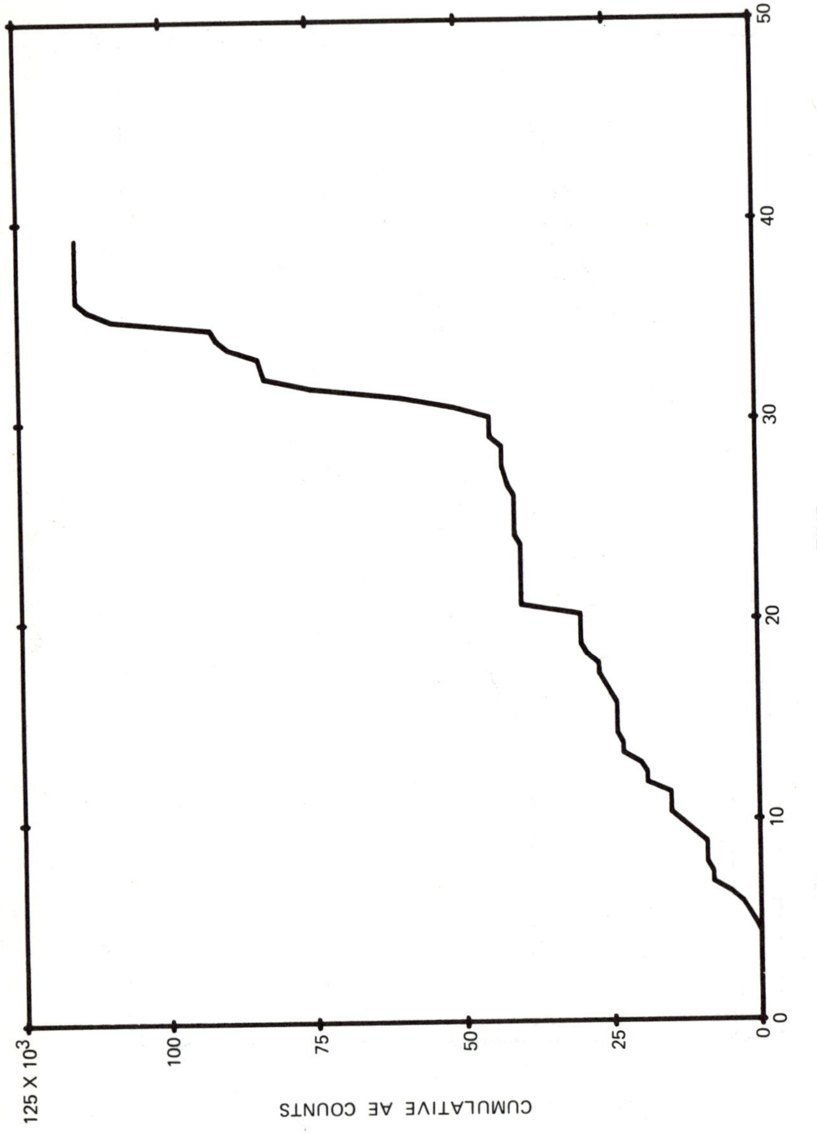

FIG. 7—*AE versus time for loading and unloading of dummy specimen in TerraTek test fixture. This test represents elastic deformation only.* (a) *AE cumulative counts, and* (b) *AE count rate.*

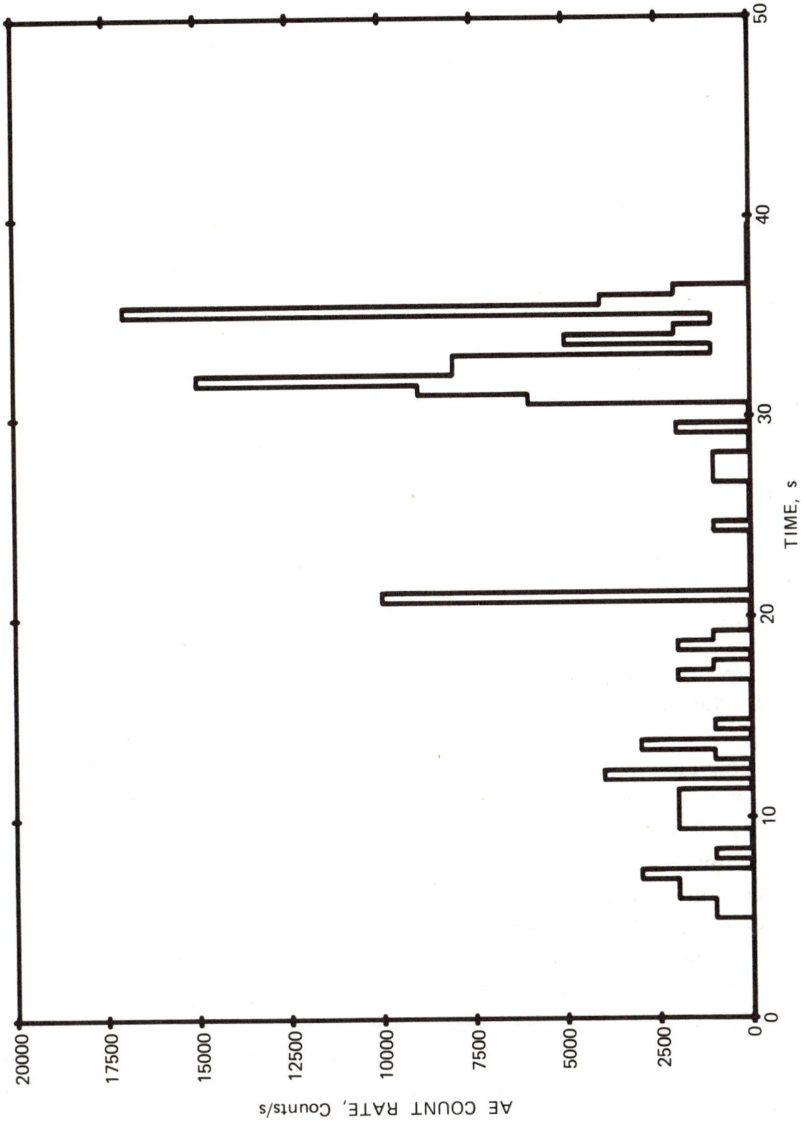

FIG. 7—*Continued.*

is shown in Fig. 7. Figure 7*a* shows total AE counts versus time, and Fig. 7*b* shows count rate versus time. These data show that noise constituted approximately 5% of the total AE activity monitored during fracture toughness testing.

Summary

When testing chevron-notched specimens, it was found that the AE count rate during loading varied significantly from one alloy to the next depending on fracture toughness. This is reflected in Table 3 by the total AE counts at the first unloading point. A comparison of fracture toughness to total counts at the first unloading point provided a relative measure of fracture toughness, that is, the lower the fracture toughness the higher the count rate. This can be attributed to microcracking processes initiated early in the first loading cycle. In all cases, AE was observed only on loading.

Log-sum amplitude distributions were used to characterize initial and final stages of crack growth. For all materials a distinct change in amplitude distribution was observed after the first unloading point. From this point on, the amplitude distributions remained essentially the same throughout the test. It appears, therefore, that in all cases the initial amplitude distribution was characteristic of plastic deformation and crack initiation, and that all subsequent distributions represented crack growth. The slopes (*b*-values) of the amplitude distributions were measured as suggested by Pollack, and, in general, the *b*-values were in agreement with Pollack's data.

Conclusions

The following conclusions can be made in regard to AE monitoring of chevron-notched fracture toughness tests:

1. AE count and count rate are directly relatable to fracture toughness values; the tougher alloys produced a relatively low total count and count rate, while more brittle alloys produced a much higher total count and count rate.
2. A distinct change in amplitude distribution occurs during the transition from crack initiation to crack growth.
3. Log-sum amplitude distribution analysis indicates high values of the *b*-parameter during the plastic deformation and crack-initiation stages of testing and low values of *b* for slow crack growth.
4. For this series of tests, *b*-values are not proportional to fracture toughness.
5. AE measurements provide a possible means for relating fracture toughness to material phenomena (plastic deformation, microcrack formation, crack growth).

Acknowledgment

This report is based on part of the results of a study of ESR casting technology sponsored by the U.S. Naval Air Systems Command, Washington, D.C.

The authors wish to express their thanks to C. Johnson for providing the ESR ingots for fracture toughness and acoustic emission evaluation.

References

[1] Barker, L. M., *Engineering Fracture Mechanics*, Vol. 9, 1977, p. 361.
[2] Barker, L. M. and Leslie, W. C. in *Fracture 1977*, Vol. 2, 4th International Conference on Fracture, Waterloo, Canada, 1977, p. 305.
[3] Barker, L. M. in *Fracture Mechanics Applied to Brittle Materials*, *ASTM STP 678*, S. W. Freiman, Ed., American Society for Testing and Materials, Philadelphia, 1979, pp. 73–82.
[4] Barker, L. M., *Fracture Mechanics: Thirteenth Conference*, *ASTM STP 743*, Richard Roberts, Ed., American Society for Testing and Materials, Philadelphia, 1981, pp. 456–475.
[5] Wadley, H. N. G., Scruby, C. B., and Speake, J. H., *International Metallurgical Reviews*, Vol. 2, 1980, p. 41.
[6] Pollack, A. A., *International Advances in Nondestructive Testing*, Vol. 7, 1981, pp. 215–239.

Fracture Toughness Measurements

Kevin R. Brown[1]

The Use of the Chevron-Notched Short-Bar Specimen for Plane-Strain Toughness Determination in Aluminum Alloys

REFERENCE: Brown, K. R., "**The Use of the Chevron-Notched Short-Bar Specimen for Plane-Strain Toughness Determination in Aluminum Alloys,**" *Chevron-Notched Specimens: Testing and Stress Analysis, ASTM STP 855,* J. H. Underwood, S. W. Freiman, and F. I. Baratta, Eds., American Society for Testing and Materials, Philadelphia, 1984, pp. 237–254.

ABSTRACT: An extensive study of the use of the chevron-notched short-bar (CNSB) specimen for measuring plane-strain fracture toughness in high-strength aluminum alloys is reported. The toughness measured in a CNSB specimen, K_{SB}, correlates closely with plane-strain fracture toughness, K_{Ic}, for toughnesses less than 35 MPa $m^{1/2}$. At higher toughnesses K_{SB} is higher than K_{Ic} for two reasons; first, sample heterogeneity causes the different sized specimens to sample different material, and second, an effect of rising crack growth resistance in the toughness range 40 to 55 MPa $m^{1/2}$; these latter test results should be considered invalid; however, within this range the K_{Ic}, K_{SB} correlation, although largely empirical, is still useful for screening purposes.

The toughness in high-strength aluminum alloy rolled plate may vary markedly through the plate thickness. Examples of these variations are presented.

KEY WORDS: fracture toughness, fracture mechanics, test methods, chevron notch, short bar, toughness inhomogeneity, mechanical properties, aluminum alloys, 2024, 7475, 7075, 7049, 7050

This paper presents the results of recent research into the use of the chevron-notched short-bar (CNSB) specimen for measuring the plane-strain fracture toughness of high-strength heat-treated aluminum alloys. These materials are used primarily for aircraft construction and are currently tested for fracture toughness by the ASTM Standard Test Method for the Plane-Strain Fracture Toughness of Metallic Materials (E 399-81), which determines K_{Ic}, or by a screening test that yields a result that has been correlated with K_{Ic}, Standard Practice for Fracture Toughness Testing of Aluminum Alloys (ASTM B 646-78).

[1]Head, Engineering Properties Section, Kaiser Aluminum & Chemical Corp., Center for Technology, Pleasanton, Calif. 94566.

The ASTM E 399 test method has two main limitations in the aluminum industry. First, it is expensive, and, because of the requirement for a fatigue precrack, specialized equipment and operators are needed. Hence, it is not appropriately carried out in a small or even an average size production plant. Second, its requirements for plane-strain conditions at the crack tip severely limit its applicability to -T6 and sometimes -T7 temper products in common thickness, and to -T3 products in heavy sections.

Currently accepted screening tests are not a panacea, as they usually require specialist machining and metrology, and they often correlate poorly with K_{Ic}. A poor correlation often requires that a high proportion of samples must be retested by ASTM E 399 to meet the required confidence levels set by the aerospace industry.

The chevron-notched sample appears attractive as

(*a*) It does not require specialist machining or testing capability, and it is potentially inexpensive, due largely to the absence of a fatigue pre-crack.

(*b*) Deep side grooves that form the chevron also increase stress triaxiality at the crack tip, and are reported [*1,2*] to retain plane-strain conditions in samples smaller than ASTM E 399 specimens. This has the potential to increase the range of products that can be tested for plane-strain fracture toughness.

This paper presents the results of a study of the use of the CNSB specimen to measure plane-strain fracture toughness in high-strength aluminum alloys. For clarity, the experimental work is divided into three sections. The first describes a statistical comparison of the results of ASTM E 399 testing, K_{Ic} or K_Q, and of short-bar testing, K_{SB}, for a range of alloys, tempers, and product forms.

The second section attempts to assess the contribution of sample heterogeneity to differences between K_{Ic} (or K_Q) and K_{SB}. The variations in fracture toughness through the thickness of plates of several aluminum alloys were determined using short-bar samples.

The third section reports a study of the effect of crack sharpness in short-bar tests of 7475-T7351 plate. It is an attempt to assess the contribution of a rising crack growth resistance [*3*] on the correlation.

In this last portion of the work, a possible systematic error of approximately 3% was found in the curve relating specimen compliance to crack length. This correction was not applied to the results of the earlier studies, as in the majority of cases it would result in a negligible change in K_{SB}.

Procedure and Materials

All short-bar specimens tested in this study were of the geometry described by Barker [*2*], with a thickness B of 25.4 mm (Fig. 1). The side grooves that form the chevron were made with a 127-mm-diameter diamond saw using a plunge cut such that the resultant chevron had curved, rather than straight sides.

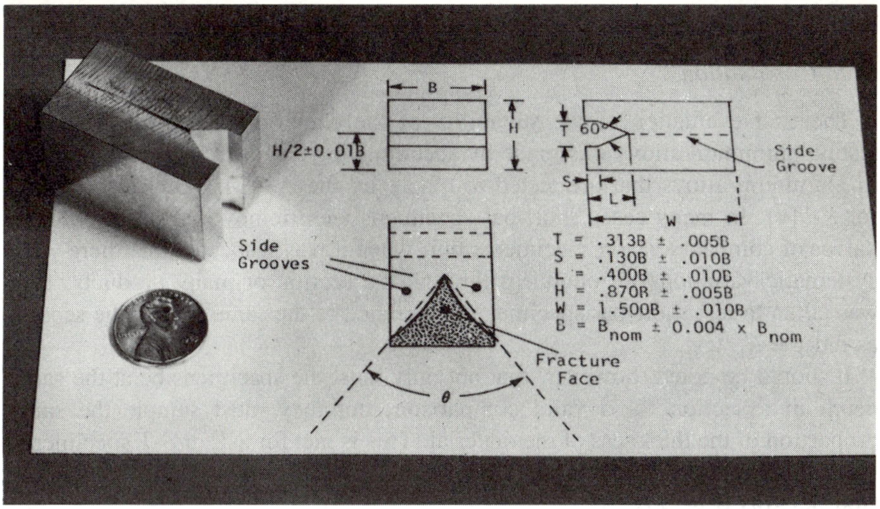

FIG. 1—*Chevron notched short-bar sample used in this study.* B = *25.4 mm.*

The saw blade tip profile was semicircular and produced a groove with a tip radius of 0.22 mm.

In the normal test method used for the majority of this work, the specimens were loaded on the inside of the grip groove in a Terratek Fractometer 2 test machine, using techniques developed by Barker [2], except that plasticity corrections were not made. The load was plotted autographically against the opening of the loading points.

In general, the loading was interrupted twice during a test to record the unloading slopes; however, in materials showing pronounced crack jump behavior, up to five unloading slopes were recorded. These slopes are presented as a ratio, *r*, to the initial elastic loading slope of the unfractured specimen.

The 25-mm rectangular section sample (short bar) was selected instead of the alternate cylindrical short-rod specimen, because the orientation of the rectangular cross section would be easier to maintain during sampling and machining in production plant conditions. Although the circular cross section of the short-rod sample lends itself to round section products, the fracture plane orientation is critical in flat products, and it can easily be marked inaccurately during the preparation of the round test specimen.

For some tests to assess the contribution of rising crack growth resistance, cracks were extended by fatigue loading the specimens in a conventional servohydraulic testing machine. Maximum fatigue loads were half the preceding maximum static load for precracked specimens, or for specimens that were fatigue cracked from the chevron tip, half the maximum load recorded on an equivalent specimen during static loading. A fatigue ratio of 0.1 was used for all fatigue cracked specimens.

All materials tested were from routine production.

Results and Discussion

Data Compilation

The first evaluation of the suitability of the chevron-notched short bar for testing aluminum alloys was made by accumulating K_{SB} data for production lots of aluminum alloys that are tested routinely by the ASTM E 399 test method for K_{Ic} [*4*]. In many cases short-bar specimens were removed from the broken halves of compact tension specimens, but, when it was suspected that there were systematic variations in toughness through the section of many products, care was taken to test short-bar specimens that related to the same part of the section as did the K_{Ic} test.

It should be noted, however, that not only must the specimens be at the same depth in a section for a valid comparison, but they must sample the same proportion of the thickness of the material. This is met for S-L or S-T specimens, but it is not met when comparing L-T or T-L specimens that have different crack front lengths (Fig. 2).

A comparison of 198 correlations of K_Q and K_{SB} is shown in Fig. 3. The data points are for plate, forgings, and extrusions of 7049, 7050, 7075, 7475, 2024, 2124, and 2419 in one or more of -T6, -T7, and -T8 tempers and are described in detail in Ref *4*. An additional set of data for 7050-T73651 plate is not included,

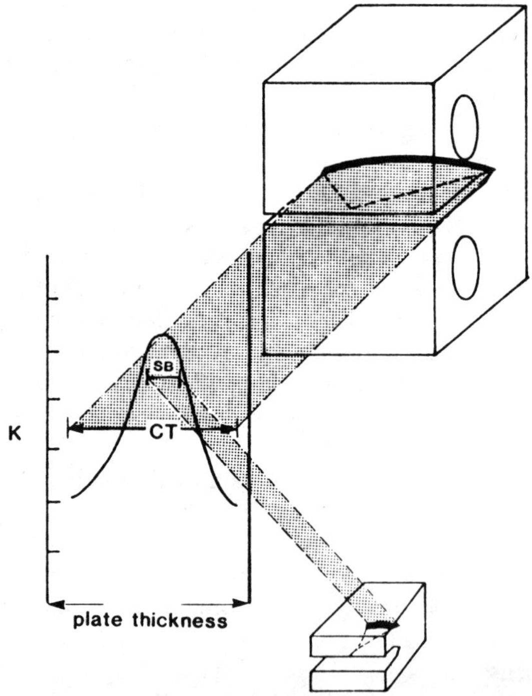

FIG. 2—*Schematic representation of the effect of systematic sample toughness variations on results of short-bar and compact-tension specimen tests.*

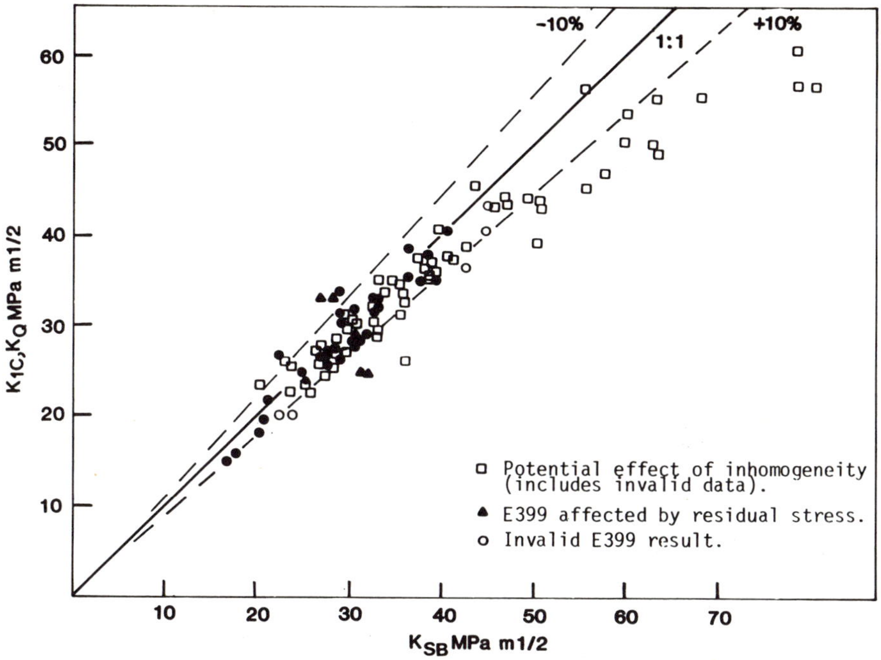

FIG. 3—*Plot of* K_{SB} *versus* K_Q *for each of 198 short-bar specimens.*

but for clarity is shown in Fig. 4. Those correlations in Fig. 3, that are potentially influenced by toughness heterogeneity, are shown as open symbols.

These data include a large proportion of invalid ASTM E 399 test results; however, the majority of these are meaningful, as defined in ASTM Plane-Strain Fracture Toughness Testing of Aluminum Alloys (B 645-78), or are acceptable to aerospace users of the material because the type of invalidity is known to yield a conservative test result less than K_{Ic}. No validity criteria were applied to the short-bar results. The eight remaining invalidities resulted from excessive crack front curvature caused by residual stress in 7049-T3 forgings that were not stress relieved. No effect of residual stress was noted during the testing of the CNSB specimens.

The results for specimens where no heterogeneity effects are expected, that is, those in S-L or S-T orientations, are replotted in Fig. 5. Other S-T results are included in Fig. 4. The screened data in Fig. 5 fit the relationship; $K_{SB} = 1.017$ (± 0.014) K_{Ic}, with a standard deviation of 1.97 and a multiple correlation coefficient of 0.998 for values of K_{Ic} up to 40.7 MPa $m^{1/2}$.

Determination of Sample Heterogeneity

To assess the degree of heterogeneity in some of these materials, CNSB specimens were machined from various depths through the thickness of a number of heat-treated aluminum alloy plates in various tempers (Figs. 6 through 10).

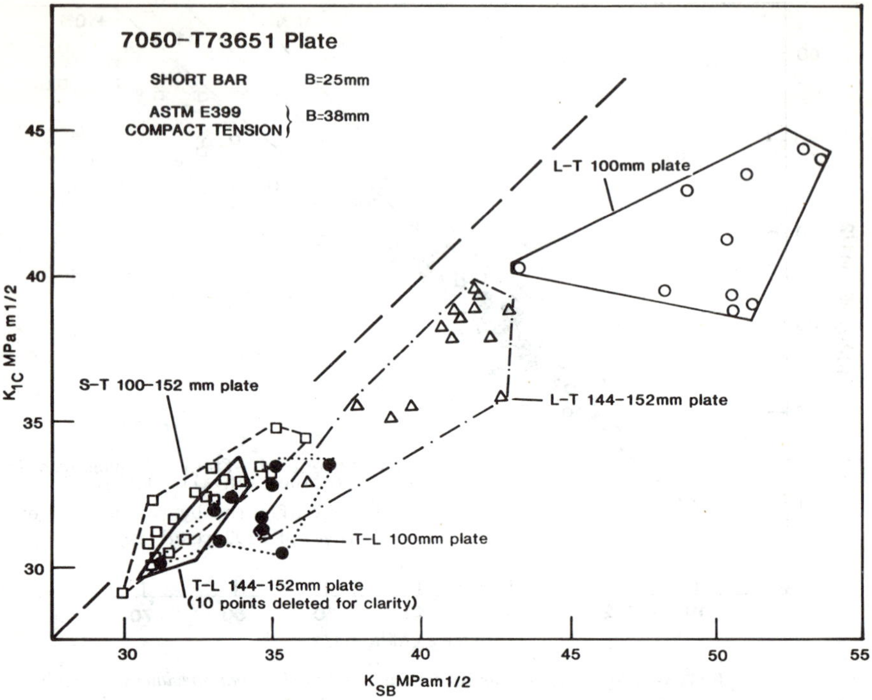

FIG. 4—K_{Ic} *versus* K_{SB} *plot for 7050-T76351 plate of thicknesses of 100 to 152 mm.*

Longitudinal mechanical property and hardness data were included where possible. Some data produced in this study have been published already [*4,5*] for other production lots.

The plate midthickness was invariably tougher than elsewhere, particularly in the 7000 series alloys, as has been reported [*6*], and, in some specimens, K_{SB}-values were 50% higher than those measured near the plate surfaces.

It can be seen that the larger ASTM E 399 compact tension specimen is in fact averaging the toughness variations measured by the CNSB specimens. The effect is emphasized in 7475-T7351 in Fig. 7*a*, where two midthickness L-T compact tension tests were run with thickness, *B*, of 32 and 48 mm. The thinner specimen gave a K_Q result that was invalid because of insufficient specimen thickness and crack length. Although this type of invalidity would be normally expected to result in a measured K_Q that was lower than K_{Ic} [*7*], it was actually higher due to the toughness inhomogeneity in the plate.

The 2024 plate shown in Fig. 6 shows the least variation in toughness through the thickness; however, these profiles may not be typical, for greater variations have been reported in similar material [*4,8*]. These two plots represent the same piece of plate; one section was further aged from -T351 to -T851 before testing. Note that in the higher strength -T851 temper, there is a close agreement between the K_{SB}-values and K_{Ic}, but in the -T351 plate there is a clear discrepancy of

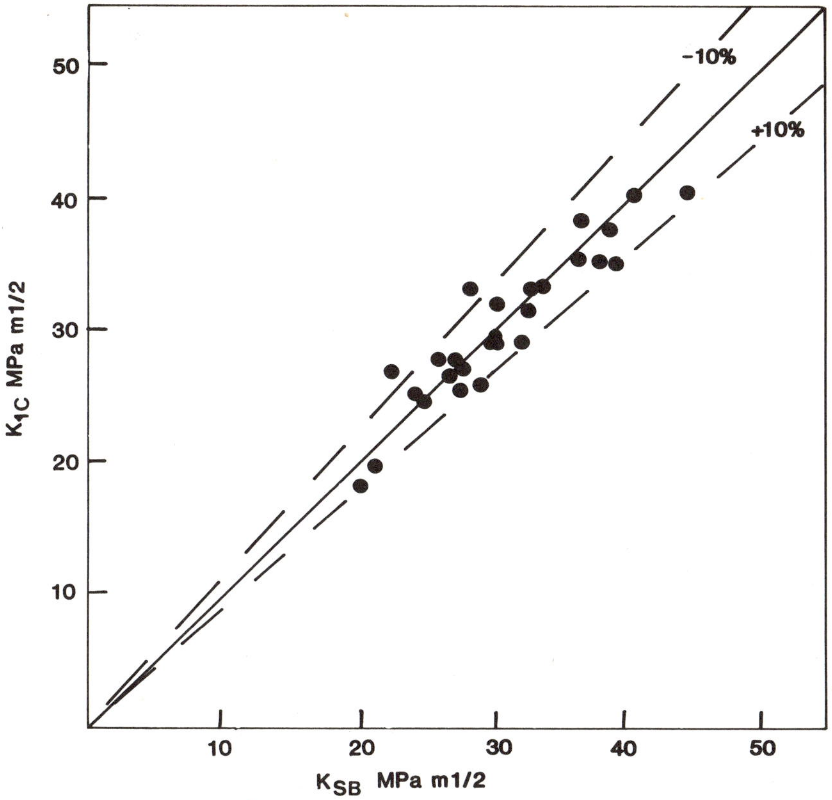

FIG. 5—*Plot of* K_{SB} *versus* K_{Ic} *using selected data.*

approximately 6% between a marginally invalid L-T compact tension test result, and four K_{SB}-values. This difference cannot be explained by sample heterogeneity and will be discussed later.

The effect of through-thickness toughness variations on correlations between K_{Ic} and K_{SB} at the midthickness of 100, 144, and 152-mm 7050-T73651 plate is also demonstrated in Fig. 4. The toughness profiles for the 100 and 152-mm plates are shown in Figs. 10*a* and *b*. Both compact tension and short-bar S-T test specimens sample the same material at the plate center, and there is a close relationship between K_{SB} and K_{Ic} (Fig. 4); however, in the L-T and T-L orientations, the smaller short-bar specimen samples a small volume of tougher material at the plate center, and K_{SB} is greater than K_{Ic}. A comparison of S-T and T-L results can be made at the same K_{Ic} level; the higher K_{SB} at the same K_{Ic} in the T-L sample cannot reasonably be attributed to a rising R-curve effect, as the transverse yield strength is approximately 5% higher than the short-transverse (thickness direction) yield strength.

A similar comparison can be made for the L-T samples.

It can also be seen in Fig. 4 that K_{SB} is relatively higher in thinner plate for

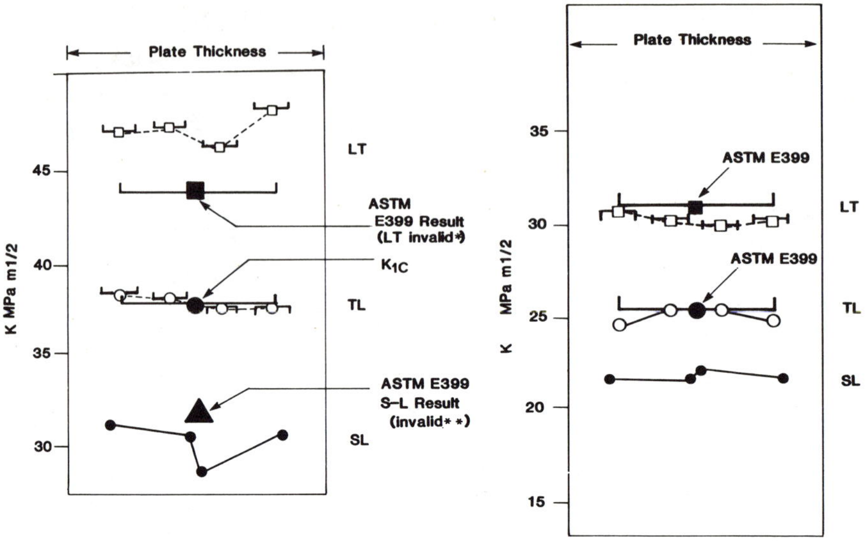

FIG. 6—*Toughness variations through the thickness of 63-mm 2024-T351 and -T851 plates.*

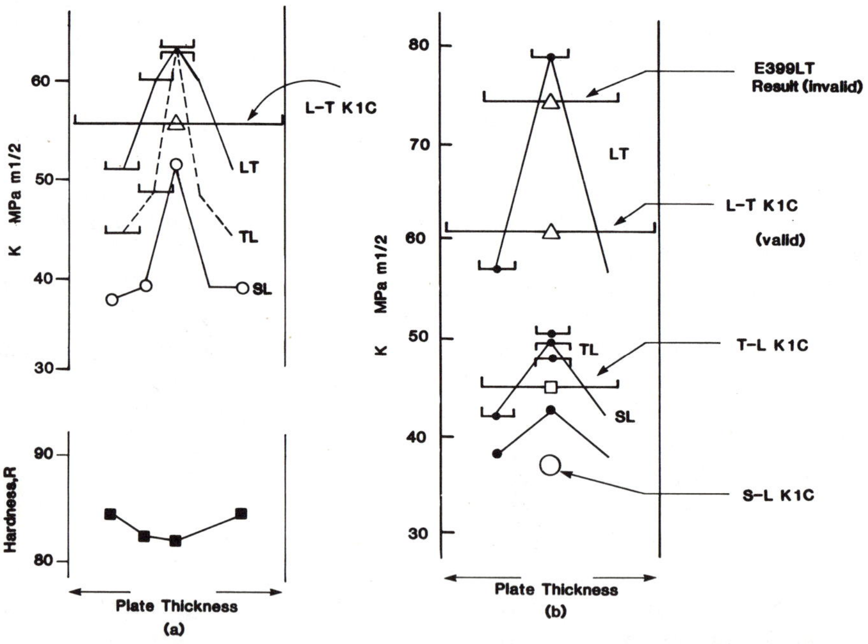

FIG. 7—*Toughness variations through the thickness of 50-mm thick 7475-T7351 plates.*

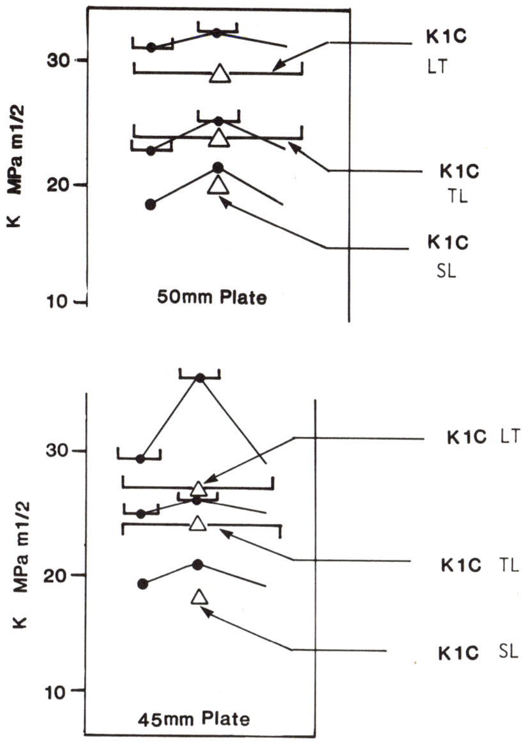

FIG. 8—*Toughness variation through the thickness of* (a) *a 50-mm- and* (b) *a 45-mm-thick plates of 7075-T6.*

both L-T and T-L specimens. Although toughness levels at the same relative depth in the plate thicknesses are similar, a steeper toughness gradient is noted in the thinner plate. The larger compact tension specimen thus samples relatively more material near the thinner plate surface, with the result that there is a relatively greater difference between K_{SB} and K_{Ic} in both orientations.

Increasing Crack Growth Resistance Effect

It is well known that crack growth in materials which show significant plasticity is accompanied by an increase in crack growth resistance. This increase is described in an R-curve which can be measured for a particular material by ASTM Standard Recommended Practice for R-Curve Determination (E 561-80). The K_{Ic}-value in the ASTM E 399 test is measured after a crack growth of an arbitrary 2% from the fatigue crack tip; however, the K_{SB}-value in the short-bar test is measured after extensive crack growth, and Munz [*3*] has suggested that a rising R-curve effect is responsible for K_{SB}-values being higher than K_{Ic} at high toughnesses. This possibility is suggested by the data for 2024-T351 shown in

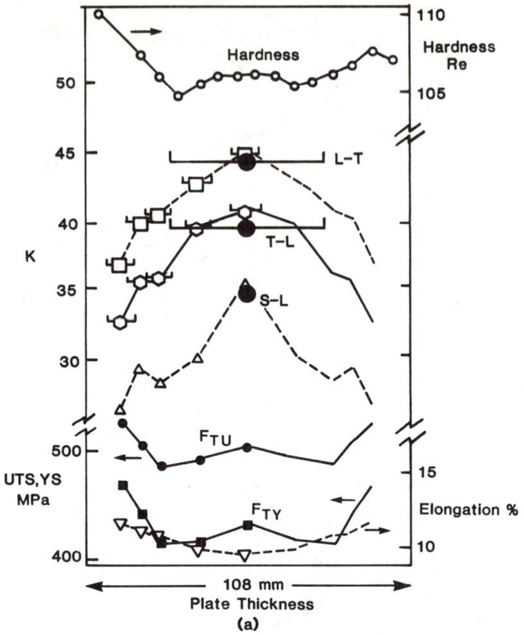

FIG. 9—*Property variations through 7049-T7351 108-mm plate.*

Fig. 6, where a series of K_{SB} exceed K_Q; however, this discrepancy could also be caused by the invalidity in K_Q [7].

This was checked on a series of 30 short-bar samples of 7475-T7351 plate. Ten identical samples were from each L-T, T-L, and S-L orientation; all samples were removed from the midthickness of the plate.

Several specimens of each orientation were tested to determine an average K_{SB}. Two further samples of each orientation were fatigue cracked approximately 8 mm from the tip of the chevron.

In the remainder of the samples, cracks were grown by static loading to various lengths. These were then unloaded, and the crack extended by fatigue at 50% of the final static load. It was intended to grow each fatigue crack through the plastic zone created by the static loading; this was achieved in most specimens (Table 1).

All fatigue cracked samples were then loaded in the usual way, and the load-elongation record determined. Secant offsets ranging from about 3% to 11% were used to determine the stress intensities at 2% crack growth, $K_{2\%}$ (Table 1).

Two problems were revealed by an examination of the fracture faces:

(*a*) The fracture was out of plane in both L-T samples used to measure K_{SB}, and in three of four L-T samples that were fatigued to sharpen a static crack (Table 1). The fracture surface tended to follow the surface of a cone whose

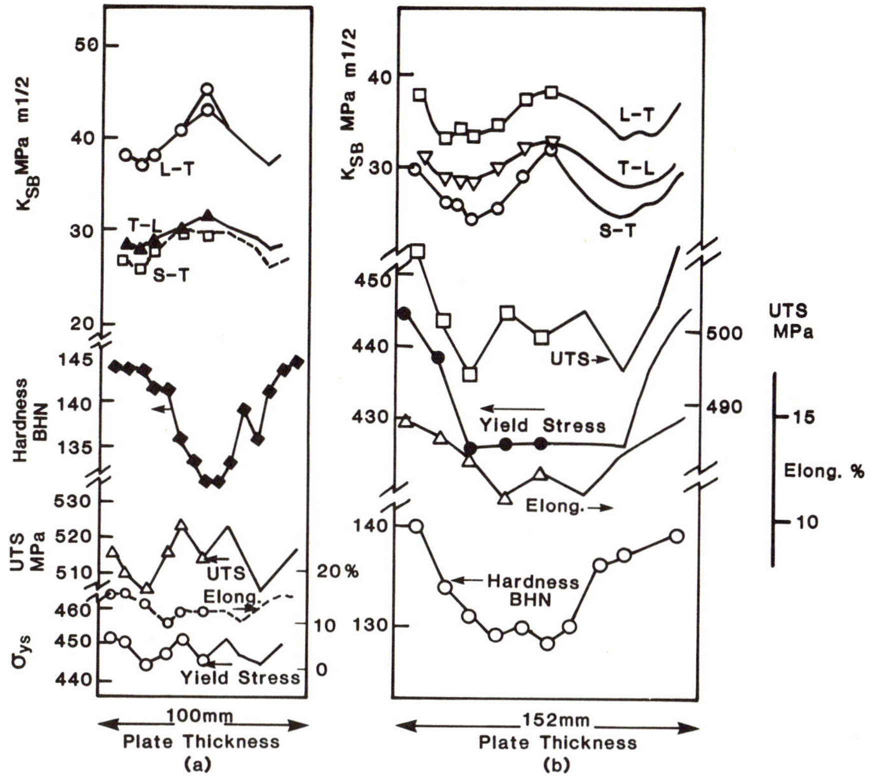

FIG. 10—*Property variations through 100-mm and 152-mm 7050-T73651 plates.*

apex corresponded to the tip of the chevron, such that the center of the crack front was several millimeters above the nominal crack plane at the crack length where K_{SB}-values were determined. Similar behavior has been reported [*9*]. In all but one of these samples, the crack made its exit through the face of the specimen, instead of the back (Fig. 11), in a manner similar to compact tension tests oriented in a tough direction in some materials. The effect of an out-of-plane crack is obvious in Table 1 and Fig. 12; K-values are about 10 MPa $m^{1/2}$ lower for a flat fracture.

(*b*) The fatigue crack started at different points of the crack front, and at different levels, in those samples precracked by static loading. Typically a fatigue crack would start at each end of the monotonic-crack front, and meet in the middle at a step (Fig. 12). This resulted in extreme crack front curvature. Somewhat surprisingly, this did not appear to alter $K_{2\%}$ significantly from that obtained with a flatter crack that had been fatigued from the chevron tip.

After all samples had been broken open, crack lengths were measured (Table 1) and the slope ratio and length data used to correct the reported specimen

TABLE 1—*Effect of crack sharpness on measured toughness of 7475-T7351 short-bar samples.*[a]

Sample[b] Orientation	Sample[b] No.	Average Crack Length, mm Static	Average Crack Length, mm Fatigue	Crack Curve, %[c] Static	Crack Curve, %[c] Fatigue	Fatigue Increase, mm	Slope Ratio Static	Slope Ratio Fatigue[d]	Load at 2% Monotonic Growth, kg	$K_{2\%}$, MPa m$^{1/2}$	Average of $K_{2\%}$, MPa m$^{1/2}$	$(K_{2\%}/\sigma_{ys})^2$ mm	Critical Intensity Stress, MPa mm$^{1/2}$ K_{Ic}	Critical Intensity Stress, MPa mm$^{1/2}$ Average K_{SB}	$\frac{K_{SB} - K_{2\%}}{K_{2\%}}$
L-T	4	none	23.52	. . .	10	23.52	. . .	0.35	1151.3	62.8					
L-T	5	none	21.51	. . .	10	21.51	. . .	0.42	1139.0	60.7	60.7 (nos. 4–6)	21.5			
L-T	6	14.40	19.02	1	22	4.57	0.80	0.59	1062.8	58.7	65.8 (nos. 4–9)	25.3 (nos. 4–9)	. . .	74.1	13
L-T	7	17.37	21.72	2	23	4.32	0.65	0.41	1362.0	72.7[e]				(2/3 tests[e])	
L-T	8	19.48	24.49	2	18	5.08	0.54	0.31	1279.4	72.2[e]	70.8 (nos. 7–9)	29.2			
L-T	9	21.29	24.69	4	20	3.30	0.46	0.30	1190.4	67.8[e]					
T-L	4	none	23.34	. . .	8	23.34	. . .	0.35	825.4	45.0					
T-L	5	none	21.18	. . .	6	21.18	. . .	0.43	860.3	45.9					
T-L	6	14.66	18.39	0	16	3.81	0.83	0.57	850.3	46.6	45.2 (nos. 4–9)	11.8	. . .	51.4	14
T-L	7	18.57	22.25	1	14	3.81	0.58	0.40	870.8	46.5					
T-L	8	20.52	24.00	0	17	3.56	0.49	0.32	769.5	43.0					
T-L	9	22.40	25.53	0	16	3.05	0.41	0.28	761.4	44.3					
T-L	10	24.61	29.41	1	8	4.83	0.34	0.17	515.7	. . .					
S-L	4	none	21.36	. . .	6	21.36	. . .	0.42	777.7	41.5					
S-L	5	none	21.92	. . .	9	21.92	. . .	0.40	759.5	40.6		10.7	40.9	45.4	10
S-L	6	13.97	17.30	2	11	3.30	0.90	0.63	759.5	42.7					
S-L	7	18.39	24.97	2	7	6.60	0.59	0.30	714.1	40.7	41.4 (nos. 4–10)				
S-L	8	21.06	29.06	0	3	8.13	0.47	0.17	470.3	. . .					
S-L	9	22.05	27.81	0	6	5.84	0.43	0.21	582.0	. . .					
S-L	10	27.00	27.66	0	1	0.76	0.26	0.22	639.7	. . .					

[a]Yield strengths (0.2% offset), Mpa: L = 414, LT = 416, ST = 400.
[b]Samples 1 to 3 in each orientation were used to generate K_{SB} data. One L-T and one S-L sample were scrapped due to a testing error.
[c]Average of mid and end points on crack front. NB differs from ASTM E 399 definition.
[d]From corrected calibration curve, Fig. 13.
[e]Fracture out of plane.

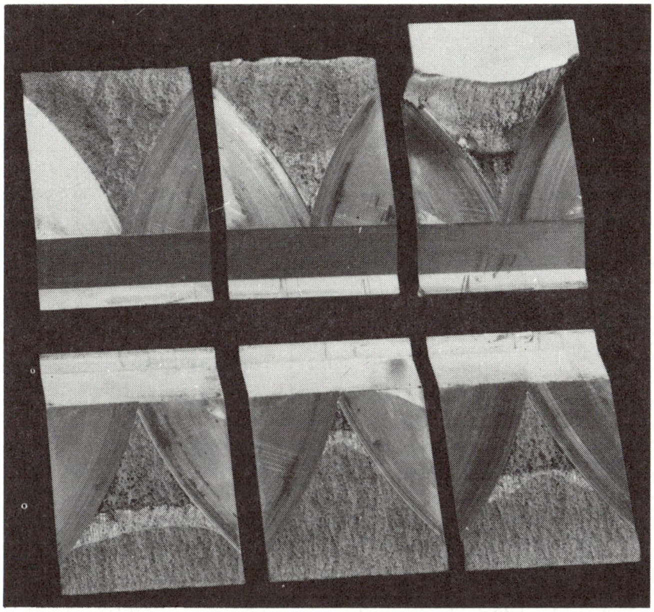

FIG. 11—*Broken short-bar specimens, showing various test features.*

calibration [2] (Fig. 13). Fatigue crack lengths were not used, as crack mouth opening calibrations (and hence slope ratios) were not maintained from pre-cracking, through fatigue cracking, to final loading for each specimen. Slope ratios for fatigue cracks were determined by reference to this corrected calibration curve.

An error of approximately 3% was found in the calibration curve; however, for measurements made near the maximum load of the load/CMOD curve, this results in an insignificant error in K_{SB}. Larger errors could be encountered in material showing extreme crack jump behavior, where K_{SB} is measured at crack lengths that are significantly longer or shorter than that at the maximum load in an ideal test.

The data relating to fatigue cracks (Table 1) is shown in relation to the curves used normally to determine K_{SB} in Figs. 12, 14, and 15; the loads corresponding to $K_{2\%}$ tend to lie beneath the normal curve.

Discussion

In agreement with the results of others [*2,3,10,11*], there appears to be a close correlation between the critical stress intensity measured in a chevron-notched short-bar test, K_{SB}, and K_{Ic} in high-strength aluminum alloys. For K_{Ic}-values

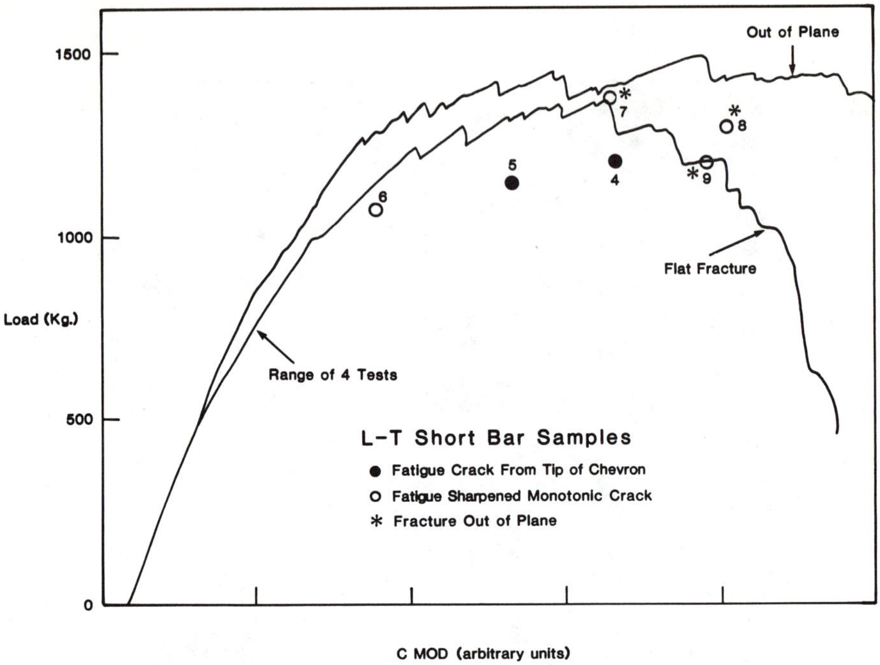

FIG. 12—*Range of scatter in tests used to determine* K_{SB} *data on L-T 7475-T7351 plate.*

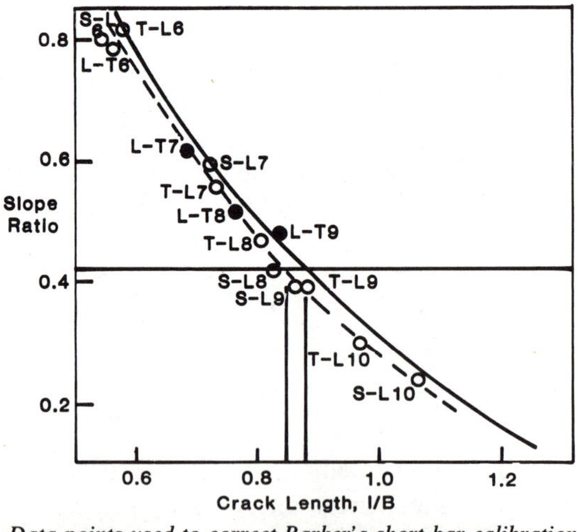

FIG. 13—*Data points used to correct Barker's short-bar calibration curve* [2].

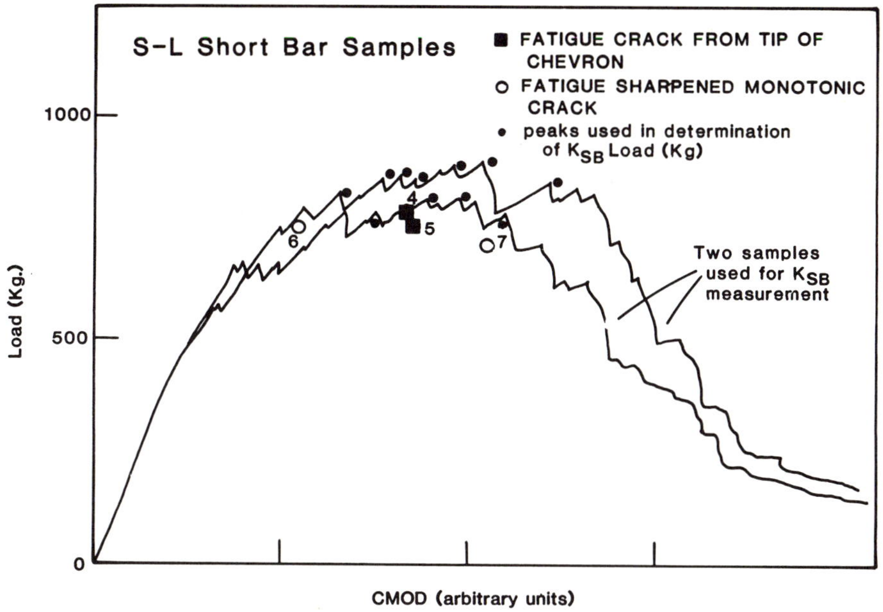

FIG. 14—*Range of scatter in tests to determine S-L* K_{SB}*-values for 7475-T7351 plate.*

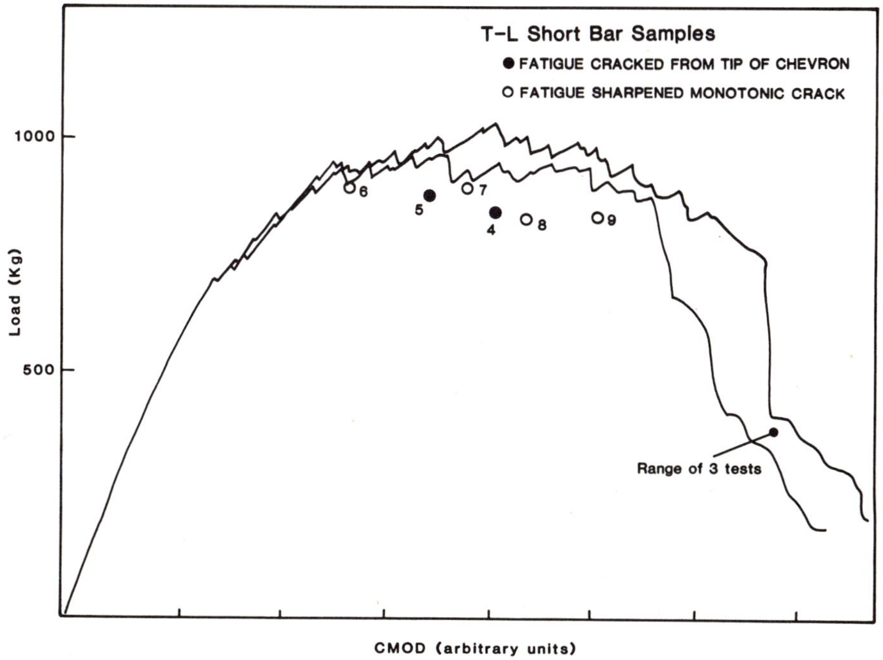

FIG. 15—*Range of scatter in tests to determine T-L* K_{SB}*-values for 7475-T7351 plate.*

below about 40 MPa $m^{1/2}$ the correlation is linear and K_{SB} approximates K_{Ic}; however, at higher toughnesses, K_{SB} exceeds K_{Ic} or K_Q.

Despite the departure from a linear correlation with K_{Ic} at higher toughnesses, the CNSB specimen appears suitable for use as a screening test, as described in ASTM B 646, provided a sufficiently large data base is developed for the product. The present correlation is closer than for other currently used tests.

Sample Inhomogeneity

The greatest part of the departure from linearity can be attributed to variations in toughness through the section of the materials studied. In the specimen used in this study, the crack front at the point at which K_{SB} is measured is about 8 mm long, and the correlations are made against L-T and T-L K_{Ic}-values determined on compact tension specimens having thicknesses, B, and hence crack front lengths of 25 to 63 mm. The different tests thus sample different volumes of material.

This problem does not occur in S-L or S-T specimens; however, the aluminum alloys examined in this program characteristically do not often have S-L or S-T toughnesses above 40 MPa $m^{1/2}$; the higher values in the correlation are obtained only in L-T or T-L orientations, and are thus confounded by the effect of toughness heterogeneity. To extend the correlation accurately, it would be necessary to use larger short-bar samples with crack fronts of the same length as the comparable compact tension sample, or to reduce the size of the compact tension specimen. In most cases this is impossible as the necessary thickness of the short-bar would exceed the product thickness, or the smaller compact tension sample would yield an invalid ASTM E 399 result.

The alternative is to test several adjacent short bars so that the tests span the crack front of the comparable compact tension specimen, as has been done here for a limited number of alloys.

Although it is a disadvantage when measuring an average toughness for the product, the smaller size of the short bar compared to the E 399 sample enables a higher resolution when testing an inhomogeneous product. Heavy sections of aluminum alloys are usually machined extensively for use in aircraft, and, for the alloys examined, the average toughness would be rarely relevant to the remaining section. The higher resolution of the CNSB specimen could make a contribution to more efficient design.

Effect of Rising Crack Growth Resistance

Because there is a good correlation between K_{SB} and K_{Ic} for S-L and S-T specimens having toughnesses below 35 to 40 MPa $m^{1/2}$ (Fig. 5), it is reasonable to conclude that there is a negligible rising R-curve effect in this range. However, the tests of short-bar samples containing fatigue cracks (Table 1) indicate clearly that at toughnesses above 40 MPa $m^{1/2}$ there was an effect of a rising crack growth resistance, K_R, as the crack propagated. For 7475-T7351 S-L short-bar samples (Table 1), there was close agreement between an average stress intensity determined for 2% crack growth, $K_{2\%}$ (41.4 MPa m$^{1/2}$) and K_{Ic} (40.9 MPa m$^{1/2}$); however, the K_{SB}-value was 10% higher than $K_{2\%}$ and 11% higher than K_{Ic}. The T-L results indicated that K_{SB} was 14% greater than $K_{2\%}$, and, in all probability, K_{Ic}. Similar comparisons for L-T samples are difficult because many cracks were out of the plane of the side groove, and the test results would be considered invalid by any reasonable criterion. This departure of the crack from the expected fracture plane is more troublesome in the short-bar sample than in an ASTM E 399 test, for, although such departures are common in compact tension tests of very anisotropic materials, the most significant data are gathered in the first 2% of crack extension, before significant departures occur. A comparison of $K_{2\%}$ values with subsequent maximum load K_{SB}-values on the same L-T specimens with flat fractures indicated that K_{SB} should not exceed $K_{2\%}$ by more than 20% at short-bar toughnesses of 74 MPa m$^{1/2}$. It would be expected that if $K_{2\%}$ is influenced by plasticity at these toughness levels, then it would be lower than a valid K_{Ic} [*7*], and consequently K_{SB} should not exceed K_{Ic} by more than 20%.

These results conflict with those of Munz [*3*], who found discrepancies between K_Q and K_{SB} of up to 65%, but who did not take sample heterogeneity into account. The whole of the difference between K_Q and K_{SB} was ascribed to a rising crack growth resistance. The use of invalid ASTM E 399 data is another possible source of disagreement, for most of the causes of test invalidity in Munz's work are known to reduce the measured toughness [*6*].

Munz tested materials that were very similar to those of this study, and marked heterogeneity would have been expected. His short-bar toughness values measured with a fatigue crack exceeded the K_{Ic} of a larger compact tension specimen, as could occur if the toughness variations were present.

Further, Munz's comparisons were between K_{Ic} or invalid ASTM E 399 test results measured after a 2% growth from the fatigue crack, and a value, (K_{IIe}), that presumably corresponds to the stress intensity at the first detectable growth of the fatigue crack in the short-bar samples. This value is expected to be relatively lower than $K_{2\%}$ measured on short bars in this study, and its use would exaggerate the effect of rising crack growth resistance, and enhance the apparent agreement with K_{Ic}. Munz's K_{IIe}-values may have been lowered further, if, as in this work, he encountered problems in nucleating a fatigue crack on a static crack front. The fracture of ligaments between independently nucleated fatigue cracks would occur at low stress intensities, and would affect K_{IIe} to a greater extent than $K_{2\%}$.

Specimen Size Limits

The present results are not sufficient to establish specimen size limits for a one-to-one correlation between K_{SB} and K_{Ic}. The data in Fig. 5 for S-L and S-T specimens correspond to values of $(K_{Ic}/\sigma_{ys})^2$ ranging from 2.8 to 11.9 mm, where the higher values correspond to toughnesses, K_{Ic}, around 40 MPa m$^{1/2}$. Unfortunately, there are insufficient data to establish with statistical confidence the agreement with K_{SB} at that toughness level. Data for T-L samples in Table 1 indicate significant departures from a linear correlation with K_{Ic} at a $(K_{2\%}/\sigma_{ys})^2$ of 11.8 mm.

These data indicate a limiting specimen size, B, of about 2.5 $(K_{Ic}/\sigma_{ys})^2$, the same as the ASTM E 399 limit; however, the absence of shear lips and crack front shortening (Fig. 12) suggests that in this toughness range the limiting parameters may actually be the side groove thickness and tip profile, rather than specimen thickness.

Conclusions

1. In a range of high-strength aluminum alloys there is a close correlation between the plane-strain fracture toughness K_{Ic} or K_Q measured by the ASTM E 399 test method and that determined on chevron-notched short-bar specimens. For toughnesses less than 35 MPa $m^{1/2}$, K_{Ic} and K_{SB} appear to measure the same material property.
2. The correlation between K_{SB} and K_{Ic} in aluminum alloys is sufficiently close for values of K_{Ic} up to 55 MPa $m^{1/2}$ to justify the use of the chevron-notched short-bar specimen as an empirical screening test to reduce the amount of testing to ASTM E 399.
3. Material heterogeneity is the most significant factor affecting correlations between K_{Ic} and K_{SB} in aluminum alloys. Toughness and other property variations through the thickness of a range of aluminum alloy plates have been determined.
4. The plane-strain toughness determined with a chevron-notched short-bar specimen can be 10 to 20% in error due to rising crack growth resistance effects at toughnesses in the range 40 to 75 MPa $m^{1/2}$ in 7475-T7351 plate; however, such results would undoubtedly be classified invalid by any future standard test procedure.
5. The short-bar test sample is less sensitive than larger compact tension samples to the effects of residual stress in aluminum alloys.

References

[1] Barker, L. M., *Engineering Fracture Mechanics*, Vol. 9, 1977, pp. 361–369.
[2] Barker, L. M., *Engineering Fracture Mechanics*, Vol. 17, No. 4, 1983, pp. 289–312.
[3] Munz, D., *Engineering Fracture Mechanics*, Vol. 15, No. 1–2, 1981, pp. 231–236.
[4] Brown, K. R., Kaiser Aluminum & Chemical Corporation Report, Center for Technology, 1 Oct. 1981.
[5] Brown, K. R., *Proceedings*, Sixth International Conference on the Strength of Metals and Alloys, ICSMA 6, Melbourne, Australia, August 1982, pp. 765–771.
[6] De Jong, H. F., *Engineering Fracture Mechanics*, Vol. 13, 1980, pp. 175–192.
[7] Kaufman, J. G. in *Developments in Fracture Mechanics Test Methods Standardization, ASTM STP 632*, American Society for Testing and Materials, Philadelphia, 1977, pp. 3–24.
[8] Barker, L. M. personal communication.
[9] Broek, D., Fracture Research, Inc., Report, Columbus, Ohio, March 1982.
[10] Barker, L. M. and Baratta, F. I., *Journal of Testing and Evaluation*, Vol. 8, 1980, p. 97.
[11] Bubsey, R. T., Munz, D., Pierce, W. S., and Shannon, J. L., *International Journal of Fracture*, Vol. 18, Feb. 1982, pp. 125–132.

Johannes Eschweiler,[1] *Gunter Marci,*[1] *and Dietrich G. Munz*[2]

Fracture Toughness of an Aluminum Alloy from Short-Bar and Compact Specimens

REFERENCE: Eschweiler, J., Marci, G., and Munz, D. G., **"Fracture Toughness of an Aluminum Alloy from Short-Bar and Compact Specimens,"** *Chevron-Notched Specimens: Testing and Stress Analysis, ASTM STP 855,* J. H. Underwood, S. W. Freiman, and F. I. Baratta, Eds., American Society for Testing and Materials, Philadelphia, 1984, pp. 255–269.

ABSTRACT: Fracture toughness of a 70-mm-thick plate of the aluminum alloy 7475-T7351 was measured with short-bar and compact specimens. The compact specimens with thicknesses of 25 and 38 mm and the short-bar specimens with thicknesses of 12.5, 25, and 50 mm had T-L orientation and were machined at different distances from the surface. Different heats of the same alloy tested in T-L and L-T orientation furnished a series of fracture-toughness values ranging from 40 to 60 MPa $m^{1/2}$ in increments of approximately 2 MPa $m^{1/2}$. For each of the selected heats K_{Ic} and K_{IcSB} was determined. Both fracture toughness test methods confirmed a fracture-toughness profile through the thickness direction of the plate. The fracture toughness was highest in the midsection and lowest in the surface region of the plate. Increasing fracture toughness with increasing specimen size was found for both testing methods, but K_{IcSB} was consistently higher than K_{Ic} for specimens of identical thickness. The difference between K_{Ic} and K_{IcSB} increases with increasing fracture toughness. From the results it is concluded that the through thickness variation of K_{Ic} is not the main reason for the higher short-bar values.

KEY WORDS: fracture toughness, chevron-notched short bar, aluminum alloy, through-thickness variation

The increasing use of the material parameter fracture toughness K_{Ic} as a quality criterion of the material brought the need for a simple, inexpensive method for estimating K_{Ic}. A candidate method for estimating K_{Ic} was proposed by Barker [*1,2*]. The test method proposed uses chevron-notched short rods or short bars. Subsequently, the compliance- and stress-intensity coefficients for chevron-notched specimens had been determined over a wide range of geometrical variables [*3–5*]. In testing aluminum alloys in accordance with ASTM Test Method for Plane-Strain Fracture Toughness of Metallic Materials (E 399-78*a*) and the method

[1]Research technician and division head, respectively, German Aerospace Research Establishment, Köln, West Germany.

[2]Professor, University of Karlsruhe, Postfach, West Germany.

proposed by Barker [*2*], Munz [*6*] found that the fracture toughness values differed considerably. An extension of this work [*7*] showed that the geometrical requirement set forth by Barker [*8*] could not be responsible for the differing results of the two test methods.

There seem to be materials which furnish results of testing chevron-notched specimens in close agreement with plane-strain fracture-toughness testing [*9,10*] (ASTM E 399-81). For some materials, in particular the aluminum alloy 7075 or 7475 in heat treatment conditions T6 and T7, testing chevron-notched specimens and testing in accordance with ASTM E 399-81 give results for fracture toughness which are not compatible [*6,7,9*]. The known variation of fracture toughness through the thickness of plate materials [*9,11,12*] and across the plate may be thought to be responsible for the disagreement. Therefore, in the first part of this work three strips of an aluminum alloy 7475-T7351 plate are tested thoroughly with respect to fracture toughness according to ASTM E 399-81, and the results are compared with fracture-toughness data obtained from chevron-notched short-bar specimens.

It was assumed that a pronounced R-curve behavior could be responsible for the differing results of the two testing methods. The second part of the investigation compares the fracture toughness data of the two testing methods for the aluminum alloy 7475-T7351 for different heats of materials in dependence on the fracture toughness measured according to ASTM E 399-81. The fracture toughness according to ASTM E 399-81 varied from 40 to 60 MPa m $^{1/2}$.

TABLE 1—*Mechanical properties of the aluminum alloy 7475-T7351 materials used in this investigation.*[a]

Plate Identification No.	L			L-T		
	$R_{p0.2}$	R_m	A_5	$R_{p0.2}$	R_m	A_5
242	406	475	12.6	420	493	10.4
236	411	478	12.8	418	496	10.2
79	395	474	12.8	406	485	10.8
246	415	482	13.0	429	503	11.0
68	430	510	11.7	422	500	11.2
15	383	461	15.2	402	478	11.0
234	450	516	11.0	451	501	10.0
54	443	516	11.8	429	507	10.8
247	417	483	12.6	429	502	10.4
63	436	510	11.2	418	497	10.4
76	399	475	13.8	414	497	11.6
28	397	470	13.0	410	493	11.6
24	391	465	13.8	401	484	11.6
201	404	480	12.5	415	498	9.7

[a] $R_{p0.2}$ = yield strength (N/mm^2).
R_m = ultimate strength (N/mm^2).
A_5 = elongation (%) 2 in. gage length.

Materials and Experimental Procedure

The aluminum alloy used in the present investigation qualifies as 7475-T7351 from the standpoint of chemical composition as well as mechanical properties, but has been declared as 7075-T7351 in some cases by the producer. The tensile properties of the different heats of this material are listed in Table 1. Compact specimens (CT-specimens) according to ASTM E 399-81 of thickness B = 25 mm or B = 38 mm or both were machined from a variety of plates. Chevron-notched short-bar specimens (SB-specimens) were machined according to Fig. 1 from the same plates. The loading fixture is shown in Fig. 2. The width of

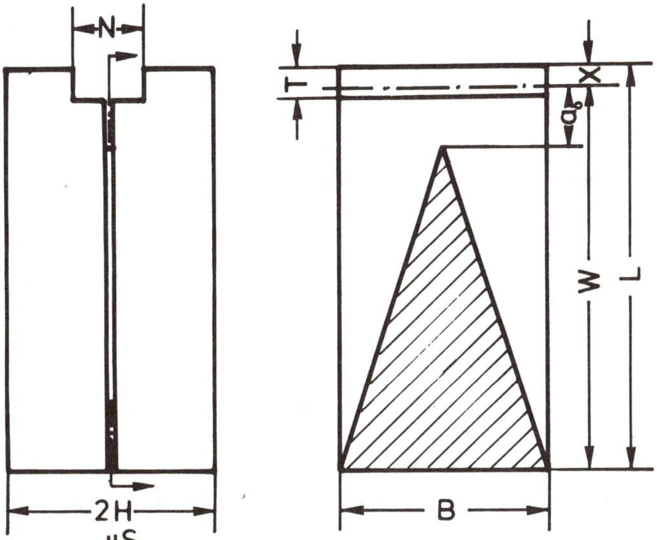

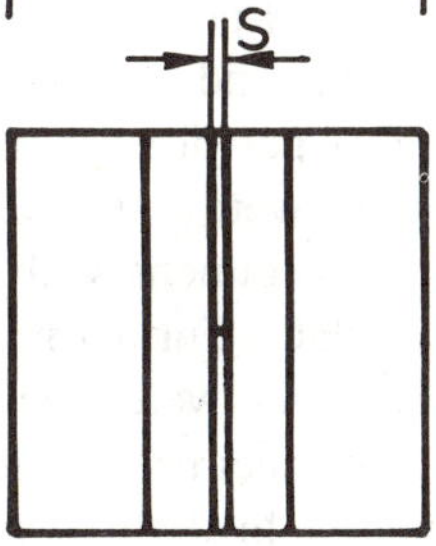

B	12.5	25	38	50
W	25	50	57	100
H	6.25	12.5	19	25
N	4.4	8.8	13.3	17.5
S	0.5	1.0	1.0	2.0
X	1.25	2.5	3.8	5.0
L	26.25	52.5	60.8	105
T	1.88	3.75	5.7	7.5
$W/H = \omega$	4	4	3	4
a/W	0.2	0.2	0.2	0.2

FIG. 1—*Dimensions of chevron-notched short-bar specimens used in this investigation.*

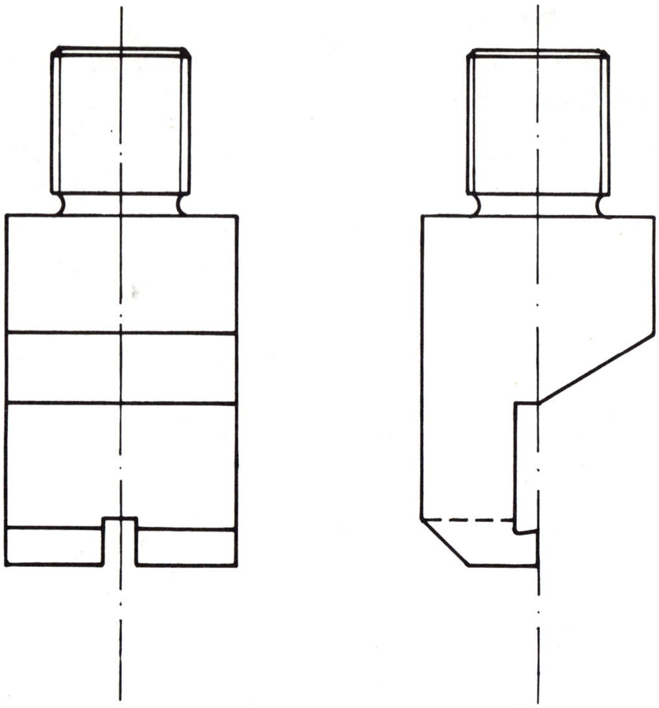

FIG. 2—*Schematic picture of the loading fixture for short-bar specimens.*

the grips and the depth of the hooks are scaled to the size B of the short-bar specimen.

For the first part of this investigation a plate (identification No. 201) of approximately 7 m by 1 m and 70 mm thickness furnished three strips of 210 mm width across the plate as shown in Fig. 3*a*. At several positions (I through V) across the width of the original plate, CT specimens and chevron-notched SB-specimens with B = 25 mm were machined out of positions A through C as shown in Fig. 3*b*. CT-specimens with B = 38 mm were machined out of positions A through C as shown in Fig. 3*c*. Four SB-specimens with B = 50 mm were machined out of the center section of the plate. Furthermore, three series of chevron-notched SB-specimens with B = 12.5 mm were machined out of the plate from positions M II (Fig. 3*a*) such that five specimens were side by side across the thickness of the plate. The crack orientation in all these specimens was in T-L orientation.

The ASTM E 399-81 fracture toughness values for these aluminum alloys 7475-T7351 vary from 38 to 50 MPa $m^{1/2}$ for the T-L orientation and from 46 to 60 MPa $m^{1/2}$ for the L-T orientation. For the second part of this investigation different heats of this material were selected, and, from specimens with T-L or L-T orientation, a sequence of materials assembled such that the fracture toughness varied from 40 to 60 MPa $m^{1/2}$ in increments of 2 MPa $m^{1/2}$. The CT-

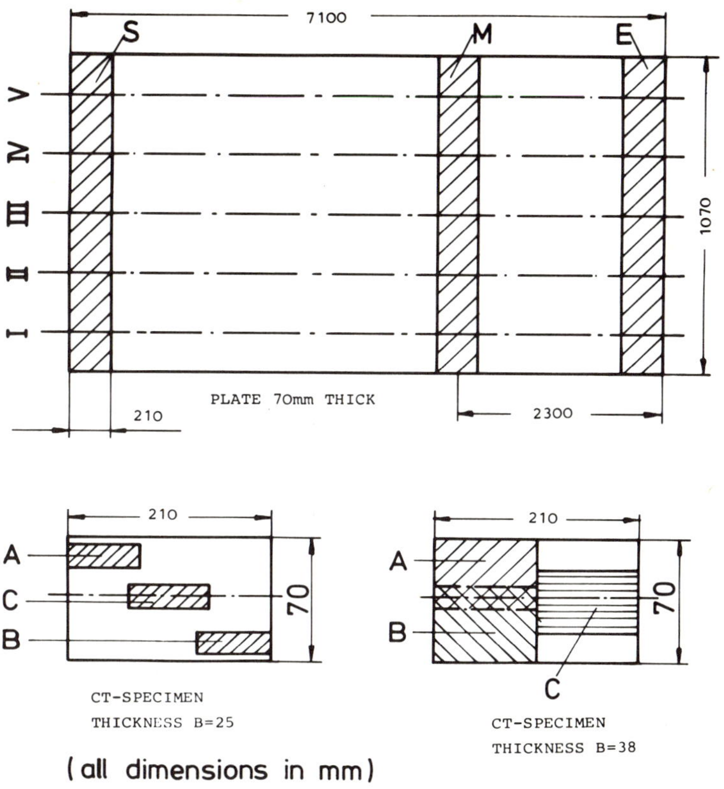

FIG. 3—*Specimen positions in the aluminum alloy plate No. 201.*

specimens with $B = 38$ mm were taken from the plate such that the midplane of the specimen coincides with the midplane of the plate. The broken halves of the tested CT-specimen were used to fabricate chevron-notched SB-specimens with $B = 38$ mm, and crack orientation identical to that of the original CT-specimen.

The fracture toughness K_{IcSB} for the chevron-notched SB-specimens was calculated from the maximum load using the following equations [5]

$$K_{IcSB} = \frac{F_{max} Y_m^*}{B\sqrt{W}} \tag{1}$$

$$Y_m^* = \{-0.36 + 5.48\omega + 0.08\omega^2 + (30.65 - 27.49\omega + 7.46\omega^2)\alpha_0 + (65.90 + 18.44\omega - 9.76\omega^2)\alpha_0^2\}\left(\frac{\alpha_1 - \alpha_0}{1 - \alpha_0}\right)^{1/2} \tag{2}$$

where Y_m^* covers $0 \leqq \alpha_0 \leqq 0.4$ and $\omega = W/H$.

Results and Discussion

The results of the ASTM E 399-81 fracture toughness determination on the three strips of plate number 201 are shown in Tables 2 and 3. Table 2 gives the K_{Ic}-values for the CT-specimen with B = 25 mm from the individual strips and a particular position across the plate thickness (Fig. 3), while Table 3 gives the K_{Ic}-values obtained from CT-specimen with B = 38 mm. The K_{IcSB}-values obtained from the plate 201 for SB-specimens with B = 12.5, 25, and 50 mm are shown in Table 4.

Figure 4 compares the fracture toughness values obtained for plate 201 from CT-specimen with B = 25 mm and SB-specimen with B = 25 mm. Similarly, Fig. 5 compares fracture toughness values for plate 201 from CT-specimen with B = 38 mm and SB-specimen with B = 50 mm. In Fig. 6 all fracture-toughness data obtained with SB-specimens from plate 201 are plotted versus the position they were machined from across the thickness of the plate (70 mm).

The results of the second part of the investigation are given in Table 5. In Table 5 the fracture toughness values K_{Ic} and K_{IcSB} for specimen with B = 38 mm are documented. Figure 7 shows these results graphically, that is, the measured K_{IcSB}-values are plotted versus the measured K_{Ic}-values. For some results

TABLE 2—*Fracture toughness* K_{Ic} *obtained with specimens of thickness* B = *25 mm at the different positions of plate No. 201.*[a]

Specimen Identification	K_{Ic}, MPa m$^{1/2}$	Specimen Identification	K_{Ic}, MPa m$^{1/2}$	Specimen Identification	K_{Ic}, MPa m$^{1/2}$
SAI	36.6	SAIII	36.7	SAV	35.2
MAI	35.3	MAIII	36.5	MAV	35.9
EAI	36.6	EAIII	36.5	EAV	35.2
MAI	36.0			MAV	36.1
MAI	37.0		$\bar{X}$ = 36.6		
					$\bar{X}$ = 35.6
	$\bar{X}$ = 36.3				
SBI	37.1	SBIII	35.7	SBV	35.7
MBI	36.3	MBIII	35.7	MBV	35.6
EBI	36.5	EBIII	35.8	EBV	36.0
MBI	36.4			MBV	36.4
MBI	35.9		$\bar{X}$ = 35.7		
					$\bar{X}$ = 35.9
	$\bar{X}$ = 36.4				
SCI	40.2	SCIII	40.5	SCV	40.5
MCI	38.3	MCIII	40.2	MCV	40.6
ECI	40.2	ECIII	41.3	ECV	40.8
	$\bar{X}$ = 39.6		$\bar{X}$ = 40.7		$\bar{X}$ = 40.6

[a]Mean value $\bar{X}$ of all A and B specimens: $\bar{X} = 36.1$ max $^{+1.0}_{-0.9}$.

[a]Mean value $\bar{X}$ of all C specimens: $\bar{X} = 40.3$ max $^{+1.0}_{-2.0}$.

TABLE 3—*Fracture toughness* K_{Ic} *obtained with specimens of thickness* B = *38 mm at the different positions of plate No. 201.*[a]

Specimen Identification	K_{Ic}, MPa $m^{1/2}$	Specimen Identification	K_{Ic}, MPa $m^{1/2}$	Specimen Identification	K_{Ic}, MPa $m^{1/2}$
SAIII	39.3	SBIII	39.1	SCIII	41.2
SAIII	38.4	SBIII	38.7	SCIII	41.2
MAIII	38.4	MBIII	38.3	MCIII	40.9
MAIII	38.4	MBIII	39.6	MCIII	41.5
EAIII	38.3	EBIII	38.3	ECIII	41.0
EAIII	39.1	EBIII	38.0	ECIII	40.5
	$\bar{X}$ = 38.6		$\bar{X}$ = 38.7		$\bar{X}$ = 41.0

[a]Mean value $\bar{X}$ of all A and B specimens: $\bar{X} = 38.6 \text{ max } {}^{+1.0}_{-0.6}$.

Mean value $\bar{X}$ of all C specimens: $\bar{X} = 41.0$ max ± 0.5.

in Table 5 the size criterion of ASTM E 399-81 for valid K_{Ic} measurements is not fulfilled. It was shown by Munz [*13*] that, by using undersized specimens, slightly smaller K_{Ic}-values are obtained, because of R-curve effects. However, these effects do not affect the main conclusion. Linear fracture mechanics still can be applied for all of the tests.

Munz [*6*] has shown that an empirical correction allows a transformation of K_{IcSB} of aluminum alloys into corrected values $\hat{K}_{IcSB}$ which agree within $\pm 10\%$ with the measured K_{Ic}-values. The equation for the correction (Eq 6 in Ref *6*) is applicable for aluminum alloy with K_{IcSB} within certain bounds, the K_{IcSB}-data in Fig. 7 exceed these bounds. Therefore, a new equation for corrected $\hat{K}_{IcSB}$-values was developed which covers the range of K_{IcSB}-values encountered with aluminum alloys, namely

$$X = \frac{K_{IcSB} + 15}{0.03} \tag{3}$$

$$K_{Ic} = 35.13 \ln X - 226 \qquad (K_{Ic} \text{ in MPa m}^{1/2}) \tag{4}$$

The corrected $\hat{K}_{IcSB}$-values corresponding to the K_{IcSB}-values measured are given in Tables 4 and 5 and graphically in Fig. 7 as crosses. It can be seen in Fig. 7 that all corrected values $\hat{K}_{IcSB}$ are within $\pm 10\%$ of the corresponding K_{Ic}-value. One should note the large scatter in K_{IcSB} for identical K_{Ic}-values or slightly different K_{Ic}-values. There seems to be a need to investigate the geometric parameters of the SB-specimens, which in our opinion are responsible for the observed scatter.

Looking at the ASTM E 399-81 fracture toughness data from plate 201 in Figs. 4 and 5, it is found, for both specimen sizes, that the fracture toughness

TABLE 4—K_{IcSB}-*values obtained with specimens of different sizes and corrected values* $\hat{K}_{IcSB}$.

SB-Specimen Thickness B, mm	Specimen Identification	K_{Ic} (average), ASTM E 399-81, MPa m$^{1/2}$	K_{IcSB}, Eqs 1 and 2, MPa m$^{1/2}$	$\hat{K}_{IcSb}$, corrected by Eqs 3 and 4, MPa m$^{1/2}$
25	MA II	36.1	43.2	39.9
25	MA II	36.1	45.2	41.1
25	MA II	36.1	43.9	40.4
25	MA II	36.1	44.7	40.8
25	MB II	36.1	43.3	40.0
25	MB II	36.1	44.1	40.5
25	MB II	36.1	44.2	40.5
25	MB II	36.1	43.3	40.0
25	MB IV	36.1	44.3	40.6
25	MB IV	36.1	46.2	41.7
25	MC IV	40.3	48.8	43.2
25	MC IV	40.3	47.4	42.4
25	MC IV	40.3	49.0	43.3
25	MC IV	40.3	49.5	43.6
50	MC II	41.0	49.5	43.6
50	MC II	41.0	49.4	43.5
50	MC II	41.0	49.5	43.6
50	MC II	41.0	49.3	43.5
12.5	M II-1		39.9	37.9
12.5	M II-2		41.1	38.6
12.5	M II-3		48.1	42.8
12.5	M II-4		40.9	38.5
12.5	M II-5		39.4	37.6
12.5	M II-1		39.6	37.7
12.5	M II-2		41.5	38.9
12.5	M II-3		46.3	41.8
12.5	M II-4		40.4	38.2
12.5	M II-5		38.7	37.1
12.5	M II-1		38.7	37.1
12.5	M II-2		41.1	38.7
12.5	M II-3		45.0	41.0
12.5	M II-4		40.5	38.3
12.5	M II-5		40.6	38.3

K_{Ic} is lowest in the surface region and highest in the midplane region of the plate. This is in agreement with the findings of other investigations [*9,11,12*]. The shaded areas in Figs. 4, 5, and 6 indicate the scatterband of the test results.

Two factors contribute to the magnitude of K_{Ic} for both specimen sizes. The first factor is the inherent fracture toughness profile across the thickness of a plate such that the specimen thickness *B* causes an experimental averaging across the position from which the specimen was machined. Because of this, the specimen with a larger thickness *B* should measure lower K_{Ic}-values in the midsection of a plate compared with specimens with smaller thickness *B*. In the surface region it should be reversed, that is, the thicker specimen should measure higher K_{Ic}-values than thinner specimens.

The second factor contributing to the magnitude of K_{Ic} is the "5% secant line"

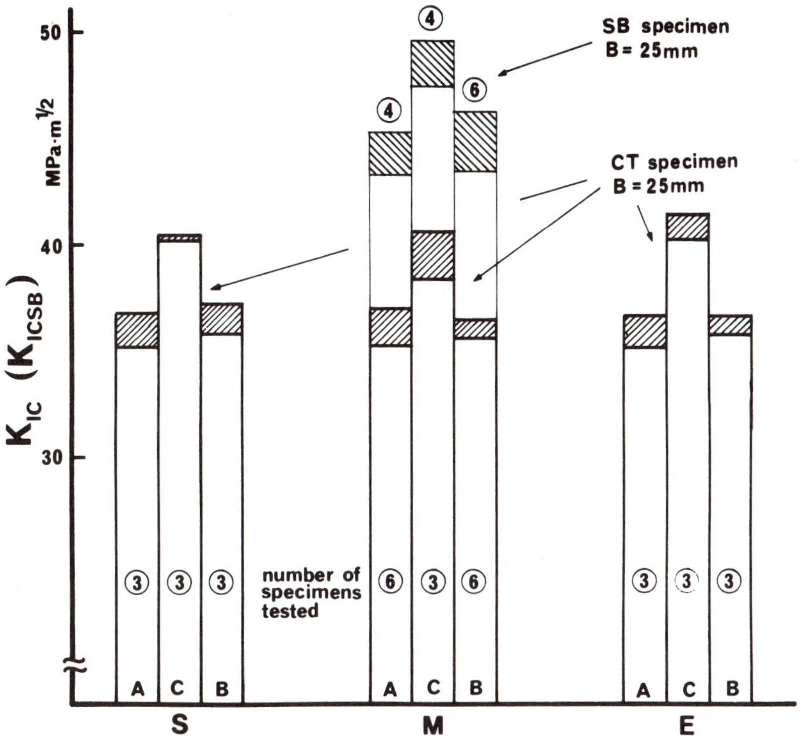

FIG. 4—*Fracture toughness* K_{Ic} *and* K_{IcSB} *for specimens with thickness* B = *25 mm taken at the different positions of plate No. 201.*

procedure as laid down in ASTM E 399-81. This criterion allows larger crack extension in larger specimens. If the material shows a pronounced R-curve behavior, as this aluminum alloy 7475-T7351 does, then higher K_{Ic}-values are measured with larger specimens [*13*].

For the larger specimen in Figs. 4 and 5, the two factors act in the opposite direction in the midsection of the plate, while they are additive in the surface region of the plate. Therefore, it is not astonishing that both specimen sizes furnish nearly identical K_{Ic}-values for the midsection of the plate. In the surface region of the plate the specimens with B = 38 mm have somewhat higher K_{Ic}-values than the specimens with B = 25 mm, as expected.

Comparison of the fracture toughness data K_{Ic} in Figs. 4 and 5 measured in the three strips of plate 201 at different locations (I through V in Fig. 3) and different positions across the plate thickness leads to the conclusion that everywhere the same fracture-toughness profile across the plate thickness is found. Furthermore, the fracture-toughness profile in plate 201 is symmetrical to the midplane of the plate, and K_{Ic} is very uniform for a given position through the thickness of the plate. Therefore, the K_{Ic}-values obtained for plate 201 furnish a reliable base to compare with the results of chevron-notched SB-specimens.

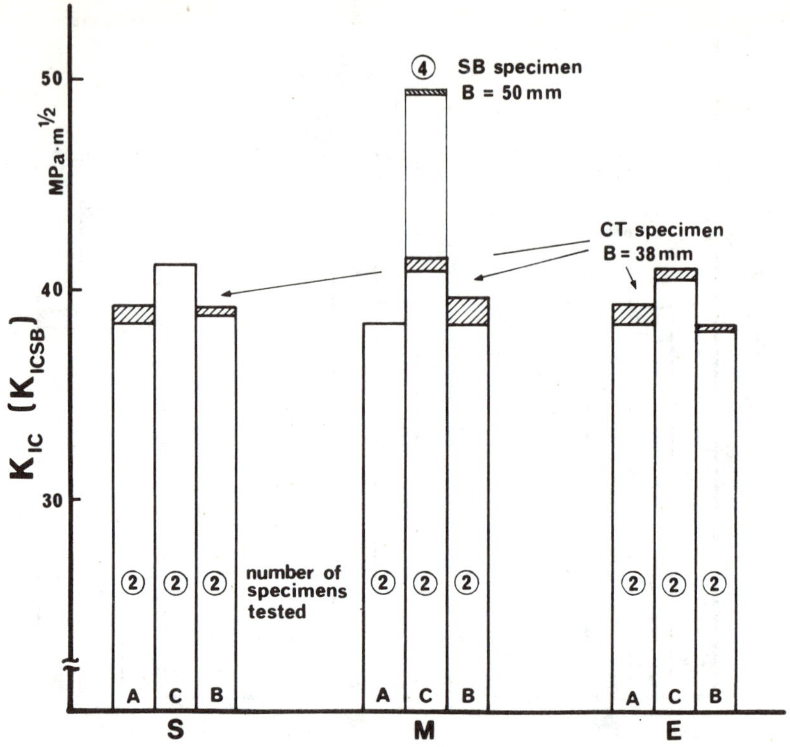

FIG. 5—*Fracture toughness* K_{Ic} *for specimens with* B = *38 mm and* K_{IcSB} *for specimens with* B = *50 mm taken at the different positions of plate No. 201.*

The K_{IcSB}-results from SB-specimens with $B = 25$ mm are compared in Fig. 4 with K_{Ic}-data obtained from CT-specimens with the same thickness B. It can be seen in Fig. 4 that K_{IcSB}-values are substantially higher than the K_{Ic}-values. The fracture toughness profile across the plate thickness agrees for both test methods. In Fig. 5 K_{IcSB}-results from SB-specimens with $B = 50$ mm are compared with K_{Ic}-results from specimens with $B = 38$ mm. Again there is a substantial difference. It is interesting to note that the K_{IcSB}-values from specimens with $B = 50$ mm agree very well with the K_{IcSB}-values from specimens with $B = 38$ mm coming from the midsection of the plate (see Fig. 6). This seems to indicate that the previously mentioned two factors might be also operative in chevron-notched SB-specimens.

Figure 6 compares the K_{IcSB}-values according to specimen thickness B and the position across the plate thickness. It is found that the fracture toughness K_{IcSB}-profile depends strongly on the specimen thickness B. All K_{IcSB}-values are higher than the corresponding K_{Ic}-values. With increasing thickness of the SB-specimen, the difference gets larger up to a limiting size as indicated by the SB-specimen with $B = 50$ mm. A reasonable explanation for the difference could be the R-

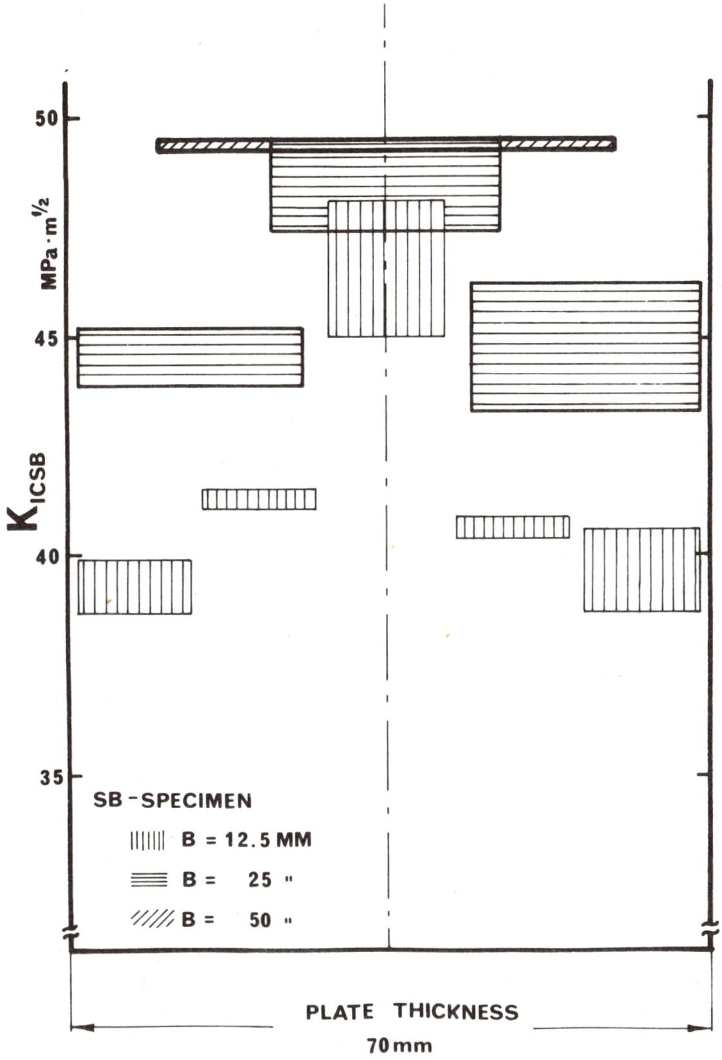

FIG. 6—*Comparison of* K_{IcSB}*-values obtained with specimen of different thickness* B.

curve behavior of the subject material. With the initiation of a crack at the tip of the chevron notch, the resistance to further crack growth will increase as growth proceeds. At a given crack length *a*, the fracture toughness, that is, the resistance to further crack growth, will be at the initiation value of the R-curve at the corner of the chevron notch. Toward the midplane of the SB-specimen, the fracture toughness increases in accordance with the crack growth at this position and the materials' R-curve. Fracture toughness will be highest at the midplane position of the specimen. The foregoing arguments assumed a uniform

TABLE 5—K_{Ic}, K_{IcSB}, *and corrected values* $\hat{K}_{IcSB}$ *for specimens with thickness* B = *38 mm in dependence on the magnitude of* K_{Ic}.

Identification	Orientation	CT-Specimen Thickness, mm	K_{Ic}, MPa m$^{1/2}$	SB-Specimen Thickness, mm	K_{IcSB}, MPa m$^{1/2}$, Eqs 1 and 2	$\hat{K}_{IcSB}$, MPa m$^{1/2}$ corrected[a] values
242	T-L	38	40.5	38	43.3	40.1
236	T-L	38	40.7	38	43.5	40.1
79	T-L	38	42.1	38	47.5	42.4
246	T-L	38	42.3	38	42.7	39.6
68	T-L	38	44.0	38	57.1	47.5
15	T-L	38	44.4	38	53.4	45.6
234	L-T	38	46.0	38	64.9	51.1
234	L-T	38	46.2	38	61.3	49.5
236	L-T	38	48.3	38	59.3	48.5
54	L-T	38	48.3	38	63.3	50.4
247	L-T	38	50.0	38	57.3	47.6
247	L-T	38	50.1	38	56.6	47.2
63	L-T	38	52.4	38	66.2	51.6
79	L-T	38	52.7	38	69.9	53.2
76	L-T	38	55.4	38	78.0	56.4
79	L-T	38	55.5	38	69.3	53.0
28	L-T	38	56.3	38	75.4	55.4
24	L-T	38	56.3	38	76.0	55.6
15	L-T	38	59.2	38	79.5	57.0
15	L-T	38	59.8	38	82.3	58.0

[a]Corrected with Eqs 3 and 4.

fracture toughness profile across the specimen thickness B. If there is a fracture-toughness profile other than uniform, then the magnitude of fracture toughness depends on the materials R-curve at the individual positions across the specimen thickness and the proceeding crack growth at this position.

The foregoing model was verified by the second part of this investigation as can be seen in Fig. 7. Figure 7 shows that with increasing K_{Ic}, the difference between K_{Ic} and K_{IcSB} is widening, and the K_{IcSB}-values are much higher than the measured K_{Ic}-values. Therefore, the aluminum alloy 7475-T7351 as used in this investigation does not lend itself to the estimation of K_{Ic} via chevron-notched SB-specimen (not directly). It is believed that the same will be true for a chevron-notched short-rod specimen, since the same arguments apply to this specimen as were cited for SB-specimen.

The problem inherent in the determination of K_{IcSB} from chevron-notched specimen results from the neglected R-curve behavior in the formulation of K_{IcSB} where it is assumed that the resistance to crack growth is constant, that is, G_{Ic}. Barker in Ref *2*, Eq 2, starts from the condition for a stable crack

$$\frac{d}{da}(b\,G_{Ic}) > \frac{d}{da}\left(\frac{F^2}{2}\frac{dc}{da}\right) \tag{5}$$

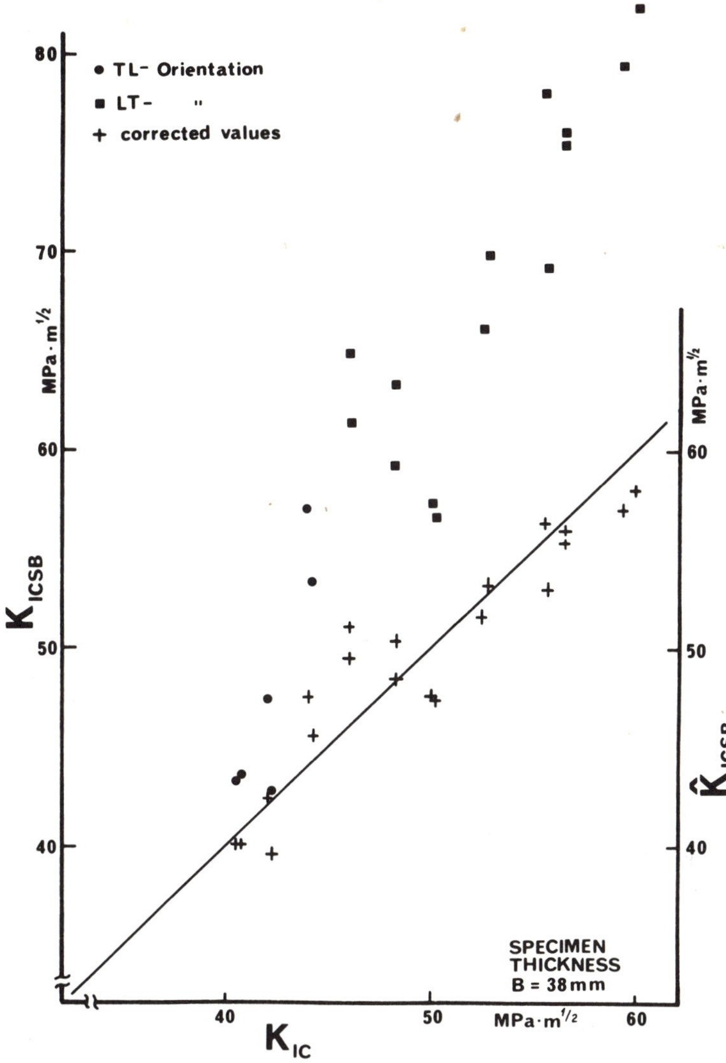

FIG. 7—*Comparison of measured* K_{IcSB} *and* K_{Ic} *data, comparison of corrected* $\hat{K}_{IcSB}$ *and* K_{Ic} *data.*

and derives

$$G_{Ic} \frac{db}{da} > \frac{F^2}{2} \frac{dc^2}{da^2} \tag{6}$$

where

b = width of the crack front,
F = applied force, and
C = specimen compliance.

If one allows for an R-curve behavior by replacing G_{Ic} by the increasing crack growth resistance R

$$\frac{d}{da}(b\,R) > \frac{d}{da}\left(\frac{F^2}{2}\frac{dc}{da}\right) \tag{7}$$

the crack stability condition becomes

$$R\frac{db}{da} + b\frac{dR}{da} > \frac{F^2}{2}\frac{dc^2}{da^2} \tag{8}$$

It is obvious that, by neglect of the second term on the left of Eq 8, the results of a K_{IcSB}-test cannot agree with "some" initiation values (5% secant criterion) K_{Ic}.

Barker [*14*] proposed an energy approach for the incorporation of plasticity effects. This approach results in even higher K_{IcSB}-values. The testing method becomes more involved and runs counter to the goal of an inexpensive, easy to handle method of estimating K_{Ic}.

Conclusions

1. The aluminum alloy 7475-T7351 as plate material shows a pronounced fracture-toughness profile through the thickness of a plate, being highest in the midsection and lowest in the surface region.
2. With both types of specimens, higher fracture toughness values were obtained for specimens from the center of the plate compared to near-surface specimens.
3. Both types of specimens showed a small increase in fracture toughness with increasing size of the specimen.
4. Fracture toughness from short-bar specimens was significantly higher than from compact specimens. The main cause is the rising R-curve of the material and not a through thickness variation of K_{Ic}.
5. The difference between K_{Ic} and K_{IcSB} increases with increasing fracture toughness K_{Ic}.
6. The K_{IcSB}-values for the aluminum alloy 7475-T7351 can be corrected by an empirical equation, such that the corrected values $\hat{K}_{IcSB}$ are within $\pm 10\%$ of the K_{Ic}-values.

References

[*1*] Barker, L. M., "Evaluation of a Simple Method of Measuring Fracture Toughness in both Brittle and Ductile Materials," *Proceedings,* ICM-II Conference, Boston, 1976, p. 1547.
[*2*] Barker, L. M., *Engineering Fracture Mechanics,* Vol. 9, 1977, p. 361.
[*3*] Munz, D., Bubsey, R. T., and Srawley, J. E., *International Journal of Fracture,* Vol. 16, 1980, pp. 359–374.

[*4*] Bubsey, R. T., Munz, D., Pierce, W. S., and Shannon, J. L., Jr., *International Journal of Fracture,* Vol. 18, 1982, pp. 125–133.

[*5*] Shannon, J. L., Jr., Bubsey, R. T., Pierce, W. S., and Munz, D., *International Journal of Fracture,* Vol. 19, No. 3, July 1982, pp. R55–R58.

[*6*] Munz, D., *Engineering Fracture Mechanics,* Vol. 15, No. 1-2, 1981, pp. 231–236.

[*7*] Eschweiler, J. and Munz, D., *International Journal of Fracture,* Vol. 20, 1982, pp. R131–R133.

[*8*] Barker, L. M., "Data Analysis Methods for Short-Rod and Short-Bar Fracture-Toughness Tests of Metallic Materials," Terra Tek Report TR 80-12, 1980.

[*9*] Brown, K. R., "Evaluation of the Chevron-Notched Short-Bar Specimen for Fracture-Toughness Testing in Aluminum Alloys," Kaiser Aluminum and Chemical Corp., Center for Technology, Information supplied by ASTM E 24.01.04 Cooperative Test Program, 1982.

[*10*] Barker, L. M. and Baratta, F. I., "Comparisons of Fracture Toughness Measurements by the Short Rod and ASTM E 399 Methods," *Journal of Testing and Evaluation,* Vol. 8, 1980, pp. 97–102.

[*11*] Munz, D., "Fracture Toughness Determination of the Aluminum Alloy 7475-T7351 with Different Specimen Sizes," Deutsche Forschungs- und Versuchsanstalt für Luft- und Raumfahrt (DFVLR), Research Report DLR-FB 77-04, 1977.

[*12*] De Jong, H. F., *Engineering Fracture Mechanics,* Vol. 13, 1980, pp. 175–192.

[*13*] Munz, D. in *Elastic-Plastic Fracture, ASTM STP 668,* American Society for Testing and Materials, Philadelphia, 1979, pp. 406–425.

[*14*] Barker, L. M., "Theory for Determining K_{Ic} from Small, Non-LEFM Specimens, Supported by Experiments on Aluminum," *International Journal of Fracture,* Vol. 15, 1979, pp. 515–536.

John L. Shannon, Jr.,[1] *and Dietrich G. Munz*[2]

Specimen Size and Geometry Effects on Fracture Toughness of Aluminum Oxide Measured with Short-Rod and Short-Bar Chevron-Notched Specimens

REFERENCE: Shannon, J. L., Jr., and Munz, D. G., **"Specimen Size and Geometry Effects on Fracture Toughness of Aluminum Oxide Measured with Short-Rod and Short-Bar Chevron-Notched Specimens,"** *Chevron-Notched Specimens: Testing and Stress Analysis, ASTM STP 855,* J. H. Underwood, S. W. Freiman, and F. I. Baratta, Eds., American Society for Testing and Materials, Philadelphia, 1984, pp. 270–280.

ABSTRACT: Plane-strain fracture toughness measurements were made on aluminum oxide using short-rod and short-bar chevron-notched specimens previously calibrated by the authors for their dimensionless stress intensity factor coefficients. The measured toughness varied systematically with variations in specimen size, proportions, and chevron notch angle apparently due to their influence on the amount of crack extension to maximum load (the measurement point). The toughness variations are explained in terms of a suspected rising R-curve for the material tested, along with a discussion of an unavoidable imprecision in the calculation of K_{Ic} for materials with rising R-curves when tested with chevron-notched specimens.

KEY WORDS: fracture, fracture strength, toughness, fracture toughness, cracks, crack strength, crack toughness, short rod specimen, short bar specimen, chevron-notched specimens, brittleness, ceramics, aluminum oxide

Nomenclature

- a Crack length
- Δa Crack extension
- a_m Crack length at minimum of Y^*
- a_{max} Crack length at P_{max}
- a_0 Initial crack length (distance from line of load application to tip of chevron)
- a_1 Length of chevron notch at specimen surface (distance from line of load application to point of chevron emergence at specimen surface)
- α a/W
- $\Delta\alpha$ $\Delta a/W$

[1]Head, Fracture Research Section, National Aeronautics and Space Administration, Lewis Research Center, Cleveland, Ohio 44135.

[2]Professor, University of Karlsruhe, D-7500 Karlsruhe, FR (West) Germany.

α_m a_m/W
α_{max} a_{max}/W
α_0 a_0/W
α_1 a_1/W
B Thickness of short-bar specimen or diameter of short-rod specimen
b Crack front length
b_m Crack front length at minimum of Y^*
K Stress intensity factor
K_{Ic} Plane-strain fracture toughness
K_{IR} Crack extension resistance
K_{IR_m} Crack extension resistance at minimum of Y^*
$K_{IR_{max}}$ Crack extension resistance at P_{max}
P Load
P_m Load at minimum of Y^*
P_{max} Maximum load
W Specimen width
Y^* Dimensionless stress-intensity factor coefficient for a trapezoidal crack, $= KB\sqrt{W}/P$
Y_m^* Minimum of Y^* as function of α
Y_{max}^* Y^* at P_{max}

The performance of short-bar and four-point-bend chevron-notched specimens in measuring the fracture toughness of aluminum oxide (Al_2O_3), using experimentally and analytically determined stress intensity factor calibrations, has been previously reported upon by the authors [*1–3*]. The measured fracture toughness varied systematically with variations in specimen size, proportions, and chevron notch angle apparently due to their influence on the amount of crack extension to maximum load (the measurement point). Similar effects would be expected for the short-rod chevron-notched specimen, first used by Barker [*4*].

Experimentally determined stress intensity factor calibrations of the short-rod specimen have been recently made by the authors to enable calculation of fracture toughness from maximum test load [*5,6*]. These calibrations were used for the short-rod specimens in this study. Fracture toughness tests were made on the same Al_2O_3 stock that was used in the investigations reported upon in Refs *1* to *3*. The short-bar specimen results of Refs *1* and *3* are combined with the short-rod specimen results newly reported here for the sake of generalizing conclusions on size and geometry effects for the chevron-notch type of fracture toughness specimen.

Experimental Material, Specimens, and Procedure

The experimental material used in this investigation of the short-rod chevron-notched specimen was from the same production of 3M Company Alsimag-614 sintered Al_2O_3 as that used in the studies of Refs *1* to *3*. The material was supplied from the manufacturer as cylindrical specimen blanks 12.7 and 25.4

mm in diameter. Its measured density was 3.736 g/cm^3. Its microstructure was a uniform mix of grains 2 to 30 μm in size. The intercept grain size, determined using ASTM Method for Estimating the Average Grain Size of Metals (E 112-81), was 10 μm.

Short-rod specimens were machined to the dimensions shown in Fig. 1 from the supplied specimen blanks. Width-to-diameter (W/B) ratios of both 1.5 and 2.0 were examined. The chevron notch length at the specimen surface (a_1) was always made equal to the specimen width (W), (that is, $\alpha_1 = 1$). The chevron angle was varied by varying the length to the chevron tip (a_0). The notches were introduced by diamond wheel slotting with kerfs (slot width, N) of either 0.40 or 1.00 mm depending on the specimen diameter.

Results for the short-bar specimen reported in Refs *1* and *3* are repeated here. Those specimens were machined to the dimensions shown in Fig. 2 from rectangular specimen blanks of 12.7 and 25.4 mm square cross section. Like the short-rod specimens, proportions (width-to-height, $W/2H$ ratios) of both 1.5 and 2.0 were produced. Chevron angles were varied by varying either a_0 or a_1. The notches were introduced either by diamond coated wire sawing or diamond wheel slotting. Slot widths in the 12.7-mm-thick specimens were 0.25 mm. Those in

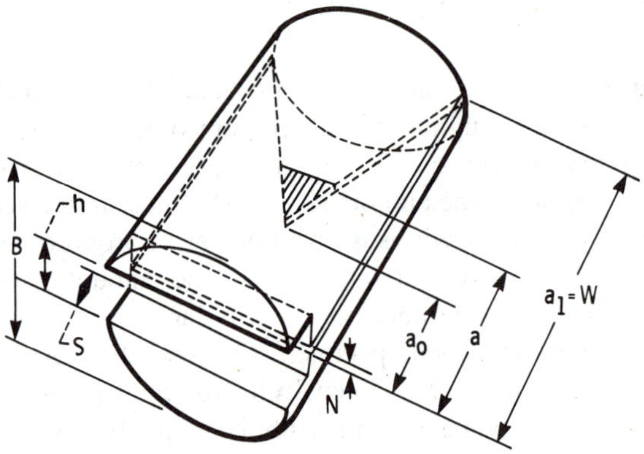

DIMENSIONS OF SPECIMENS TESTED IN THIS INVESTIGATION							
B	W	W/B	h	S	N	a_0	a_1
25.4	50.8	2.0	6.4	3.8	1.0	12.5-22.5 (SIX SPECIMENS)	50.8
25.4	38.1	1.5	6.4	3.8	1.0	7.5-16.0 (TEN SPECIMENS)	38.1
12.7	25.4	2.0	6.4	3.8	0.4	4.5-10.0 (EIGHT SPECIMENS)	25.4
12.7	19.1	1.5	6.4	3.8	0.4	3.4-7.9 (NINE SPECIMENS)	19.1

ALL DIMENSIONS IN mm.

FIG. 1—*Chevron-notched short-rod fracture toughness test specimen.*

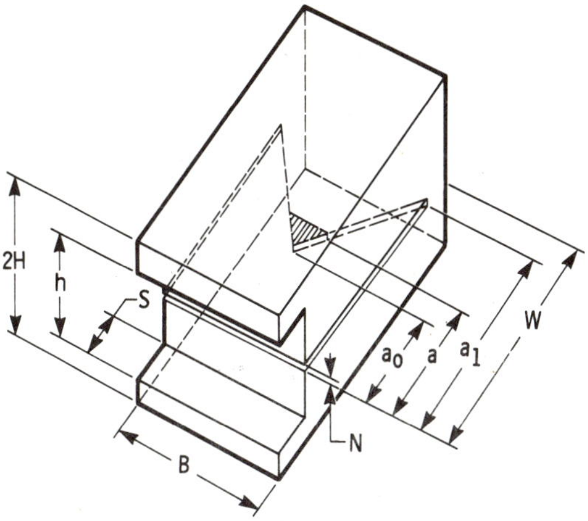

DIMENSIONS OF SPECIMENS TESTED IN INVESTIGATIONS OF REFERENCES (1) AND (3)								
B	W	2H	W/2H	h	S	N	a_0	a_1
25.4	50.8	25.4	2.0	12.7	3.8	0.25 or 0.70	10.6-22.2 (NINE SPECIMENS)	50.8
25.4	38.1	25.4	1.5	12.7	3.8	0.25 or 0.70	8.6-17.6 (NINE SPECIMENS)	38.1
12.7	25.4	12.7	2.0	6.3	3.8	0.25	4.8-11.5 (TEN SPECIMENS)	25.4
12.7	25.4	12.7	2.0	6.3	3.8	0.25	5.1 (FIFTEEN SPECIMENS)	10.2-25.4
12.7	19.1	12.7	1.5	6.3	3.8	0.25	1.7-6.9 (EIGHT SPECIMENS)	19.1

ALL DIMENSIONS IN mm.

FIG. 2—*Chevron-notched short-bar fracture toughness test specimen.*

the 25.4-mm-thick specimens were either 0.25 or 0.70 mm. As pointed out in Refs *1* and *3*, no effect of slot width was observed for these 25.4-mm-thick specimens.

The test setup, shown schematically in Fig. 3, was the same as that used in Refs *1* and *3*. Care was exercised in aligning the loading rods according to a procedure previously used by the authors for compliance calibrations of the short-bar [*7*] and short-rod [*5*] specimens. A double-cantilever displacement gage [ASTM Method of Test for Plane-Strain Fracture Toughness of Metallic Materials (E 399-83)] was inserted into knife edges integral with the loading rods, as shown in Fig. 3. The displacement gage force was tared from the load measurement, and the specimen was installed by pressing it firmly against the loading rods to seat the loading knife edges in the corners of the recessed notch mouth of the specimen.

Specimen load was applied at a constant test machine crosshead speed of 0.05 mm/min. Relative humidity of the laboratory air ranged from 54 to 82% during the period of testing. Ancillary experiments showed no difference in the results

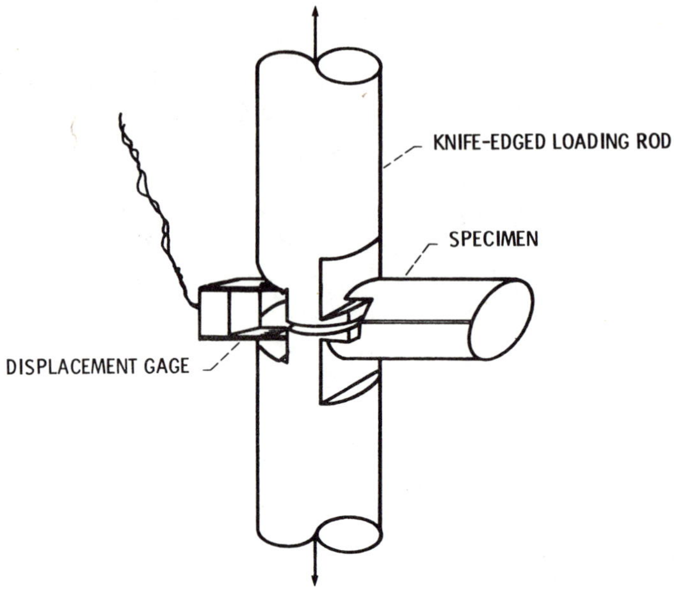

FIG. 3—*Test setup.*

of tests performed in the laboratory environment and those performed in dry nitrogen, indicating that the test loading rate employed was sufficiently rapid to not be affected by atmospheric moisture. A typical load versus displacement test record is shown in Fig. 4. The slope of the record trace is initially depressed because of local surface damage to the specimen at the loading knife edge line of contact, but increases continuously to become linear, and thereafter decreases with stable crack extension. After passing through its maximum, the trace suddenly drops because of rapid specimen fracture.

For materials with a flat crack growth resistance curve (R-curve),[3] the chevron-notched specimen fracture toughness is proportional to the maximum load as expressed in the following

$$K_{Ic} = \frac{P_{max}}{B\sqrt{W}} Y_m^* \qquad \text{(see footnote 4)} \tag{1}$$

[3]Plot of crack extension resistance K_{IR} versus crack extension Δa.

[4]The authors recognize that the designation K_{Ic} is customarily reserved for the value of plane-strain fracture toughness determined in strict accordance with ASTM E 399. For materials with flat crack growth resistance curves, we believe the chevron-notched specimen will yield a plane-strain fracture toughness value fully equivalent to K_{Ic} of the E 399 test and without the encumbrances of posttest crack length measurement and secant-line construction on the test record. For materials with nonflat crack growth resistance curves, as suspicioned for the Al_2O_3 material tested here, the measured K_{Ic} will be different from the E 399 value, but is nevertheless designated K_{Ic} in this paper for convenience.

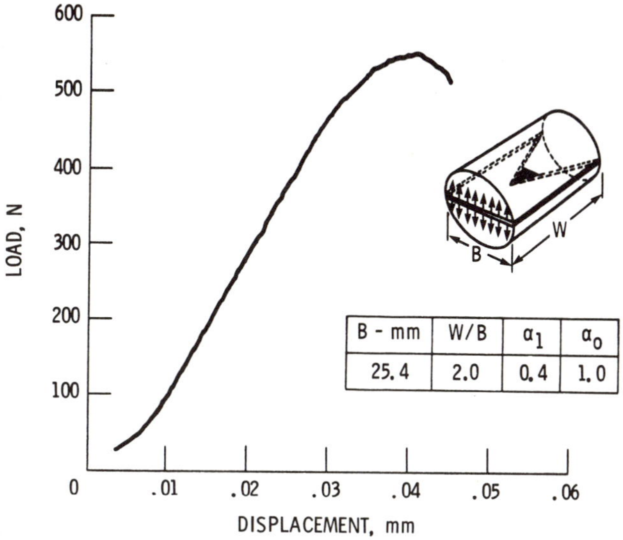

FIG. 4—*Typical load versus displacement record for chevron-notched short-rod specimen test of sintered aluminum oxide (Alsimag-614).*

where Y_m^* is the minimum value of the dimensionless stress intensity factor coefficient as a function of relative crack length for the particular specimen used (that is, for the particular specimen proportions W/B for the short-rod specimen and $W/2H$ for the short-bar specimen, and chevron notch parameters α_0 and α_1). From the experimental compliance calibrations of Refs *5* and *6*, the following polynomial expression for Y_m^* for short-rod chevron-notched specimens with $\alpha_1 = 1$ was developed for use in the present investigation

$$Y_m^* = 19.98 - 9.54(W/B) + 6.80(W/B)^2 + [-118.7 + 125.1(W/B) - 22.08(W/B)^2]\alpha_0 + [379.4 - 363.6(W/B) + 84.4(W/B)^2]\alpha_0^2 \quad (2)$$

Results and Discussion

As expected, the short-rod specimen results of this investigation (Fig. 5) closely resemble the short-bar specimen results previously obtained by the authors [*1*]. There is little if any influence of α_0 on K_{Ic} within the range investigated, but a notable effect of specimen size and proportions, B and W/B. As explained previously for the short-bar specimen [*1*], these effects can be ascribed to a rising R-curve for the aluminum oxide material tested.[5] Specimen size, proportion, and chevron notch angle all affect the amount of crack extension to maximum load (the measurement point), and therefore K_{Ic} as dictated by the shape of the R-curve.

[5]A rising R-curve has been previously determined for pure alumina (Degussit AL 23) by Hübner and Jillek [*8*] using straight-through notched, four-point-bend specimens.

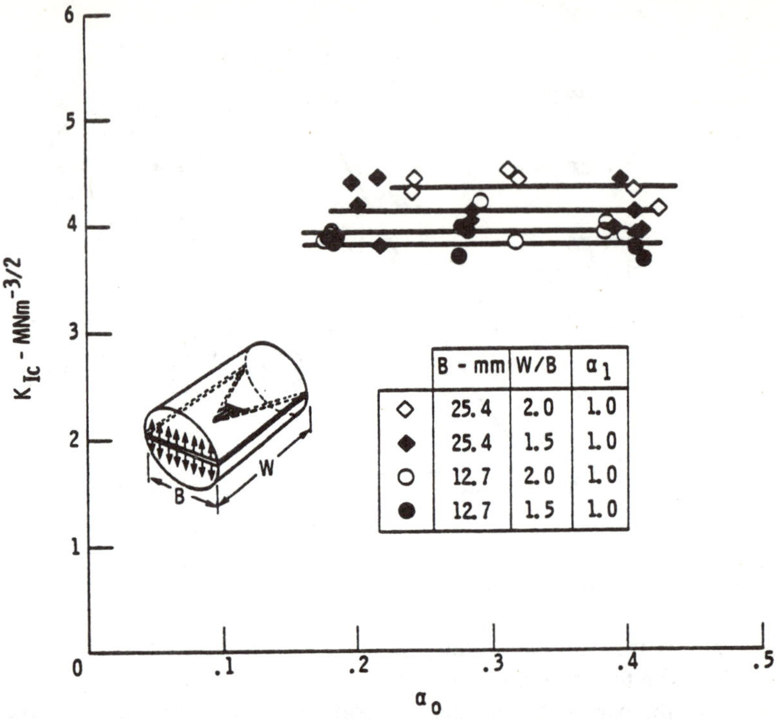

FIG. 5—*Effect of* α_0 *on* K_{Ic} *of sintered aluminum oxide (Alsimag-614) determined with chevron-notched short-rod specimens.*

As the crack proceeds down the chevron shaped ligament, the test load passes through a smooth maximum. For materials with flat R-curves, this maximum occurs at the same relative crack length where the corresponding dimensionless stress intensity factor calibration curve (Y^* versus α) exhibits a minimum, namely, α_m. This value of Y^* is designated Y_m^* and is used in Eq 1 to compute K_{Ic} directly from maximum load.

For any flat R-curve material, the crack extension to maximum load is $(a_m - a_0)$. a_m is not measured on the specimen but is computed as W times α_m, where α_m is read from the Y^* versus α curve for that specimen at Y_m^*.

For materials with rising R-curves, maximum load and Y_m^* do not occur coincidently at α_m. The load peaks, instead, at a relative crack length greater than α_m. This results in some error in the calculation of K_{Ic} at maximum load because the corresponding value of crack length is unknown. Nevertheless, a plot of K_{Ic} versus $(a_m - a_0)$ should be a reasonable approximation of the basic trend of the fracture resistance versus crack extension to maximum load curve and serve as an indication of whether the material has a flat R-curve or not.

Figure 6 is the K_{Ic} versus $(a_m - a_0)$ curve for the short-rod specimen, and Fig. 7 is the corresponding curve for the short-bar specimen reported upon previously [1]. These curves are not R-curves, but rather the loci of points lying on a family of R-curves, with each R-curve being specific to the particular

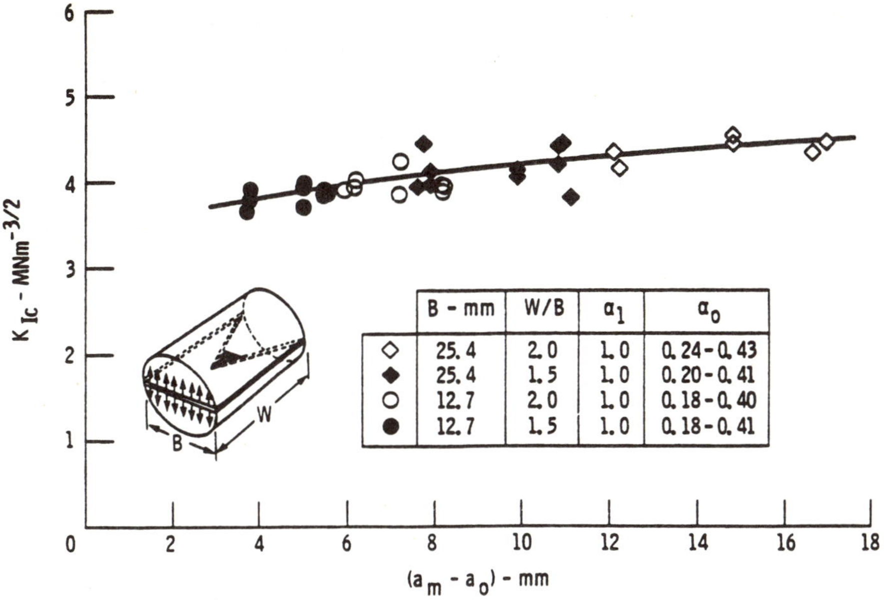

FIG. 6—*Variation in* K_{Ic} *of sintered aluminum oxide* (*Alsimag-614*) *with amount of crack extension to maximum load for short-rod chevron-notched specimens of* $\alpha_1 = 1$ *in two diameters and* W/B *proportions, and variable chevron-notch angle as obtained by varying* α_0.

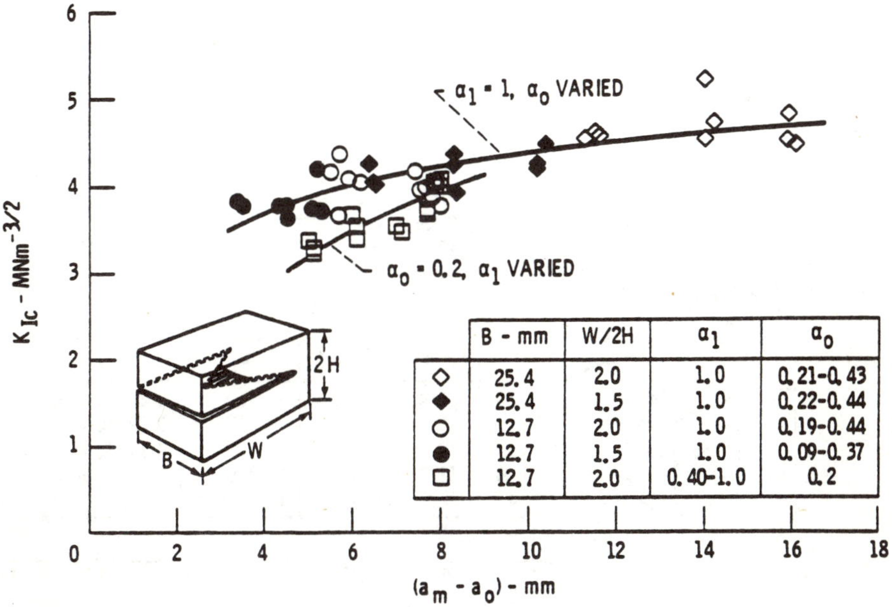

FIG. 7—*Variation in* K_{Ic} *of sintered aluminum oxide* (*Alsimag-614*) *with amount of crack extension to maximum load for short-bar chevron-notched specimen in two thicknesses and* W/2H *proportions, and variable chevron-notch angle obtained by selectively varying either* α_0 *or* α_1.

chevron-notched specimen involved. In effect, the chevron-notched specimen is variably side-grooved. As the crack proceeds down the chevron, the constraint provided by the notch varies, and how it varies will depend on the particular geometry of the chevron. Thus we can imagine that the data in Fig. 7, for example, are specific points on a family of R-curves as shown schematically in Fig. 8.

For each of the two curves in Fig. 7, one chevron notch parameter (either α_0 or α_1) was fixed while the other was varied. K_{Ic} at a given value of crack extension to maximum load is different in the region of lesser extensions due to the different constraint afforded by the two manners of adjusting the chevron angle. The data can be normalized to account for the variable side grooving effect by multiplying the crack extension by $(B - b_m)/B$ as shown in Fig. 9, where all the data falls on a common curve.

The difficulty in calculating K_{Ic} for materials with rising R-curves can be understood as follows. From an energy consideration, crack extension in the chevron-notched specimen occurs so as to satisfy the following relation [7]

$$K_{IR} = \frac{P}{B\sqrt{W}} Y^* \tag{3}$$

For a given material the K_{IR} versus Δa curve is fixed, and for a given specimen

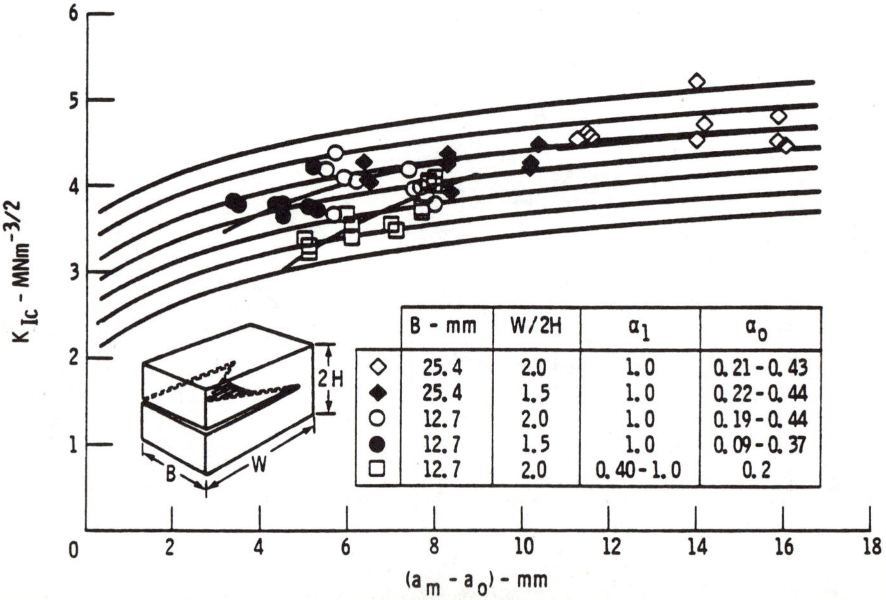

FIG. 8—*Same as Fig. 7 but with family of R-curves schematically overlain.*

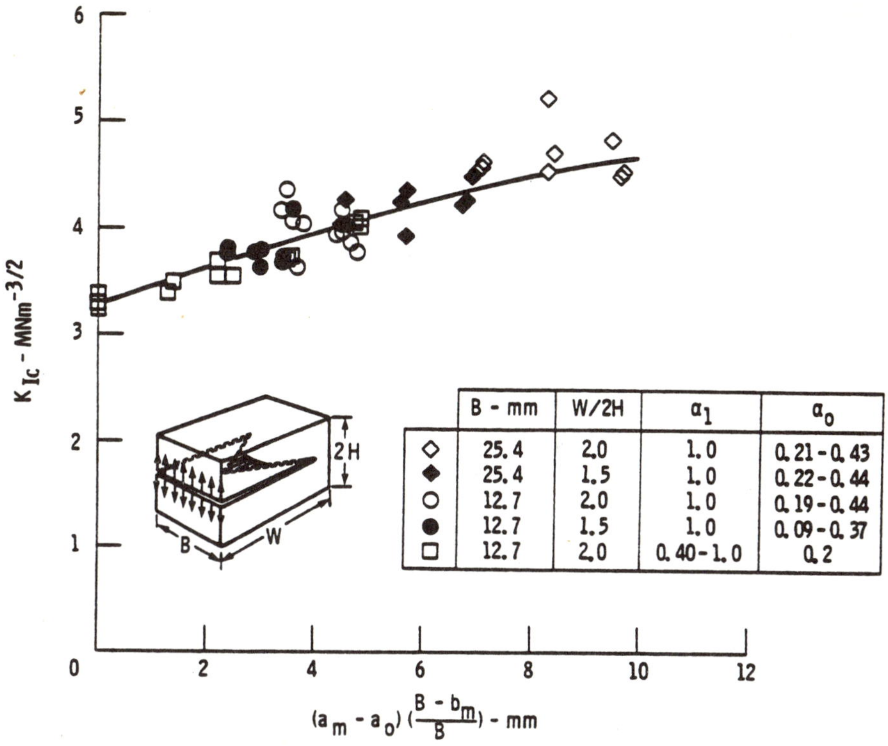

FIG. 9—*Data of Fig. 7 replotted with abscissa normalized to account for variable side grooving effect of the chevron notch.*

geometry the Y^* versus Δa curve is fixed. The P versus Δa curve then follows directly from Eq 3.

The consequence of a rising R-curve is not just to raise the load for continued crack extension compared to that dictated by a flat R-curve, but also to shift the load maximum with respect to the Y^* curve minimum as shown schematically in Fig. 10. As seen for a hypothetical rising R-curve, and a typically shaped Y^* curve, P_{max} occurs at a relative crack length α_{max} greater than α_m corresponding to the Y^* curve minimum. Referring still to Fig. 10, K_{IR_m} and $K_{IR_{max}}$ are computed using the following combinations of load and stress intensity factor coefficients

$$K_{IR_m} = \frac{P_m \cdot Y_m^*}{B\sqrt{W}} \quad \text{and} \quad K_{IR_{max}} = \frac{P_{max} \cdot Y_{max}^*}{B\sqrt{W}}$$

Calculation of K_{Ic} is conventionally done on the assumption of a flat R-curve using the combination of P_{max} and Y_m^* in Eq 1. K_{Ic} calculated in this way yields a value which lies on the R-curve between K_{IR_m} and $K_{IR_{max}}$, the uncertainty being greater the steeper the R-curve. Since it is reasonable to assume that K_{IR} depends

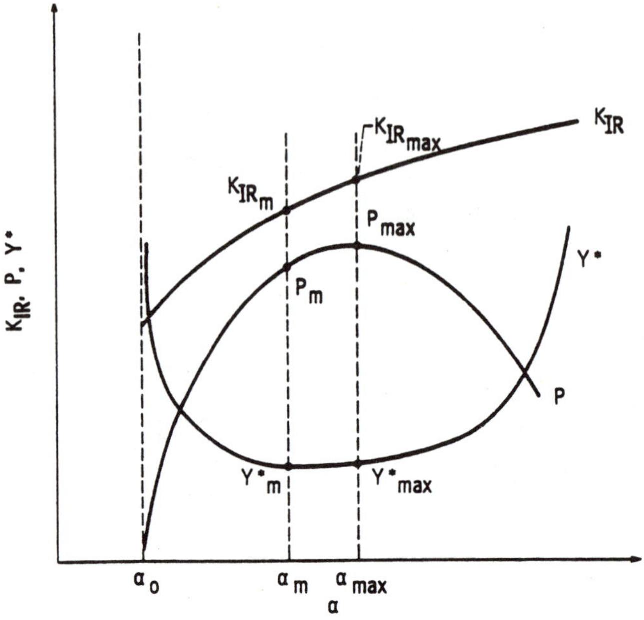

FIG. 10—*Lack of coincidence between the load maximum and the stress intensity factor coefficient minimum values as functions of relative crack length for rising R-curve materials.*

on the absolute amount of crack extension Δa and not the relative amount $\Delta\alpha$, K_{Ic} from proportionate chevron-notched specimens can be expected to increase with increasing specimen size.

References

[1] Munz, D., Bubsey, R. T., and Shannon, J. L., Jr., *Journal of Testing and Evaluation,* Vol. 8, No. 3, May 1980, pp. 103–107.
[2] Munz, D., Bubsey, R. T., and Shannon, J. L., Jr., *Journal of the American Ceramics Society,* Vol. 63, No. 5-6, May-June 1980, pp. 300–305.
[3] Shannon, J. L., Jr., Bubsey, R. T., Munz, D., and Pierce, W. S., *In Advances in Fracture Research,* D. Francois, Ed., Pergamon Press, New York, 1982, pp. 1127–1141.
[4] Barker, L. M., *Engineering Fracture Mechanics,* Vol. 9, No. 2, 1977, pp. 361–369.
[5] Bubsey, R. T., Munz, D., Pierce, W. S., and Shannon, J. L., Jr., *International Journal of Fracture,* Vol. 18, No. 2, Feb. 1982, pp. 125–133.
[6] Shannon, J. L., Jr., Bubsey, R. T., Pierce, W. S., and Munz, D., *International Journal of Fracture,* Vol. 19, No. 3, July 1982, pp. R55–R58.
[7] Munz, D., Bubsey, R. T., and Srawley, J. E., *International Journal of Fracture,* Vol. 16, No. 4, Aug. 1980, pp. 359–374.
[8] Hübner, H. and Jillek, W., *Journal of Materials Science,* Vol. 12, 1977, pp. 117–125.

James R. Tingle,[1] Carl A. Shumaker, Jr.,[1] David P. Jones,[1] and Raymond A. Cutler[2]

The Effect of Binder Chemistry on the Fracture Toughness of Cemented Tungsten Carbides

REFERENCE: Tingle, J. R., Shumaker, C. A., Jr., Jones, D. P., and Cutler, R. A., **"The Effect of Binder Chemistry on the Fracture Toughness of Cemented Tungsten Carbides,"** *Chevron-Notched Specimens: Testing and Stress Analysis, ASTM STP 855,* J. H. Underwood, S. W. Freiman, and F. I. Baratta, Eds., American Society for Testing and Materials, Philadelphia, 1984, pp. 281–296.

ABSTRACT: The fracture toughness of WC-10Co compositions with varying carbon concentrations were investigated. Both the material property, $K_{IcSR,}$ and the apparent toughness, K_Q, decreased rapidly upon the appearance of eta carbides in the microstructure. The amount of tungsten in the cobalt binder, as well as the presence of free carbon or eta carbides in the microstructure, affected the magnitude of the macroscopic residual stresses in the cermet.

The fracture toughness of compositions in which varying amounts of nickel had been substituted for cobalt were also investigated. It was found that up to 25% nickel could be substituted for cobalt without adversely affecting the fracture toughness or the hardness of the cermet.

KEY WORDS: fracture toughness, cemented carbides, binder chemistry, macroscopic residual stresses

Substantial progress in understanding the dependence of mechanical properties on microstructural parameters and deformation characteristics of hard metals has been made in recent years [*1*]. The chemistry of the binder phase in the tungsten-carbon-cobalt (W-C-Co) system is extremely important as evidenced by the narrow two-phase region in which tungsten carbide (WC) and cobalt coexist [*2*]. Excess carbon results in the presence of graphite in the binder or at WC-binder interfaces, while carbon deficiency causes the eta carbide $(Co,W)_6C$ to form. Both graphite and eta phase generally have a very deleterious effect on the mechanical properties of cemented carbides. Uhrenius et al [*2*] found that $(Co,W)_6C$ nucleates more readily than WC and identified additional carbon deficient phases

[1]Manager, Production Engineering, corporate metallurgist, and research associate, respectively, Vermont American Corporation, Louisville, Ky. 40201.

[2]Formerly, ceramic engineer, Terra Tek, Inc. Salt Lake City, Utah 84108; presently, Ceramatec, Inc., Salt Lake City, Utah 84115.

as Co_6W_6C and Co_7W_6 (μ-phase). Westin and Franzen [*3*] showed that the amount of tungsten retained in the binder phase is related to the cooling rate. Suzuki and Kubota [*4*] used lattice parameter measurements to show that the amount of tungsten in solid solution with the binder of a WC-10Co alloy varied from 2 to 3% near the carbon-cobalt-tungsten carbide (C-Co-WC) boundary to a maximum of 9 to 10% tungsten near the $(Co,W)_6C$-Co-WC boundary. Pope and Knight [*5*] used electrolytic leaching and wet chemistry to determine the amount of tungsten in the binder phase of a WC-6Co alloy. They found that the minimum tungsten content in the binder (1.9% tungsten) occurred near the carbon-porosity boundary, and a maximum of 23% tungsten occurred at the eta phase boundary. Since the amount of tungsten dissolved in the cobalt is a function of sintering temperature and cooling rate, it is not surprising to find conflicting reports regarding the amount of tungsten in the binder [*4–7*]. Ghandahari et al [*8*] have suggested that macroscopic compositional gradients may exist in sintered carbides.

Suzuki and Kubuta [*4*] noted that mechanical properties (that is, strength and hardness) vary considerably within the two-phase region (WC-Co) and are controlled by the tungsten content in the binder. They showed that transverse rupture strength was highest near the carbon porosity boundary and decreased rapidly with both increasing and decreasing carbon content. Hardness, however, decreased linearly with increasing carbon content throughout the carbon levels studied (5.9 to 6.4% carbon based on WC). Data on the effect of binder chemistry on the fracture toughness of tungsten carbide-cobalt (WC-Co) alloys is sparse [*9,10*]. Bolton and Keely [*10*] showed that a highly carbon deficient WC-5Co alloy showed significantly lower toughness as compared to near stoichiometric WC-5Co alloys. The purpose of the present research was to provide quantitative information on the effect of binder chemistry on the fracture toughness of WC-Co hard metals. In addition, the effect of nickel substitution for cobalt on the fracture toughness of WC-Co, nickel hard metals was also investigated.

Experimental Procedure

Specimen Preparation

Carbon Level Series—Nine compositions, as shown in Table 1, were milled under identical conditions to provide samples with sintered microstructures encompassing the free carbon to eta carbide range. Percentage total carbon and total carbon based on WC (assuming all of the carbon is in the WC phase and 90 weight % WC in all compositions) are calculated for the compositions in Table 1. Two groups of six samples of each composition (A-I in Table 1) were sintered in a laboratory furnace under different conditions. Group 1 was sintered at 1390°C for 1 h. The samples were cooled at a rate of approximately 500°C/h under vacuum to 1000°C, and then the vacuum furnace was backfilled with helium and cooled to room temperature. Group 2 was sintered at 1425°C for 1

TABLE 1—*Nominal compositions of carbon level powders.*

Code	Composition, weight %: WC	Co	C	W	Total Carbon, %	% Carbon (based on WC)
A	89.88	10.00	0.12	0.00	5.63	6.26
B	89.93	10.00	0.07	0.00	5.58	6.20
C	89.98	10.00	0.02	0.00	5.54	6.16
D	90.00	10.00	0.00	0.00	5.52	6.13
E	89.63	10.00	0.00	0.37	5.49	6.10
F	88.90	10.00	0.00	1.10	5.45	6.06
G	88.16	10.00	0.00	1.84	5.40	6.00
H	87.43	10.00	0.00	2.57	5.36	5.96
I	86.69	10.00	0.00	3.31	5.31	5.90

h and cooled under vacuum at a temperature drop of 100°C/h. The samples were randomly placed within the sintering furnace to eliminate the effect of any temperature gradients within the furnace. Sintered cylinders were ground to obtain 12.7-mm-diameter by 19.1-mm-long short-rod specimens. The cylinders were cut with a diamond saw, leaving a V-shaped ligament in the intended crack plane [*11*].

Nickel Substitution Series—Thirteen compositions, as shown in Table 2, were milled under identical conditions to obtain binders containing a wide range of nickel levels. The 6.2 weight % binder samples (J-M in Table 2) were sintered at 1500°C for 1 h. The 10 weight % binder cylinders (N-V in Table 2) were sintered at 1420°C for 1 h and then hot isostatically pressed (HIP) at approxi-

TABLE 2—*Nominal compositions of nickel substitution powders.*

Code	Composition, weight %: WC	TaC	TiC	Co	Ni	% Nickel in Binder
J[a]	93.55	0.15	0.10	6.20	0.00	0.0
K	93.55	0.15	0.10	5.58	0.62	10.0
L	93.55	0.15	0.10	5.27	0.93	15.0
M	93.55	0.15	0.10	4.65	1.55	25.0
N[b]	90.00	0.00	0.00	10.00	0.00	0.0
O	90.00	0.00	0.00	9.50	0.50	5.0
P	90.00	0.00	0.00	9.00	1.00	10.0
Q	90.00	0.00	0.00	8.75	1.25	12.5
R	90.00	0.00	0.00	8.50	1.50	15.0
S	90.00	0.00	0.00	8.25	1.75	17.5
T	90.00	0.00	0.00	8.00	2.00	20.0
U	90.00	0.00	0.00	7.50	2.50	25.0
V	90.00	0.00	0.00	5.00	5.00	50.0

[a]Compositions J-M were made by ball milling 30-μm WC which resulted in a mean WC grain size of 4 to 6 μm after sintering.

[b]Compositions N-V were made by ball milling 6.5-μm WC which resulted in a mean WC grain size of 1 to 3 μm after sintering.

mately 1320°C and 103 MPa (15 000 psi) pressure before centerless grinding to make 12.7-mm-diameter short-rod specimens.

Characterization

The microstructural characterization of the carbon level series after sintering is shown in Table 3. These data include density and hardness measurements on fractured short-rod specimens, as well as microstructural and magnetic property results for these same specimens. Sintered samples were pulverized, and the carbon concentrations were measured using a Leco carbon analyzer. Two to four measurements were made on each composition. Standard deviations were generally lower than 0.03 when four measurements were made, but standard deviations as high as 0.07 were recorded. Transverse rupture strength (TRS) data were obtained from bars sintered under identical conditions as fracture toughness cylinders. TRS data showed no clear trend as a function of carbon content. The large scatter in the data (coefficients of variation were usually higher than 10%, and above 30% in one instance) limit the usefulness of this mechanical property and show its dependence on crack initiation. A modified TRS test using a chamfered specimen [*12*] may have provided more meaningful data.

The polished microstructures of all specimens showed minimal porosity (less than A02, B02 per ASTM B 276-79) with samples A and B being the only sintered microstructures exhibiting carbon porosity. Eta phase in relatively large quantities was found in sample I, with somewhat smaller amounts present in sample H. Eta phase was found to be present in the core of fracture toughness cylinders in sample G-1. No eta carbides were detected in the G-2 samples sintered at the lower temperature. In specimens sintered at both temperatures, the mean WC grain size and mean free path of the binder phase became smaller with decreasing carbon concentration.

The carbon concentration decreases systematically between composition A and I and inconsistencies in the total carbon measurements appear to be related to carbon gradients, or perhaps the general difficulty of determining carbon levels in sintered specimens. Both metallography and carbon analyses of case and core specimens suggest that carbon gradients exist, although these data do not permit the quantification of these gradients.

In these specimens, density and magnetic saturation appear to be better indicators of carbon concentration. It has been shown that density increases with decreasing carbon content in a linear manner [*4*]. Suzuki and Kubota [*4*] showed that the magnetic saturation was constant in WC-10Co samples having carbon porosity and attributed this to a constant binder phase composition. Tillwick and Joffe [*13*] showed that magnetic saturation decreased linearly in the two phase region (WC + Co) due to the increase in the amount of tungsten in solid solution in the cobalt binder. The magnetic saturation decreases more rapidly upon the precipitation of eta carbides due to the removal of cobalt from the binder phase.

Both density and magnetic saturation results in this study indicate that the

TABLE 3—*Microstructural characterization of sintered carbon level series.*

Code	Group	Density, g/cm[c]	Hardness[a]	TRS[b], MPa $\bar{x}$	TRS[b], MPa s	Coercive Force, orsteads	Magnetic Saturation, electromagnetic units/g	Total Carbon, %	Micro-structure	d[c], μm
A	1	14.42	88.7	2089	37	103	15.2	5.66	free carbon	1.7
	2	14.43	88.6	2117	121	101	15.2	5.60	free carbon	1.6
B	1	14.47	89.1	1999	331	97	15.2	5.60	free carbon	. . .
	2	14.50	88.8	2192	338	95	15.2	5.59	free carbon	. . .
C	1	14.54	88.9	2255	221	109	14.8	5.44	. . .	. . .
	2	14.55	89.0	2530	207	109	14.6	5.49	. . .	. . .
D	1	14.55	89.1	1930	314	113	14.6	5.47	. . .	. . .
	2	14.57	89.0	2241	363	110	14.5	5.45	. . .	. . .
E	1	14.56	89.1	2241	335	120	14.0	5.44	. . .	1.6
	2	14.57	89.1	1731	536	117	13.6	5.47	. . .	1.6
F	1	14.60	89.2	2579	333	129	12.8	5.36	. . .	. . .
	2	14.61	89.3	2461	401	127	12.4	5.29	. . .	. . .
G	1	14.62	89.5	2661	273	143	11.7	5.36	. . .	. . .
	2	14.64	89.5	2089	282	154	11.5	5.37	. . .	. . .
H	1	14.65	89.9	2227	175	154	11.0	5.28	moderate eta	. . .
	2	14.67	89.9	1793	438	164	10.8	5.30	heavy eta	. . .
I	1	14.68	90.1	1779	164	145	10.7	5.26	heavy eta	1.4
	2	14.68	89.9	1806	271	143	10.5	5.25	heavy eta	1.5

[a]Rockwell A hardness.
[b]Transverse rupture strength, multiply by 0.145 to convert to psi ($\bar{x}$ = mean, s = standard deviation).
[c]WC grain size assuming 83.87 volume % WC.

more quickly cooled inserts (group 1) have higher carbon contents than the group 2 samples. Rapid cooling shifts the eta phase boundary to lower carbon contents because more tungsten can be retained in solid solution with the cobalt. The higher sintering temperature of group 2 samples in the decarburizing atmosphere of the sintering furnace may be responsible for the lower magnetic saturation of compositions within the two phase region.

Both coercive force and hardness of sintered samples increased gradually as the amount of tungsten in the cobalt binder increased (compositions C to F) and rapidly upon the formation of eta phase. The combination of magnetic saturation, density, coercive force, total carbon, and metallography in Table 3 show that compositions A and B have free carbon, compositions C-F are in the two phase region and compositions H-I have eta carbides. The carbon concentrations decrease, without exception, between compositions A and I.

Limited characterization of the nickel series specimens was performed, with results as shown in Table 4. The average sintered grain size for composition J-M was 4 to 6 μm, while compositions N-V had an average grain size of approximately 2 μm. All sintered microstructures were within the two phase region and no porosity was detected. Hardness readings were made by taking 12 measurements on three samples from each composition in a random fashion. Both the 6.2 weight % binder and 10 weight % binder materials showed that nickel could be substituted for up to 25 weight % of the cobalt without adversely affecting the hardness.

Testing and Data Reduction

The theory for the short-rod method for measuring plane-strain fracture toughness has been reported elsewhere for brittle materials [*11,14–15*]. One advantage

TABLE 4—*Microstructural characterization of nickel substitution series.*

		Hardness[a]	
Code	Density, g/cm^3	$\bar{x}$	s
J	14.85	88.5	0.1
K	14.89	88.4	0.1
L	14.87	88.7	0.2
M	14.86	88.5	0.1
N	14.53	87.9	0.3
O	14.59	88.0	0.2
P	14.57	87.9	0.1
Q	14.57	88.1	0.1
R	14.57	87.9	0.2
S	14.57	87.9	0.3
T	14.55	87.8	0.3
U	14.55	87.8	0.1
V	14.55	87.5	0.2

[a]Rockwell A hardness.

of the short-rod technique is that the effect of macroscopic residual stresses can be corrected so that the fracture toughness is not a function of the residual stresses and is truly a material property [15]. While microscopic residual stresses fluctuate from maximum compression to maximum tension over distances which are on the same order as the microstructure of the material, macroscopic residual stresses can affect the measurement of fracture toughness by causing an entire crack front of macroscopic dimensions to be in a state of net tension or compression before the application of any external load. Residual stress fields in which the periphery of the specimen is either in longitudinal tension or longitudinal compression affect the short-rod technique the most [15].

The effect of residual stress can be determined by measuring the load applied to the specimen as a function of mouth opening [15]. The magnitude and direction of the residual stress, p, is determined graphically by analyzing two unloading-reloading cycles performed at different points during the test [15]. The evaluation of p is simple from an experimental standpoint since it requires no calibration of the specimen mouth opening displacement transducer or measurement of crack length. A positive value of p indicates tensile longitudinal residual stresses near the periphery of the specimen while a negative value of p indicates compressive residual stresses. It has been shown [15] that the short-rod fracture toughness, K_{IcSR}, corrected for residual stresses, is given by

$$K_{IcSR} = \left(\frac{AF}{B^{3/2}}\right)\left(\frac{1 + p}{1 - p}\right)^{1/2} \tag{1}$$

where

A = a dimensionless function of the scaled crack length,
F = the mouth opening load required to advance the crack, and
B = specimen diameter.

Equation 1 simplifies to

$$K_{IcSR} = AF\,(1 + p)/B^{3/2} \tag{2}$$

when p is small.

Most cemented carbides are affected by macroscopic residual stress fields due to nonequilibrium cooling after sintering or HIP, surface grinding, binder migration or surface treatments. It is therefore important to determine the apparent toughness (K_Q), the toughness uncorrected for macroscopic residual stresses, as well as the material property K_{IcSR}.

The apparent toughness is given by

$$K_Q = AF/B^{3/2} \tag{3}$$

and is equal to K_{IcSR} when there are no net macroscopic residual stresses. Four to six specimens of each composition were tested in order to determine both K_{IcSR} and K_Q.

Results and Discussion

Carbon Level Series

Fracture Toughness—The fracture toughness results for the carbon level series are listed in Table 5. The effect of carbon content on the material property, K_{Ic}, is shown for groups 1 and 2 in Fig. 1.

The general trend observed from the data in Fig. 1 is an essentially constant fracture toughness, independent of carbon content, between 6.25 and 6.05% carbon. The fracture toughness then drops rapidly for carbon deficient samples (below 6.00% carbon) in which eta carbides were detected. The two sintering temperatures and cooling rates had only a small effect on toughness, as the data for a given composition are within one standard deviation of each other, except for composition B.

The reason for the low toughness of the first group of composition B appears to be related to the distribution of carbon porosity in these specimens. The carbon porosity in composition A specimens and the slowly cooled composition B

TABLE 5—*Fracture toughness results for carbon level series.*

Code	Group	Number of Specimens Tested	p^a		Apparent Toughness, K_Q (MPa$\sqrt{m}$)		Fracture Toughness, K_{IcSR} (MPa$\sqrt{m}$)	
			$\bar{x}$	s	$\bar{x}$	s	$\bar{x}$	s
A	1	4	−0.029	0.008	14.03	0.20	13.62	0.25
	2	6	−0.044	0.011	14.70	0.27	14.06	0.28
B	1	5	+0.042	0.024	12.09	0.25	12.60	0.31
	2	5	−0.044	0.027	14.17	0.37	13.55	0.42
C	1	6	−0.067	0.006	14.33	0.14	13.38	0.21
	2	4	−0.068	0.025	13.95	0.31	13.00	0.36
D	1	6	−0.086	0.029	14.73	0.29	13.46	0.61
	2	5	−0.081	0.017	14.02	0.31	12.88	0.51
E	1	6	−0.075	0.027	15.46	0.45	14.30	0.71
	2	5	−0.098	0.036	14.51	0.39	13.09	0.55
F	1	6	−0.130	0.036	15.31	0.43	13.31	0.81
	2	4	−0.112	0.053	14.10	0.14	12.53	0.83
G	1	6	−0.147	0.056	14.84	0.36	12.67	1.04
	2	6	−0.084	0.011	12.59	0.21	11.52	0.26
H	1	6	−0.048	0.019	11.36	0.15	10.82	0.09
	2	5	−0.057	0.015	11.90	0.28	11.22	0.39
I	1	6	−0.025	0.015	11.68	0.43	11.39	0.38
	2	5	−0.054	0.030	11.88	0.17	11.22	0.22

[a]*p* is a measure of the magnitude of the macroscopic residual stresses.
[b]Uncorrected for macroscopic residual stresses.

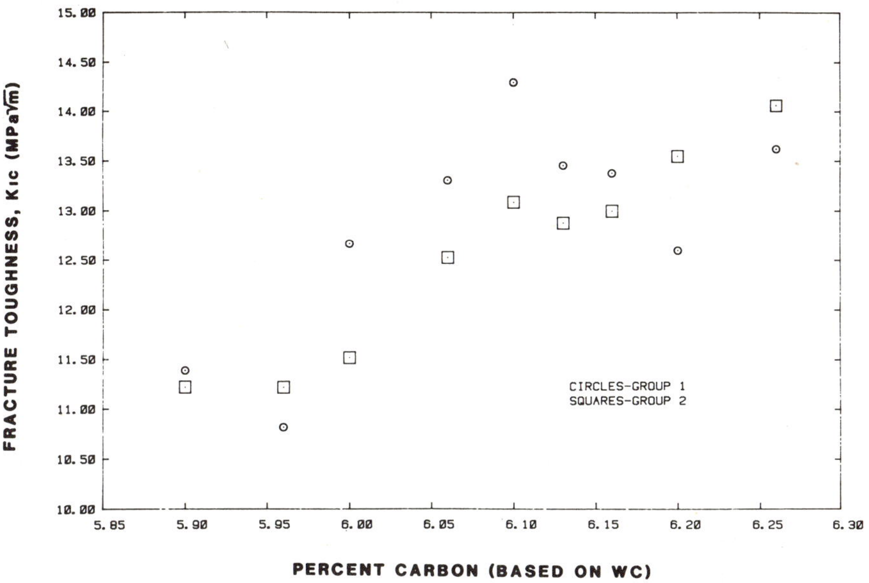

FIG. 1—*Effect of carbon content on the fracture toughness of WC-10Co.*

cermets is very fine and uniformly distributed in the microstructure. The quickly cooled composition B specimens contain massive regions of segregated carbon porosity separated by material which is only slightly affected by the presence of carbon porosity.

It is important to note that neither the apparent toughness nor the fracture toughness decreased significantly in three of the four groups which contained carbon porosity, whereas most mechanical properties rapidly deteriorate when excess carbon is noted in the microstructure [*4,16*]. The fracture toughness does not appear to decrease due to carbon porosity until large regions of segregated carbon are observed or higher carbon concentrations than those investigated in this study occur. The much lower toughness of the specimens with large regions of carbon porosity (quickly cooled composition B) as compared to the other three groups containing carbon porosity indicates that the morphology and distribution of the carbon porosity can affect the fracture toughness.

The effect of binder chemistry on fracture toughness is better viewed by examining K_{IcSR} as a function of magnetic saturation (see Fig. 2). The free carbon and eta phase boundaries in Fig. 2 are based on metallographic observations. The sharp decrease in fracture toughness which occurs when eta carbides are observed is shown in Fig. 2. It is well known that eta carbides are brittle and lower the strength of cemented carbides [*4,17*]. What is not well known, however, is the effect of the formation of eta carbides on the volume of binder in the cemented carbide. There are several eta carbides of general form $(W,Co)_6C$, but if one assumes the composition W_3Co_3C, one possible mechanism for its

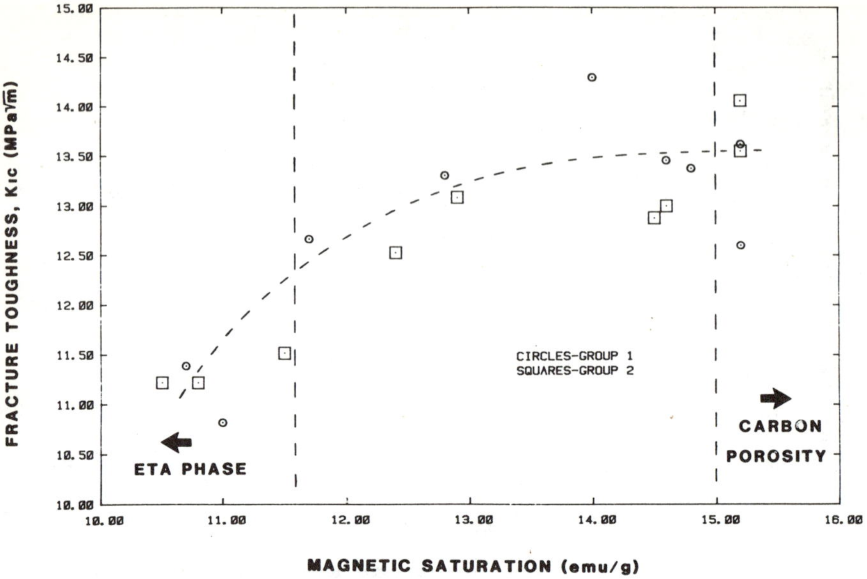

FIG. 2—*Variation in fracture toughness as a function of magnetic saturation.*

formation is

$$WC + 2W + 3Co \rightleftharpoons W_3Co_3C \quad (4)$$

If then the composition of the binder phase does not change once it is saturated with tungsten [*4*], then additional tungsten and WC must combine with tungsten and cobalt from the saturated binder to form the eta carbide. Assuming 16 weight % tungsten in the cobalt binder, a value intermediate between those reported in the literature [*4,5*], the weight ratio of tungsten added to binder removed (weight tungsten added/weight cobalt-tungsten binder removed) in forming W_3Co_3C according to Eq (4) is 1.59. It follows that if compositon G has a saturated binder, then the increased tungsten in composition I would remove 8 weight % of the cobalt-tungsten binder in order to form the W_3Co_3C. Using data from the literature [*17*] as a guide in predicting the decrease in K_{Ic} with decreasing volume percent cobalt, the decrease in binder content does not account for the entire decrease in fracture toughness between compositions G and I, but it is certainly an important factor. The sharp decrease in the K_{IcSR} at the onset of metallographic observation of eta carbides suggests that the precipitation process not only decreases the volume of binder but also makes crack propagation easier.

The variation in apparent toughness (uncorrected for macroscopic residual stresses) with carbon content is shown in Figs. 3 and 4. The apparent toughness, like K_{IcSR}, also decreases sharply upon the formation of eta carbides and appears to reach a maximum in the middle of the two phase (WC + βCo) region. When the comparison of the apparent toughness to the fracture toughness is made, the

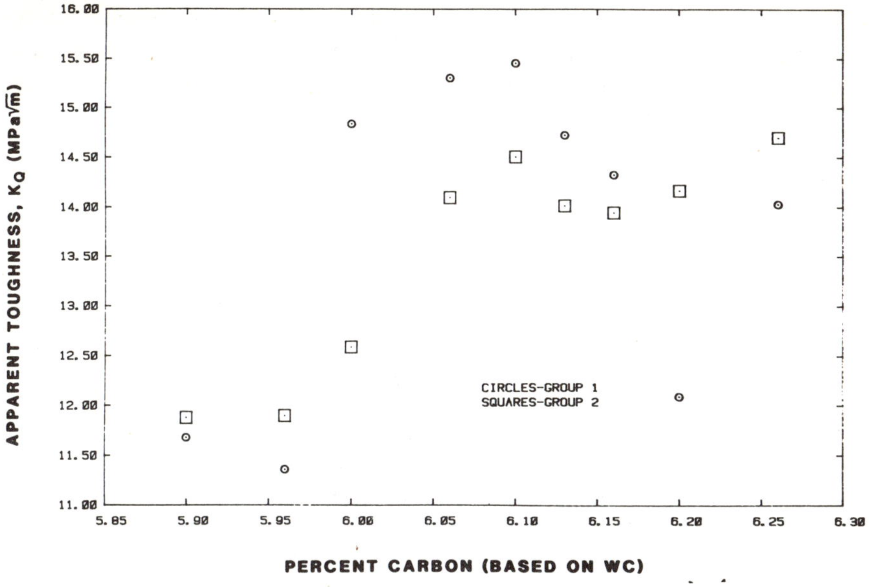

FIG. 3—*Effect of carbon content on the apparent toughness of WC-10Co.*

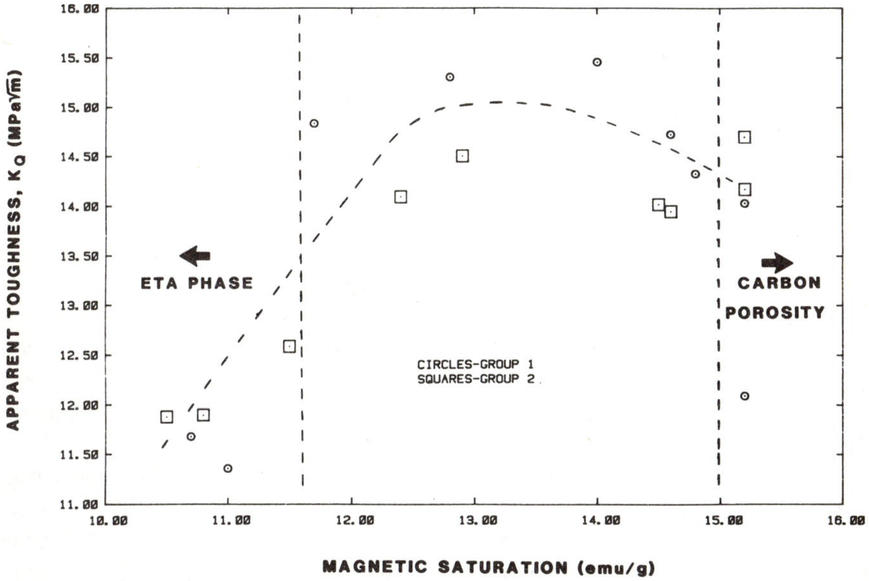

FIG. 4—*Variation in apparent toughness as a function of magnetic saturation.*

influence of macroscopic compressive residual stresses near the periphery of the short rod samples is readily seen. These compressive stresses, which are discussed below, are highest at the carbon deficient edge of the two phase region and can cause the apparent toughness to be as much as 17% higher than K_{IcSR}. Since cemented carbides are much stronger in compression than tension, they normally fail due to tensile stresses. It is therefore advantageous to have macroscopic compressive stresses near the surface. Processing which produces cemented carbides which have lower apparent toughness than fracture toughness values may not result in optimum carbide performance for most applications.

Residual Stresses—The determination of surface residual stresses resulting from both mechanical and thermal treatments on WC-Co materials has been carefully investigated using X-ray diffraction techniques [*18*]. French showed that the WC particles are always in a state of compression regardless of the cobalt content. His results showed that these compressive stresses increase as cobalt content decreases. French emphasized that the determination of residual stress by the two exposure X-ray diffraction technique measures only stresses within 4 μm of the surface, and therefore only biaxial stress is measured [*18*]. The measurement of residual stresses by any method is very difficult.

The determination of residual stresses using the short-rod technique is not quantitative, but certainly reflects macroscopic stresses at much greater depths into the surface than is possible with X-ray diffraction techniques. Previous results showed that large compressive mechanical residual stresses could be introduced by tumbling sintered WC-Co specimens before testing [*19*]. The variation in macroscopic residual stresses as a function of carbon content is shown in Fig. 5. The difference in cooling rates between the two groups was not large enough to cause a systematic variation in temperature related residual stresses. All specimens tested, except the five quickly cooled composition B specimens, had residual compressive stresses. It should be recalled that the distribution of carbon porosity in specimens with a positive value of *p* (composition B, group 1), was very different from the morphology and distribution of the other three groups which contained carbon porosity. The magnitude of the compressive stresses increased with decreasing carbon content and then decreased as eta carbides were formed. The compressive stresses were therefore minimized by the formation of free carbon or eta phase.

The reason for the variation of residual stresses with carbon content is unknown but the change in thermal expansion of the composite, binder volume, and binder migration as a function of carbon content could all contribute to the observed variation. As the carbon content decreases from free carbon towards eta phase, the weight percent tungsten in the binder increases from approximately 2 to 16%. The weight percent binder, therefore, increases from 10% (composition E) to nearly 12% (composition G) as one saturates the binder phase with tungsten prior to forming eta phase. As the eta carbides form, the weight percent binder decreases to 11% for composition I (assuming Eq 4 for eta carbide formation). The variation in macroscopic residual stresses, as shown in Fig. 5, correlates

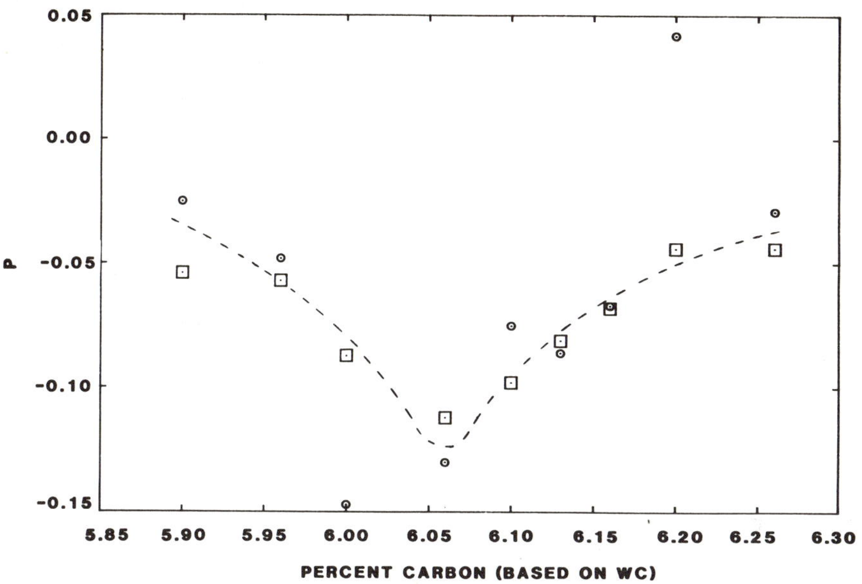

FIG. 5—*Effect of carbon content on macroscopic residual stresses.*

well with the predicted change in weight percent binder as a function of carbon. If the increase in binder content corresponded to an increase in the thermal expansion coefficient of the composite, then the compressive stresses would also increase upon nonequilibrium cooling.

The problem with this explanation is that the addition of tungsten to the binder probably lowers the thermal expansion coefficient of the binder because the expansion coefficient of cobalt is three times that of tungsten. In addition, the volume percent change in binder is much lower than the weight percent change because of the difference in atomic weights between tungsten and cobalt. If the triaxial stress state can be modeled after the work of French for biaxial stresses [*18*], then a small decrease in the magnitude of the compressive stresses with increasing binder contents would be predicted. It is possible that binder migration or the distribution of the eta carbides affects the thermal expansion of the composite in such a way as to correspond with the experimentally observed trend in macroscopic residual stresses. The important observation which can be made from Fig. 5 is that WC-Co compositions with large amounts of tungsten in solid solution with the cobalt binder have higher apparent toughness and are predicted to perform better in applications where tensile failure predominates than compositions with low amounts of tungsten in the binder (near stoichiometric carbon).

Nickel Substitution Series

Nickel additions of up to 25% of the binder had no adverse effect on fracture toughness (see Table 6 and Fig. 6) or hardness (see Table 4) for the two cermet

TABLE 6—*Fracture toughness results for nickel level cylinders.*

Code	% Nickel in Binder	Number of Specimens Tested	Apparent Toughness[a], K_a (MPa$\sqrt{m}$)			Fracture Toughness, K_{IcSR} (MPa$\sqrt{m}$)		
			$\bar{x}$	s	% COV[b]	$\bar{x}$	s	% COV
J	0.0	10	13.79	0.23	1.68	13.44	0.28	2.05
K	10.0	5	14.21	0.15	1.08	13.30	0.41	3.12
L	15.0	5	14.39	0.34	2.33	13.71	0.15	1.10
M	25.0	5	14.51	0.50	3.42	14.09	0.19	1.37
N	0.0	5	13.94	0.26	1.84	14.36	0.05	0.36
O	5.0	5	13.47	0.39	2.93	13.86	0.33	2.38
P	10.0	4	13.09	0.63	5.01	13.64	0.34	2.47
Q	12.5	5	14.21	0.72	5.07	14.27	0.09	0.64
R	15.0	5	13.83	0.71	5.12	14.01	0.09	0.64
S	17.5	5	13.61	0.44	3.25	13.81	0.22	1.58
T	20.0	5	13.90	0.53	3.81	14.02	0.29	2.04
U	25.0	5	13.66	0.80	5.86	14.02	0.48	3.42
V	50.0	5	14.29	0.22	1.53	13.83	0.19	1.38

[a]Uncorrected for macroscopic residual stresses.
[b]Percent coefficient of variation (COV) (that is, $(s/\bar{x})100$).

types tested (see Table 2). Previous results showed that the fracture toughness, hardness, and abrasion resistance of WC-Co,Ni cermets all decrease at nickel additions (to the binder) above 50% but are favorable below 25% levels [*19*]. Precht et al [*20*] showed that nickel additions of up to 15% of the cobalt increased the abrasion resistance of WC-Co,Ni cermets as compared to WC-Co cermets. They demonstrated that this was due to the stabilization of the relatively ductile face centered cubic (fcc) cobalt phase by nickel, thus preventing the transformation to the less ductile hexagonal-close-packed (hcp) cobalt structure [*20*]. X-ray diffraction studies of fractured surfaces of compositions N-V detected predominately face-centered-cubic (fcc) cobalt and only a slight amount of hcp cobalt in all specimens analyzed.

The conclusion that nickel can be partially substituted for cobalt without adversely affecting mechanical properties is important to cemented carbide producers since the price and availability of nickel is much more stable than cobalt. What is needed ultimately, however, are compositions which are not strongly dependent upon any critical materials.

The nickel substition fracture toughness cylinders were HIP before grinding, and widely different residual stress values were measured for a given composition. The values of p, however, were not dependent on the composition of the binder but rather appeared to depend on specimen placement in the HIP furnace which affected cooling rates. Barker [*15*] showed that the scatter in apparent toughness could be minimized when fracture toughness was calculated. The much lower coefficients of variation for K_{IcSR} as compared to K_Q, for a given composition (see Table 6), supports the data of Barker and his conclusion that macroscopic residual stresses can be corrected so that K_{IcSR} is a material property.

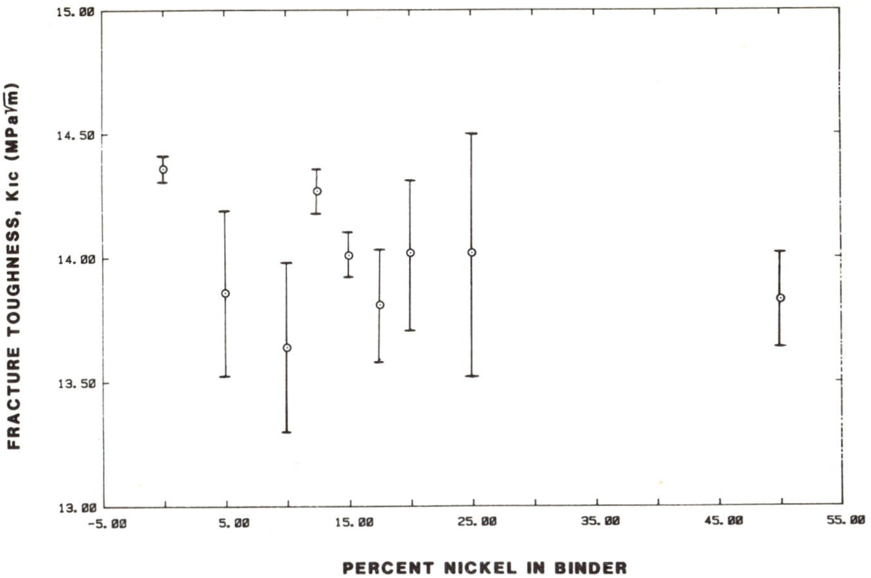

FIG. 6—*Effect of nickel substitution for cobalt in WC-10Ni,Co on fracture toughness.*

Conclusions

1. The fracture toughness of the WC-Co compositions investigated was marginally affected by the presence of free carbon in the microstructure. Fracture toughness decreased slowly with increasing tungsten in the binder within the two phase region (WC + βCo), and decreased rapidly when eta carbides were formed.

2. In WC-Co compositions containing carbon porosity, the fracture toughness was dependent on the morphology and distribution of the carbon porosity. Specimens containing fine, uniformly distributed carbon porosity were tougher than specimens which contained large, segregated agglomerates of free carbon.

3. The magnitude of macroscopic residual compressive stresses increased with higher solubility of tungsten in the binder and decreased as eta carbides were formed. The apparent toughness was therefore highest for compositions in the middle of the two phase region.

4. Nickel substitutions up to 25% of the cobalt binder had no adverse effect on fracture toughness or hardness.

Acknowledgments

The authors wish to thank L. E. Pope and L. M. Barker of Sandia National Laboratories for many helpful discussions.

References

[*1*] Almond, E. A. in *Science of Hard Materials*, R. K. Viswanadham, D. J. Rowcliffe, and J. Gurland, Eds., Plenum Press, New York and London, 1983, pp. 517–561.

[*2*] Uhrenius, B., Carlsson, B., and Franzén, T., *Scandinavian Journal of Metallurgy*, Vol. 5, 1976, pp. 49–56.

[*3*] Westin, L. and Franzén, T., *Scandanavian Journal of Metallurgy*, Vol. 8, 1979, pp. 205–215.

[*4*] Suzuki, H. and Kobota, H., *Planseeberichte für Pulver metallurgie*, Vol. 14, 1966, pp. 96–109.

[*5*] Pope, L. E. and Knight, D., to be published.

[*6*] Dawihl, W. and Frisch, B., *Cobalt*, Vol. 22, 1964, pp. 22–30.

[*7*] Nishiyama, A. and Ishida, R., *Transactions of the Japanese Institute of Metals*, Vol. 3, 1962, pp. 185–190.

[*8*] Ghandehari, M. H., Pope, L. E., and Pierce, F. D., *International Journal of Power Metallurgy and Power Technology*, Vol. 15, No. 1, 1979, pp. 55–65.

[*9*] Farb, J. E., "The Fracture Toughness of Cobalt Cemented Tungsten Carbides," M. S. thesis, University of Utah, 1978, pp. 39–42.

[*10*] Bolton, D. and Keeley, R. J., *Refractory and Hard Metals*, Vol. 9, 1982, pp. 103–111.

[*11*] Barker, L. M., *Engineering Fracture Mechanics*, Vol. 9, No. 2, 1977, pp. 361–369.

[*12*] Roebuck, B., *Journal of Materials Sciences*, Vol. 14, 1979, pp. 2837–2844.

[*13*] Tillwick, D. L. and Joffe, I., *Scripta Metallurgics*, Vol. 7, 1973, pp. 479–484.

[*14*] Barker, L. M. in *Fracture Mechanics of Ceramics*, Vol. 3, R. C. Bradt, D. P. H. Hasselman, and F. F. Lange, Eds., Plenum Publishing Co., New York, 1978, pp. 483–494.

[*15*] Barker, L. M., *Advances in Fracture Research* (*ICF5*), D. Francois, Ed., Vol. 5, 1981, p. 2563.

[*16*] Exner, H. E. and Gurland, J., *Powder Metallurgy*., Vol. 13, 1970, pp. 13–31.

[*17*] Chermant, J. L. and Osterstock, F., *Journal of Material Sciences*, Vol. 11, 1976, pp. 1939–1951.

[*18*] French, D. N., *Journal of the American Ceramic Society*, Vol. 52, 1969, pp. 267–275.

[*19*] Cutler, R. A., Precht, W., and Austin, J. A., "Correlation of the Mechanical Properties of WC-Co with Drilling Performance in Hard Formations," Terra Tek Report TR 82–21, 1982, pp. 77–79.

[*20*] Precht, W., Viswanadham, R. K., and Venables, J. D. in *Science of Hard Materials*, R. K. Viswanadham, D. J. Rowcliffe, and J. Gurland, Eds., Plenum Press, New York and London. 1983, pp. 815–828.

Joonpyo Hong[1] *and Peter Schwarzkopf*[1]

A Comparison Study of Fracture Toughness Measurement for Tungsten Carbide-Cobalt Hard Metals

REFERENCE: Hong, J. and Schwarzkopf, P., "**A Comparison Study of Fracture Toughness Measurement for Tungsten Carbide-Cobalt Hard Metals,**" *Chevron-Notched Specimens: Testing and Stress Analysis, ASTM STP 855,* J. H. Underwood, S. W. Freiman, and R. I. Baratta, Eds., American Society for Testing and Materials, Philadelphia, 1984, pp. 297–308.

ABSTRACT: K_{Ic}-values of tungsten carbide-cobalt (WC-Co) hard metals were measured using the short-rod (chevron-notched) specimen geometry. Nine alloys of various cobalt contents and tungsten carbide particle sizes were tested using modified self-aligning tensile grips. A compliance method was adopted for the test.

For comparison, the same WC-Co alloys were tested using single-edge-notched beam (SENB) specimen geometry in four-point bending. A sharp notch was introduced into the SENB specimen using a thin copper foil blade in an electrical discharge machine (EDM).

The same alloys were also tested using Barker's flatjack method with short-rod specimen geometry.

Microstructural parameters, mean free path of cobalt, mean intercept length, and contiguity of WC, were measured using a semiautomatic image analyzing method for all the alloys. Optical, scanning electron, and transmission electron micrography were used for quantitative image analysis. From these microstructural parameters, K_{Ic}-values were calculated using a theoretical equation derived by Hong and Gurland.

The four independent experimental K_{Ic}-values for nine alloys are discussed with respect to the accuracy and convenience of each experimental measuring technique.

KEY WORDS: cemented carbide, fracture toughness, WC-Co alloy, microstructural parameters, compliance, short-rod specimen, four-point bending, single-edge-notched-beam (SENB) specimen, flatjack, K_{Ic}, pre-crack, mean free path, contiguity, true mean free path

Cemented carbides are used for metal cutting, metal working, rock drilling, and a great variety of parts which require exceptional wear resistance. Cemented carbides possess an excellent combination of mechanical properties: high hardness or wear resistance with good toughness at both room and high temperatures. Although cemented carbides have relatively high toughness compared with materials with equivalent hardness, such as ceramics, usage is often limited by low fracture toughness compared to other materials. In the literature [*1–15*], the fracture toughness values of tungsten carbide-cobalt (Wc-Co) alloys containing

[1]Sii Tungsten Carbide Mfg., Tustin, Calif. 92680.

a variety of cobalt compositions and different microstructural parameters have been measured and reported for various specimen geometries. Although accurate fracture toughness values for cemented carbide are important, a standard method of measuring fracture toughness has not yet been established.

The practical aspect of fracture toughness determination demands not only precision in measurement but also a convenient and easily reproducible test method. Of the various methods of fracture toughness measurement, three independent methods using two different specimen geometries were tried in this study for comparison purposes. Nine WC-Co grades were tested covering a range of carbide grain size and cobalt composition. Microstructural parameters for all these alloys were measured and the fracture toughness values calculated using the equation developed by Hong and Gurland [*16,27*].

Experimental Work

Three different methods of measuring K_{Ic}-values were used on the nine WC-Co grades. K_{Ic}-values were also calculated using microstructural parameters. The nine alloys had cobalt contents ranging from 6 to 16 weight percent and carbide mean free paths ranging from 0.7 to 2 μm. None of the specimens exhibited any eta-phase, and they all had porosity levels less than A02, B02, with no C-type porosity [ASTM Test Method for Apparent Porosity in Cemented Carbides (B 276)].

The three experimental and the single calculated K_{Ic} determinations were the following:

1. Compliance method using short-rod specimen geometry.
2. Four-point bending method using single-edge-notched-beam (SENB) specimen.
3. Barker's flatjack method, using short-rod specimen geometry.
4. K_{Ic}-values calculated from microstructural parameters.

These four methods will be discussed in order.

Compliance Method Using Short-Rod Specimen

A short-rod specimen shown in Fig. 1 was prepared with the following dimensions: B = 12.7 mm, W = 19.05 mm, a_o = 6.35 mm, τ = 0.381 mm, V-notch angle = 58°, radius of curvature of slot = 62.38 mm, C = 3.96 mm, and b = 3.3 mm.

A self-aligning grip with specially prepared attachments (Fig. 2) was used for specimen loading in an MTS servohydraulic testing machine. The grip with the specimen in the test position is shown in Fig. 3.

During loading, a graph of mouth opening displacement versus load was produced on the X-Y recorder (Fig. 4). After a certain load, the specimen was unloaded, and from the shape of this plot the compliance was obtained.

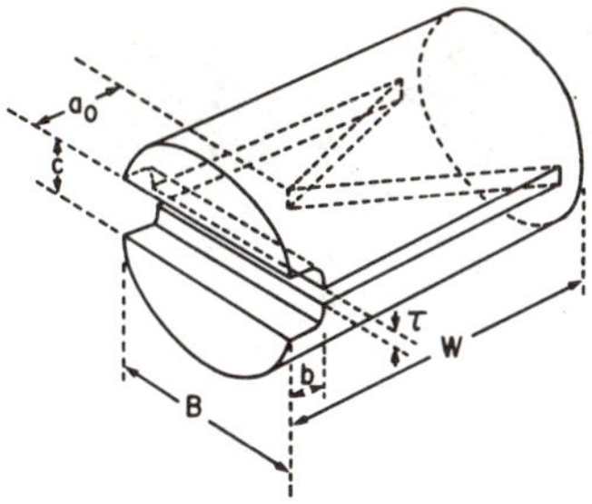

FIG. 1—*Short-rod specimen geometry.*

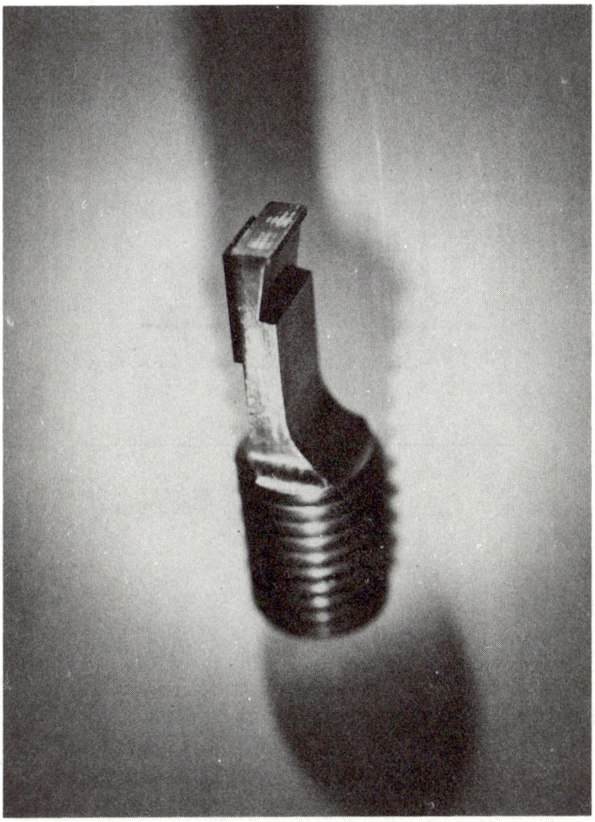

FIG. 2—*Grip attachment for short-rod specimen.*

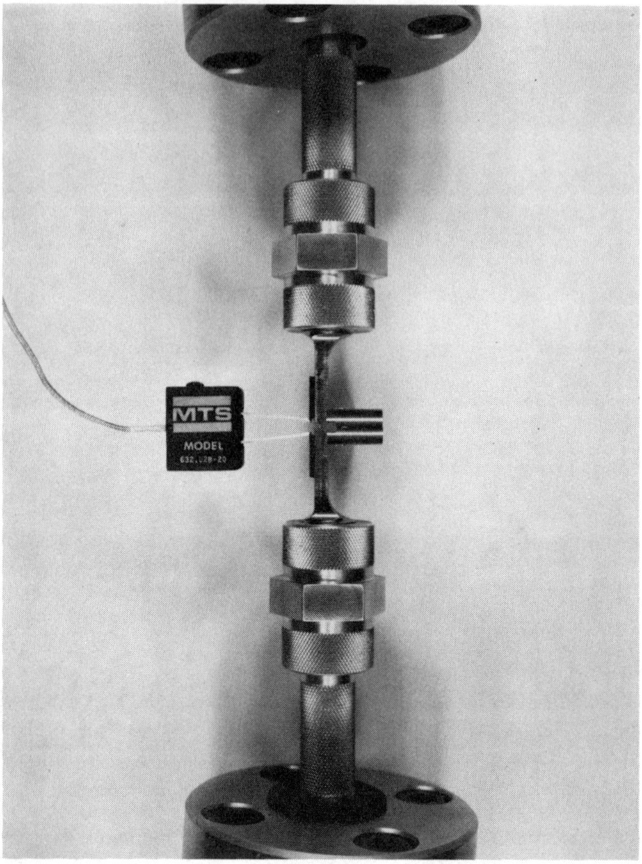

FIG. 3—*Short-rod specimen setup for testing.*

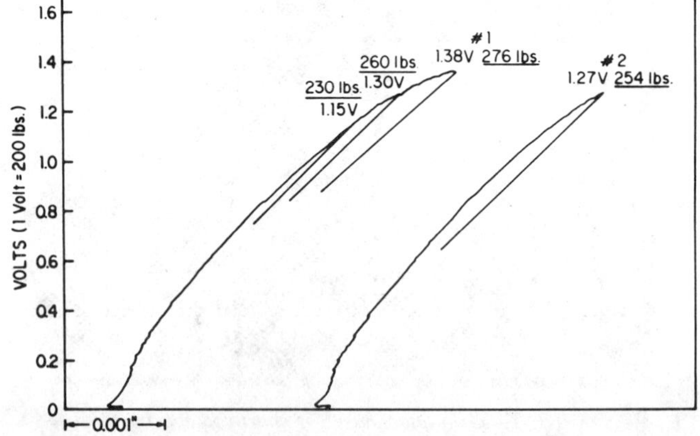

FIG. 4—*Mouth opening displacement versus load (1 lb = 4.448 N, 1 in. = 2.54 cm).*

In order to measure crack length and width, the specimen was placed in dye penetrant overnight, heated to dry, then broken apart. Measurements were taken on photos of the crack surface (Fig. 5).

For alloys of the various cobalt contents, plots of crack length versus compliance were drawn. Two of these are shown in Fig. 6. The plots were assumed to be of the form [*19*]

$$C = \alpha + \beta \cdot a + \gamma e^{\delta \cdot a} \tag{1}$$

Here α, β, and δ are constants while a is crack length and C is compliance. Least square fits were calculated (Fig. 6) and the slope (dC/da) was obtained. K_{Ic}-values were obtained from the equations

$$G_{Ic} = \tfrac{1}{2} P_c^2 \, (\partial C/\partial a)_c / B \tag{2}$$

$$G_{Ic} = \frac{1 - \nu^2}{E} K_{Ic}^2 \tag{3}$$

FIG. 5—*Fracture surface view of short-rod specimen (note infiltrated dye penetrant).*

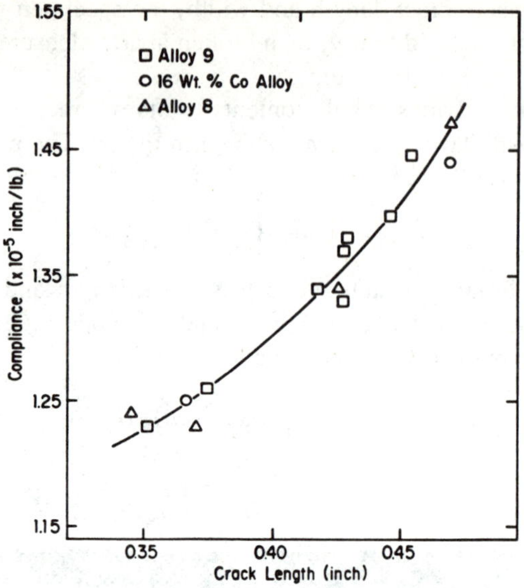

FIG. 6*a*—*Crack length versus compliance (1 in. = 2.54 cm, 1 in./lb = 5.71 × 10^{-3} m/N).*

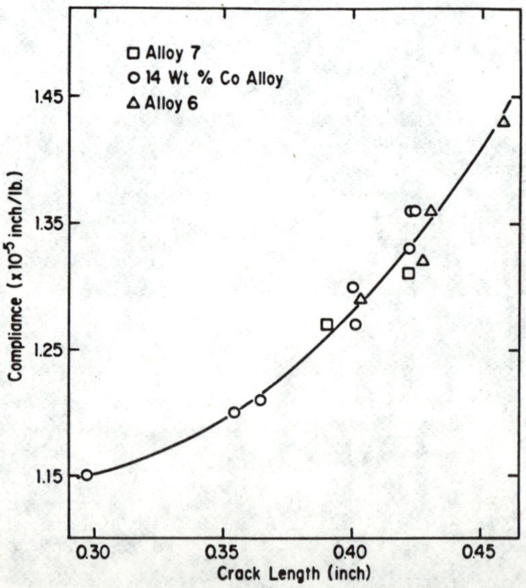

FIG. 6*b*—*Crack length versus compliance (1 in. = 2.54 cm, 1 in./lb = 5.71 × 10^{-3} m/N).*

where

P_c = peak load,
$(\partial C/\partial a)_c$ = $(\partial C/\partial a)$ value at peak,
E = elastic modulus, and
ν = Poisson's ratio.

Yeh's [*21*] theory was used to calculate E- and ν-values. The values for all the alloys are reported in Table 1 and Fig. 7.

SENB Specimen

The SENB specimen shown in Fig. 8 was used with dimensions l = 5.83 mm, b = 2.97 mm, and t = 33.5 mm.

Electrical discharge machining (EDM) with a work-hardened copper foil blade of 18 μm thickness was used to produce the precrack. The machining was carried out at a lower voltage and finished off with a higher voltage, "zapping procedure" [*22*]. At least ten specimens of each alloy were fractured in four-point bending. Lower and upper span was 14.5 and 4.83 mm, respectively. The fixture with specimen is shown in Fig. 9. It was used in a self-aligning compression grip in the MTS machine.

Pre-crack size was measured after fracture directly on the microscope. The stress-intensity factor K_{Ic} was calculated using the following equation [*23*]

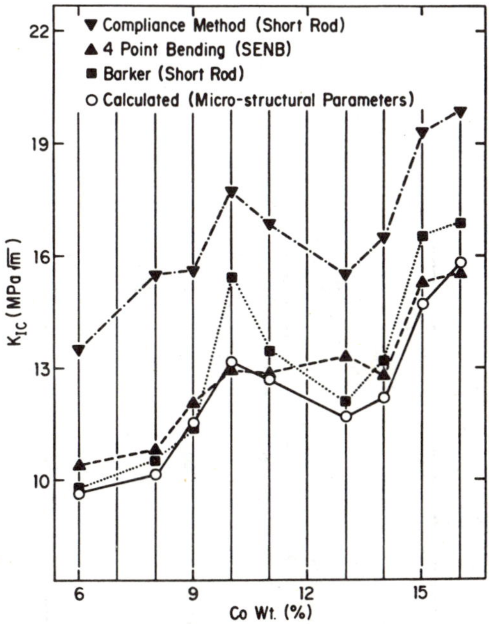

FIG. 7—K_{Ic} *values of nine alloys.*

TABLE 1—K_{Ic}-values and microstructural parameters.

Alloy	Cobalt, weight %	K_{Ic}, MPA $\sqrt{m}$				Microstructural Parameters, µm		
		Compliance Method	SENB	Short Rod	Calculated	d	λ	C
1	6	13.57	10.39	9.83	9.67	0.723	0.09	0.395
2	8	15.46	10.71	10.63	10.15	0.767	0.139	0.219
3	9	15.65	12.05	11.39	11.55	0.753	0.055	0.736
4	10	17.70	12.91	14.37	13.12	1.98	0.38	0.198
5	11	16.88	12.83	13.48	12.69	0.939	0.232	0.314
6	13	15.47	13.27	12.08	11.62	0.793	0.209	0.174
7	14	16.47	12.69	13.97	12.17	0.797	0.209	0.272
8	15	19.26	15.27	16.50	14.65	1.36	0.396	0.281
9	16	19.80	15.53	16.86	15.79	1.75	0.583	0.179

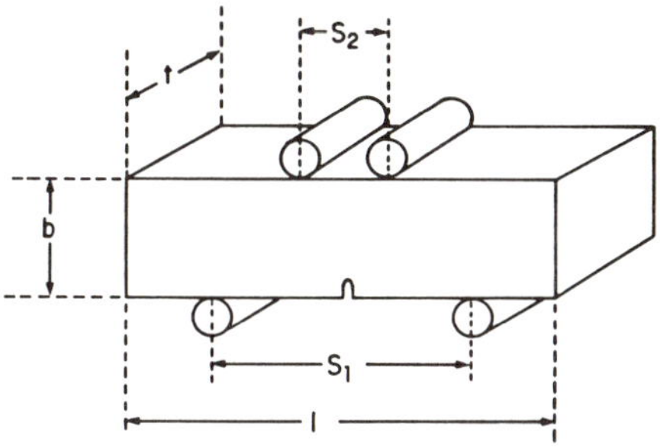

FIG. 8—*SENB specimen and 4-point bending.*

$$K_{Ic} = \frac{3\ P_s}{2\ b^2 t} \cdot \sqrt{\pi a}\ \{1.090 - 1.735\ (a/b) + 8.20\ (a/b)^2 - 14.57\ (a/b)^3 + 14.57\ (a/b)^4\} \quad (4)$$

where

P = load,
s = loading span,
t = thickness of the specimen,
a = initial notch length, and
b = width of the specimen.

The values obtained by this method show little scatter within the same alloy. The average values are reported in Table 1 and Fig. 7.

Flatjack Method Using a Short-Rod Specimen

Barker [*18*] developed an inflatable bladder, ''flatjack'' loading method on a short-rod specimen. The flatjack is placed inside the thin slot of the specimen, and the specimen is loaded by pressurizing the flatjack.

Barker showed that pressure becomes a maximum at certain crack length a_c and that a_c depends only on the specimen geometry, not on the material. There-

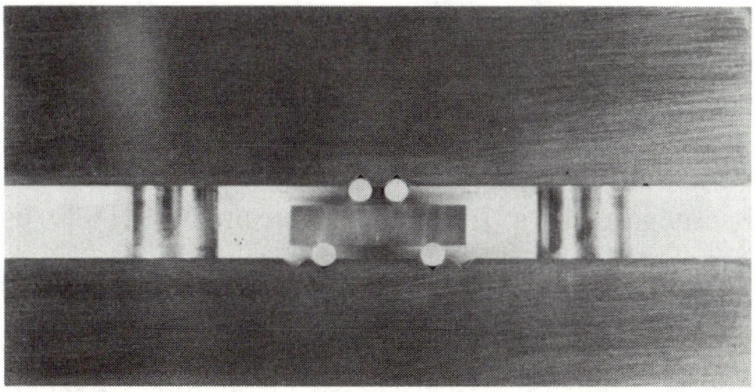

FIG. 9—*SENB specimen in 4-point bending fixture.*

fore, crack location at the peak load is constant for certain geometry. Barker developed the equation

$$K_{Ic} = A_F P_c \sqrt{B} \tag{5}$$

where

P_c = peak pressure,
B = specimen diameter, and
A_F = a constant which depends on specimen geometry and loading geometry.

Using the TerraTek Fractometer (Model #3401 and #2401), which utilizes Barker's method, K_{Ic} was measured for all the specimens and reported in Table 1 and Fig. 7. We have a high degree of confidence in the values obtained from the same alloy tested using Barker's method.

Microstructural Parameters and K_{Ic} Values

The microstructural parameters [*17*], mean free path of carbide (d), mean free path of cobalt (λ), and contiguity (C), have been measured using a semi-automatic image analysis method [*24*]. Optical, scanning electron, and transmission electron micrographs were analyzed. The results are given in Table 1.

Hong and Gurland [*16*] derived the following equation of K_{Ic} as a function of microstructural parameters

$$K_{Ic} = \sqrt{c\sigma_y \frac{\lambda_c}{d + \lambda_c} \lambda_c} \tag{6}$$

with

c = 5.03×10^{-5} MN/m,
λ_c = true mean free path of cobalt [*17*], and
σ_y = in situ yield stress.

For σ_y-values, Doi's 0.2% offset yield stress [*25*] was used

$$\sigma_y = 154 \text{ MN/m}^2 + \frac{1804}{\sqrt{\lambda}} \text{ MN}\sqrt{\mu\text{m}}/\text{m}^2 \tag{7}$$

This equation assumes crack tip opening displacement (CTOD) is comparable to the true mean free path (λ_c).

K_{Ic}-values were calculated by the Hong and Gurland equation using measured microstructural parameters. The results are given in Table 1 and Fig. 7. Among the earlier reported K_{Ic}-values [*1–15*] of Viswanadham [*2*], Nakamura [*3*], and Pickens [*5*] included microstructural parameters, and these data were used as reference when the equation was derived. This equation fits the reported data well [*16,26*]. Therefore, this equation can be used as a method of comparing the current data with former work [*2,3,5*].

Discussions and Conclusions

The four different K_{Ic}-values for the nine alloys are reported in Table 1 and Fig. 7. Microstructural parameters of these alloys are also given in Table 1 and Fig. 7.

The following observations were made:

1. K_{Ic}-values from the compliance method (short-rod specimen) are consistently higher than from the other three methods.
2. K_{Ic}-values from SENB specimens show higher scatter among each of the alloys than do those from Barker's short-rod method.
3. Fracture toughness values from Barker's flatjack method are very consistent among each of the alloys. However, the values from Barker's method have a tendency to give higher values than do those from the other methods for more ductile materials.
4. K_{Ic}-values calculated from microstructural parameters are fairly consistent with those from SENB results and from Barker's method.

A probable reason for the higher values of the compliance method is smaller values of the measured length than actual length. The crack tip in cemented carbide is usually discontinuous [*27*]. Even if it is continuous, it might be difficult for the dye penetrant to reach to the very tip. This smaller measured value could result in larger $(\partial C/\partial a)_c$ and smaller B-values, which cause larger G_{Ic} and K_{Ic} in Eqs 2 and 3. Residual stress resulting from the repeated loading and unloading of the compliance method might be another possibility for discrepancy in the result. The data trends are very similar for the compliance technique and for the other three methods (Fig. 7), even though absolute values are shifted. This shows a certain degree of confidence in this method as well as in the others.

As can be seen from Fig. 7, the values obtained by three of the methods—SENB specimen, flatjack method, and calculated values—are similar. The consistency of the calculated values with the other two methods indirectly confirms that these data are also consistent with previous experimental work [*2–4*] as discussed in the section on Microstructural Parameters and K_{Ic} Values.

Barker's flatjack method and four-point bending of SENB specimens are convenient experimentally because only peak values are counted. The data do not confirm which method is most accurate. However, for practical purposes, Barker's flatjack method seems to be a good method even though it has a tendency to give higher values for ductile grades compared with other methods. This method is easy and gives reproducible data.

Acknowledgment

The authors appreciate Chuck Lennon for his help in experimental work.

References

[1] Nidikan, B. and Davies, T. J., *Planseeberichte für Pulvermetallurgie,* Vol. 28, 1980, p. 29.
[2] Viswanadham, P. K., Sun, T. S., Drake, E. F., and Peck, J. A., *Journal of Materials Sciences,* Vol. 16, 1981, p. 1029.
[3] Nakamura, M. and Gurland, J., *Metallurgical Transactions,* Vol. 11A, 1980, p. 141.
[4] Peters, C. T., *Journal of Materials Sciences,* Vol. 14, 1979, p. 1619.
[5] Pickens, J. R. and Gurland, J., *Materials Sciences and Engineering,* Vol. 33, 1978, p. 135.
[6] Murray, M. J. and Perrott, C. M., *Proceedings,* 1976 International Conference on Hard Materials Tool Technology, 1976, p. 266.
[7] Chermant, J. L. and Osterstock, F., *Journal of Materials Sciences,* Vol. 11, 1976, p. 1939.
[8] Berry, G., *International Conference of New Tool Materials and Cutting Techniques,* London, 25 April 1975.
[9] Chermant, J. L., Deschanvres, A., and Iost, A., *Fracture Mechanics and Ceramics,* Vol. 1, R. C. Bradt et al, Eds., Plenum, New York, 1974, p. 346.
[10] Inglestrom, N. and Nordberg, H., *Engineering Fracture Mechanics,* Vol. 6, 1974, p. 597.
[11] Kenny, P., *Powder Metallurgy,* Vol. 14, 1971, p. 22.
[12] Lueth, R. C., *Fracture Mechanics and Ceramics,* Vol. 2, R. C. Bradt et al, Eds., Plenum, New York, 1974, p. 791.
[13] Yen, S. S., M.S. thesis, Lehigh University, Bethlehem, Pa., 1971.
[14] Lindau, L., *Fracture 1977,* Vol. 2, Waterloo, Canada, June 1977, p. 19.
[15] Johanson, T., *4th European Symposium for Powder Metallurgy,* Grenoble, 1975, 5-11-1.
[16] Hong, J. and Gurland, J., to be published.
[17] Lee, H. C., Ph.D. thesis, Brown University, Providence, R.I., June 1977, p. 51.
[18] Barker, L. M., *Fracture Mechanics of Ceramics,* Vol. 3, 1978, p. 483.
[19] Barker, L. M., TerraTek Report TR 81-97, Feb. 1981.
[20] Knott, J. F., *Fundamentals of Fracture Mechanics,* Wiley, New York, pp. 105 and 108.
[21] Yeh, R. H. T., *Journal of Applied Physics,* Vol. 42, No. 3, 1971, p. 1101.
[22] Pickens, J. R. and Gurland, J., *Materials Sciences and Engineering,* Vol. 33, 1978, p. 135.
[23] Brown, W. F., Jr. and Srawley, J. E., *Plane-Strain Crack Fracture Toughness Testing of High-Strength Metallic Materials, ASTM STP 410,* American Society for Testing and Materials, Philadelphia, 1967, p. 1.
[24] Hong, J., Gurland, J., and Schwarzkopf, P., to be published.
[25] Doi, H., *Elastic and Plastic Properties of WC-Co Composite Alloys,* Scientific Publication Division, Freund Publishing House Ltd., Tel-Aviv, Israel, 1974.
[26] Hong, J., Ph.D. thesis, Brown University, Providence, R.I., 1981, p. 73.
[27] Hong, J. and Gurland, J., *Proceedings,* International Conference on Science of Hard Materials, Jackson Lake Lodge, Wyoming, Aug. 1981, p. 649.

Ralph F. Krause, Jr.,[1] *and Edwin R. Fuller, Jr.*[1]

Fracture Toughness of Polymer Concrete Materials Using Various Chevron-Notched Configurations

REFERENCE: Krause, R. F., Jr., and Fuller, E. R., Jr., **"Fracture Toughness of Polymer Concrete Materials Using Various Chevron-Notched Configurations,"** *Chevron-Notched Specimens: Testing and Stress Analysis, ASTM STP 855,* J. H. Underwood, S. W. Freiman, and F. I. Baratta, Eds., American Society for Testing and Materials, Philadelphia, 1984, pp. 309–323.

ABSTRACT: The fracture toughness of two similar polymer concrete materials was determined using several fracture mechanics configurations to show any influence of crack geometry on resistance to fracture in these materials. The testing configurations included a conventional straight-through notch in a flexure bar and various chevron-notched geometries in both flexure-bar and short-rod specimens. The materials were polymerized mixtures of monomers, anhydrous Type III portland cement, and silica sand. In one composition the monomers were styrene and trimethylolpropane-trimethacrylate; whereas, in the other composition, acrylonitrile was added as well. The fracture toughness was calculated from published stress-intensity coefficients for the straight-through notch which were adapted for use with a chevron notch by assuming that the derivative of the compliance with respect to crack length was the same for both notch types. Effects of varying chevron-notched angle, chevron-vertex position, and width of specimen in the crack plane were examined. Variation of these parameters to produce a wide variation in the width of the crack front shows minor if any influence on the fracture toughness measurements. In contrast, however, the calculated amount of crack extension for all the various specimen configurations indicates a trend of fracture toughness which is consistent with the concept of a rising crack-growth resistance curve for these materials.

KEY WORDS: chevron notch, crack area extended, flexure-bar specimen, fracture toughness, polymer concrete, short-rod specimen, straight-through notch, stress-intensity coefficient

Nomenclature

- a Crack length (Fig. 1)
- a_m Crack length at minimum Y
- a_0 Initial notch length (distance from crack mouth to chevron vertex)
- a_1 Distance from crack mouth to intersection of chevron notch and specimen edge

[1]Research chemist and physicist, respectively, Inorganic Materials Division, Center for Materials Science, National Bureau of Standards, Gaithersburg, Md. 20899.

ΔA Crack area extended beyond the initial notch (Eq 9)
ΔA_m Crack area extended at minimum Y
b Width of crack front perpendicular to the crack propagation direction
b_m Width of crack front at minimum Y
B Width of specimen in the crack plane
$C = D/P$ Linear-elastic compliance
$C' = E'BC$ Dimensionless compliance
D Load-point displacement
E Young's modulus of elasticity for a given material
$E' = E$ For plane stress
$E' = E/(1 - \nu^2)$ For plane strain
G Crack extension energy per unit crack area (strain energy release rate)
K Plane strain, stress-intensity factor (Mode I)
K_m Value of stress-intensity factor at maximum P and minimum Y
K_R Crack growth resistance or fracture toughness of a given material
n Thickness of the notch gap
P Applied load
S_1 Support span for loading a flexure bar specimen
S_2 Minor span for loading a four-point flexure bar specimen
W Length of specimen in the crack propagation direction
Y Dimensionless stress-intensity coefficient (Eq 3)
ν Poisson's ratio
χ Chevron-notched angle (Fig. 1).

Polymer concrete materials have been proposed as alternate materials in various geothermal applications because of their excellent resistance to corrosion from geothermal brines and because of their tendency to inhibit the buildup of surface corrosion layers [*1*]. Proposed applications, which replace or augment metallic structures, include pressurized pipes, heat exchanger tubing, and pressure vessel liners. One of the parameters needed in the design of these structures is fracture toughness, a measure of the resistance of a flawed material to fracture. The purpose of this paper is to evaluate the fracture toughness of two similar polymer concrete materials using various fracture mechanics configurations to show their influence on the measurements.

Chevron-notched specimens are potentially ideal for application in measuring fracture toughness of ceramic and concrete materials. A sharp crack is produced during the early stage of test loading so that no precracking is required; and the maximum load corresponds to a particular crack length, which is dependent on the initial notch geometry, so that no crack length measurement is required [*2*]. In contrast, when the conventional straight-through notch is applied to ceramic and concrete materials, it is virtually impossible to initiate crack growth by fatigue precracking and subsequently even difficult to measure the critical crack length. A recourse is to estimate fracture toughness using the initial notch length, but the thickness of the notch gap must be less than some limit to avoid a measured value which increases with increasing thickness of the notch gap [*3*].

Consideration of the chevron notch raises the question of whether its geometry influences fracture toughness measurements. To explore this issue, we investigated a number of chevron-notched configurations. Besides the nature in which loading is applied to a particular specimen type, the size of specimen in the crack plane, B and W, the chevron-notched angle, χ, and its vertex position along the centerline of the specimen, a_0, have to be specified (Fig. 1). We varied χ, and the relative parameters, a_0/W and B/W, to produce a wide variation in the width of the crack front, b, to examine its influence on the fracture toughness measurements.

Crack-growth resistance or fracture toughness of most materials is generally considered independent of crack extension. This is approximately true for cracks in metallic materials under conditions of plane strain; but, in the case of plane stress which occurs in thin sheets, crack resistance has been shown to rise with the amount of crack extension [*4*]. Similar behavior has been shown to occur in fiber reinforced concrete beams [*5*] although the explanation is most likely

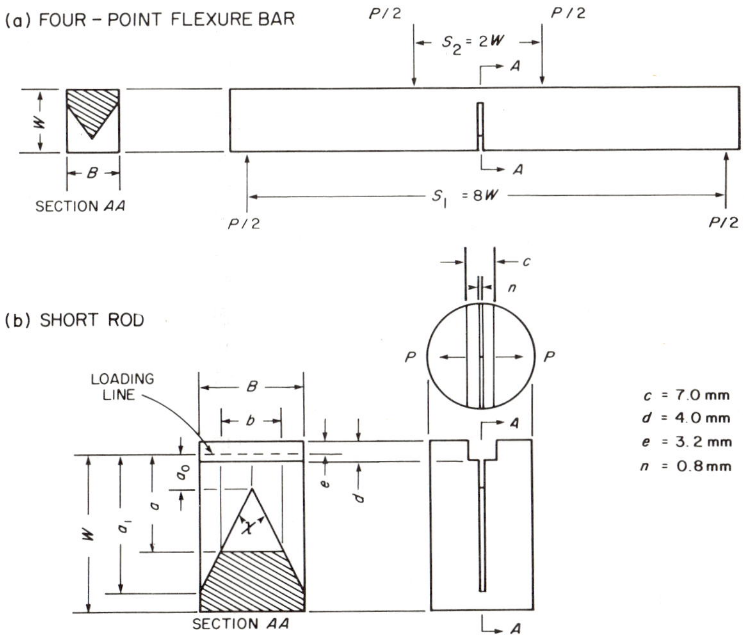

FIG. 1—*Chevron-notched specimen configurations. (The shaded area in the cross section is the uncracked region. The chevron notch is symmetrical and is centered between the loading lines.)*

not related to plane stress. To investigate this possibility in the polymer concretes, we examined various specimen configurations to determine the dependence of crack-growth resistance on crack extension.

Experimental Procedure

The composition of two similar polymer concrete materials, which were fabricated at Brookhaven National Laboratory for use in the present study, are shown in Table 1. These materials are polymerized mixtures of monomers, anhydrous Type III portland cement, and silica sand. They involve calcium silicate and polymer complexes [*6*], not to be confused with polymer-impregnated hydraulic cements. The liquid monomer phase was blended with the fine aggregate phase, approximately 1 to 0.1 mm particle size, and vibrated 3 min to remove air bubbles. The mortar was cast both in upright tubes, nominally 25 mm inside diameter, and in rectangular box forms, 50 mm by 150 mm inside base, producing finished rods 125 mm high and finished billets 150 mm high, respectively. The mortar was cured in the molds for about 4 h, demolded, then cured at 190°C for about 18 h more. Besides the irregular-sized aggregate, cavities about 0.5 mm or less in size were observed scattered throughout the finished product. Thus, the microstructure of these materials varied in texture.

Two types of specimen configurations were machined from the stock materials: short-rod specimens from the rod stock and both flexure and short-rod specimens from the billet stock. The configurations for the four-point flexure specimen and the short-rod specimen are shown in Fig. 1. A three-point flexure specimen with $S/W = 4$ was also used. The flexure bars were first diamond-sawed roughly from the stock billets; then their sides were finished by diamond-grinding under a water rinse with an automatic surface grinder. Approximately 1 mm of material was removed to produce sides which were square and parallel within 0.02 mm. The three-point flexure specimens were sawed two abreast, and the four-point flexure specimens were sawed three abreast in each layer of the billet stock. One set of short-rod specimens was diamond core-drilled from a billet, while

TABLE 1—*Composition of polymer concrete materials.*

Identity	Monomer Mixture[a]	Parts by Weight of Monomer	Aggregate Mixture	Parts by Weight of Aggregate
A	styrene	55.5	silica sand	65.5
	polystyrene	4.5	silica flour	8
	TMP-TMA[b]	40.0	Type III portland cement	24
	additives	8.5	additives	2.5
B	styrene	55	silica sand	70
	acrylonitrile	36	Type III portland cement	30
	TMP-TMA[b]	9		
	additives	8.5		

[a]The monomer mixture was 10 to 12% by weight of the total.
[b]Trimethylolpropane-trimethacrylate.

the other short rod specimens needed only to be diamond-sawed to length from the cast rods, each rod providing generally 3 specimens. All the notches were diamond-sawed with a blade 0.8 mm wide. The orientation of the specimens from the billet stock was such that the crack plane was normal to the long dimension of the billet cross section, and that the crack propagated in the direction of the short dimension of its cross section. The chevron notches were completed by rotating the specimens to produce overlapped slices within 0.02 mm. To avoid any bias in the results from variation in density along the height of the billet, due to gravity during the cure, specimens for the various chevron-notched geometries were selected alternately along the height of the billet.

The fracture tests were conducted on a gear-driven, universal testing machine. The machine and load train were shown to have negligible anelastic relaxation of fixed extension for the load range used in the present experiments. The maximum load sustained by each specimen was measured within 0.2% full scale (100, 200, 500, 1000 N) by a load cell which was calibrated periodically during the tests. A four-point flexure specimen and a short-rod specimen were centered, respectively, in the loading fixtures shown in Fig. 2. A three-point flexure specimen was centered on the 10.16 cm (4 in.) span of a commercial fixture which functioned with the machine in the tension mode. A specimen was preloaded to about 10 N, then loading was applied at a constant displacement rate of 0.05 cm/min. The test was continued until the load returned to near zero, there being no sudden failure, but a long unloading tail as indicated in Fig. 3. The slopes of the curves in Fig. 3 begin to decrease just prior to the peak load, indicating stable crack growth. It is believed that the machine compliance had little effect on the crack growth stability in these experiments; for example, the machine compliance was observed to be about 6, 10, and 22% of the specimen compliance for data sets 8, 9, and 10, respectively. (See Table 3 for definition of data.)

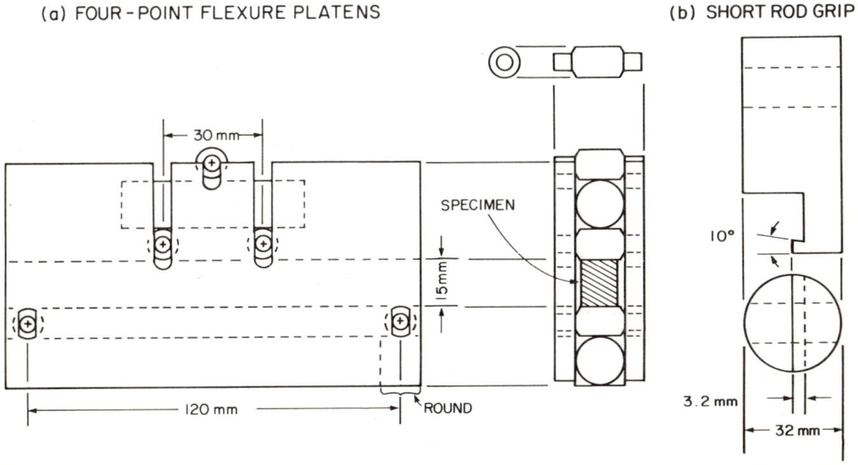

FIG. 2—*Specimen loading fixtures.*

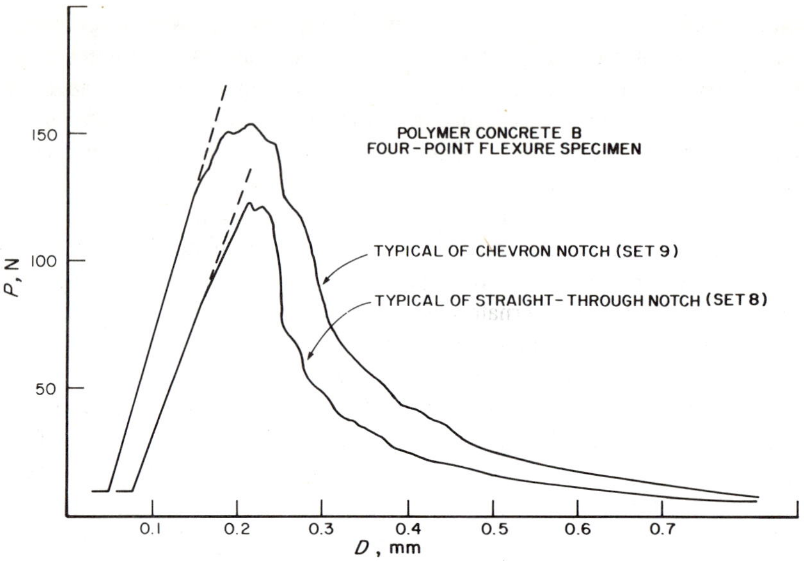

FIG. 3—*Typical specimen loading* P *versus displacement* D *of the specimen and machine at constant crosshead speed of 0.05 cm/min. (Data sets are identified in Table 3. Zero displacement is offset for purpose of illustration.)*

While the dimensions B and W were measured before load testing, the dimensions a_0 and a_1 were determined afterwards from the fractured specimen, using a traveling microscope with rotatable cross hairs that could be aligned with the chevron and specimen sides to locate their intersections within 0.01 mm.

Method of Analysis

The Griffith energy criterion for fracture states that crack extension can occur if the energy to form an incremental crack can just be delivered by the system, which includes the energy stored both in the specimen and in the machine. Using the compliance relation, $D = CP$, an evaluation of the crack extension energy can be made [*4*] which yields

$$G = K^2/E' = P^2(\partial C/\partial A)/2 \tag{1}$$

The incremental crack area for the chevron-notched specimen (Fig. 1) is defined as $dA = b\,da$ where b can be deduced from

$$2 \tan (\chi/2) = b/(a - a_0) = B/(a_1 - a_0) \tag{2}$$

A customary rearrangement of Eq 1 for the chevron-notched specimen is

$$K = P\,Y/B\,W^{1/2} \tag{3}$$

where

$$Y = [(B/b)(\partial C'/\partial a)W/2]^{1/2} \tag{4}$$

with $C' = E'BC$ as the dimensionless compliance.

When $\chi = 180°$, the special case of the straight-through notch exists such that $a_0 = a_1$ and $b = B$. For the purpose of the present investigation we assumed that $(\partial C'/\partial a)$ was the same for the chevron notch as for the straight-through notch, commonly known as the straight-through crack assumption (STCA) [*2,3*]. Using this assumption the Y coefficient for a chevron-notched specimen is the same as that for a straight-through notched specimen except the former is multiplied by the factor $[B/b]^{1/2}$. We adopted the straight-through crack assumption because the stress-intensity coefficient for a chevron-notched flexure specimen has not been calibrated.

For straight-through-notched flexure specimens, the stress-intensity coefficient Y is expressed as follows

$$Y = (3/2)(S/W)(a/W)^{1/2}\,\Gamma/(1 - a/W)^{3/2} \tag{5}$$

where $S = S_1$ for three-point loading and $S = (S_1 - S_2)$ for four-point loading. For the three-point flexure specimen with $S/W = (4 \pm 0.001)$, the parameter Γ has been formulated [*7*] as follows

$$\Gamma = \{1.99 - (a/W)(1 - a/W)[2.15 - 3.93\,a/W + 2.7\,(a/W)^2]\}/(1 + 2\,a/W) \tag{6}$$

to fit the collocation results of analysis within $\pm 0.2\%$ over the range $0 < a/W < 1$. For pure bending, which the four-point flexure specimen approximates, the parameter Γ has been also formulated [*8*] as follows

$$\Gamma = 1.9887 - 1.326\,a/W - [3.49 - 0.68\,a/W + 1.35\,(a/W)^2](a/W)(1 - a/W)/(1 + a/W)^2 \tag{7}$$

to fit the collocation results of analysis within $\pm 0.5\%$ deviation over the range $0 < a/W < 1$, except at $a/W = 0.1$ where the deviation is 1%.

Since we applied the STCA for analyzing the chevron-notched flexure specimens, we also applied this assumption to the short-rod specimen for consistency. For the straight-through notched short rod with $W/B = 1.5$, the stress-intensity coefficient has been formulated [*9*] as follows

$$Y = -3.851 + 73.63\,a/W - 65.85\,(a/W)^2 - 94.43\,(a/W)^3 + 179.8\,(a/W)^4 \tag{8}$$

to fit the compliance data over the range $0.3 < a/W < 0.75$. While compliance calibration measurements on the chevron-notched short-rod specimens have been published elsewhere [*9,10*], the minimum Y-values reported do not cover all the geometries selected for the present work.

Using the straight-through crack assumption, Fig. 4 shows Y generated as a function of a/W for selected geometries of the chevron-notched, four-point flexure specimen with $S_1/W = 8$ and $S_2/W = 2$. Thus we were able to anticipate the nature of the Y function, especially in the vicinity of its minimum, for a particular geometry as characterized by the parameters χ, B/W, and a_0/W. Note that a_1/W is completely determined from these parameters through Eq 2. Sets of selected geometries are compared in Table 2 to illustrate the effect that variation of each of the characterizing parameters has upon the relative crack length, a_m/W, and the relative width of the crack front, b_m/W, at the minimum value of Y. (The subscript m denotes the value of various parameters at the point of minimum Y). Note that a_m/W depends only on a_0/W. A discontinuous slope, which is indicated in curve (d) of Fig. 4, occurs in the curves at a_1 where the crack front reaches full specimen width, B. The Y curves for the three-point flexure and short-rod specimens were also characterized in a similar manner.

Comparison of Results

Values of K_m determined from the various sets of specimen configurations for polymer concretes A and B are shown in Table 3. The individual K_m-values were calculated from Eq 3, using as assumed in Ref *2* the maximum P and the minimum value of the appropriate Y function, that is, $K_m = P_{max}Y_{min}/B\sqrt{W}$. The Y function

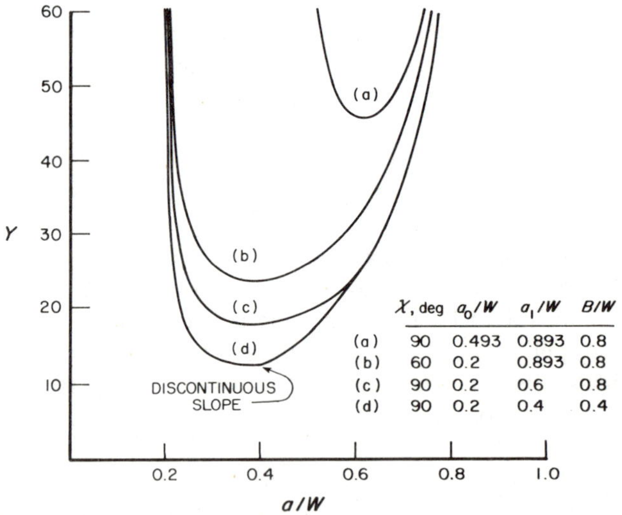

FIG. 4—*Stress-intensity coefficient* Y *versus relative crack length* a/W *for various chevron-notched geometries of a four-point flexure bar with* $S_1/W = 8$ *and* $S_2/W = 2$. (*The discontinuous slope in curve* (d) *shows where the crack front reaches full specimen thickness.*)

TABLE 2—*STCA calculations of* a_m/W, b_m/W *and the minimum* Y *for various chevron-notched geometries applied to the four-point flexure specimen with* $S_1/W = 8$ *and* $S_2/W = 2$.

B/W	χ, deg	a_0/W	a_1/W	a_m/W	b_m/W	Y_{min}
	60	0.307		0.477	0.196	31.84
0.8	90	0.600	1.0	0.700	0.200	72.72
	120	0.769		0.827	0.201	165.10
	60		0.893		0.218	23.64
0.8	90	0.2	0.600	0.389	0.378	17.96
	120		0.431		0.655	13.65
		0.2	0.6	0.389	0.378	17.96
0.8	90	0.4	0.8	0.549	0.298	32.36
		0.6	1.0	0.700	0.200	72.72
0.4			0.4			12.70
0.6	90	0.2	0.5	0.389	0.378	15.55
0.8			0.6			17.96

of Eq 4 was evaluated from Eq 2 and the respective STCA: Eqs 5 and 6 for the three-point flexure specimen, Eqs 5 and 7 for the four-point flexure specimen, and Eq 8 for the short-rod specimen. The maximum *P*-values shown in Table 3 correspond to the nominal dimensions and the mean K_m given. From a cursory view of the results it is obvious that the short-rod values are considerably larger

TABLE 3—*Fracture toughness of polymer concrete materials,* K_m, *as calculated from maximum* P *and the minimum* Y *function for various specimen configurations.*

Set	Nominal Dimensions a_0/W	χ, deg	B/W	W, mm	P_{max}, N	Replicates	K_m[c], MPa$\sqrt{m}$
			POLYMER CONCRETE A: THREE-POINT FLEXURE BAR				
1	0.33	60	0.51	24.5	228	4	2.06 ± 0.17
2	0.50	180	0.51	24.5	299	5	1.67 ± 0.09
			POLYMER CONCRETE B: FOUR-POINT FLEXURE BAR				
3	0.2	60	0.8	15	112	5	1.80 ± 0.13
4	0.2	90	0.8	15	164	6	2.00 ± 0.22
5	0.49	90	0.8	15	53	5	1.63 ± 0.23
6	0.2	120	0.8	15	209	5	1.94 ± 0.23
7	0.2	90	0.4	15	97	5	1.68 ± 0.24
8	0.5	180	0.8	15	116	5	1.33 ± 0.16
9[a]	0.2	90	0.8	15	140	6	1.71 ± 0.14
			POLYMER CONCRETE B: SHORT ROD				
10[b]	0.34	60	0.68	33.2	350	5	2.48 ± 0.13
11	0.34	60	0.68	38.1	502	6	2.85 ± 0.14
12	0.34	90	0.70	37.4	642	6	2.86 ± 0.08
13	0.60	90	0.70	37.5	249	6	2.28 ± 0.10

[a]Specimens of set 9 were diamond sawed from a separate billet while the specimens of sets 3 through 8 were diamond sawed from the same billet.

[b]Specimens of set 10 were diamond core drilled from the same billet as set 9 while the short-rod specimens of sets 11, 12, and 13 were diamond sawed from cast rods.

[c]The values listed are the mean and standard deviation of the replicates given.

than the flexure values, and that for a given material the chevron-notched values are considerably larger than the straight-through notched values ($\chi = 180°$). Whether these differences are significant depends on the accuracy of the respective results. The STCA approximation of the Y functions for the chevron-notched short-rod and flexure specimens is probably the major suspect for error.

The STCA approximation of the minimum Y-value for the short-rod specimens of data sets 10 and 11 can be compared with the results of more exact analyses and generally they agree within the precision of measurements. The chevron-notched compliance calibration of Shannon, Bubsey, Pierce, and Munz [*10*] yields a minimum Y which is about 3% less than that obtained by STCA; the three-dimensional finite element calibration factor of Beech and Ingraffea [*11*] is about the same as the STCA value; and the compliance calibration factor from Barker [*12*] is about 5% less than the STCA value. Two other assumptions were made involving the short-rod specimens. It was assumed that Eq 8 which represents Y for $W/B = 1.5$ was applied without correction for the actual values of W/B measured. Whereas, interpolation of the compliance calibration data [*9*] would have given Y-values for sets 10 and 11 about 1% lower than those obtained through Eq 8, and values for sets 12 and 13 about 2% lower. Finally, it was assumed that any variation in the position and length of the loading line for the short-rod specimens had negligible effect on the results. Whereas, in fact the relative channel width, c/B, which defines the relative chord length along which the load is applied (Fig. 1), varied from 0.27 to 0.31 in the present work, but was 0.37 in the calibration study [*9*].

The STCA results for the chevron-notched, four-point flexure specimens can be compared with those from the Bluhm slice model, BSM [*13*]. The STCA results in Table 3 for the 32 measurements in sets 3 through 9, excluding the straight-through notched data of set 8, give an overall mean K_m and standard deviation of 1.80 ± 0.23 MPa$\sqrt{\text{m}}$, which is not distinguishable from the mean values of K_m for the respective sets within the precision of measurements. If the BSM method were applied instead of the STCA for these specimens, the results would more closely group around the mean. The BSM values for data sets 3 and 5 are about 1 to 5% less than the corresponding STCA values, and the BSM values for data sets 4 and 6 are about 16 to 18% less than the STCA values, using the calculations of Refs *2* and *3*. Thus, the BSM method would tend to reduce the differences and to give better agreement among the respective K_m-values of these data sets, but the overall mean would be slightly lower. The difference between the mean K_m-values of sets 4 and 9, which have the same chevron-notched geometry, is probably due to variation in the two stock billets used.

The accuracy of the STCA evaluation for the chevron-notched three-point flexure specimen is more difficult to verify, but it can be compared to the results of Shih. He determined [*14*] a calibration coefficient for this specimen configuration from load measurements on an aluminum alloy and an evaluation of its fracture toughness by a standard compact tension test. He reported [*14*] that in a cooperative testing program this calibration coefficient yielded a value of the fracture toughness for Pyroceram which agreed very closely with that obtained from double-torsion and short-rod specimens. This coefficient, however, is about 23% less than that from his earlier work [*15*] which was based on an alternative, literature value of the fracture toughness for the alloy and which is only 3% less than the STCA evaluation.

Both the specimen width, B, and the thickness of the notch gap, n, that were used in the present work are believed to have minor influence on the observed K_m. For metals the observed values of K tend to increase with decrease in B [*4*]. Whereas, just the opposite was observed for data sets 4 and 7 in an examination of this effect. With regards the thickness of the notch gap, Barker [*12*] recommended $n/B < 0.03$. Indeed, such was used for the short-rod specimens in the present work; but n/B for the flexure-bar specimens was about 0.07. Because observed values of K tend to increase with increasing n [*12*], a smaller notch gap for the flexure-bar specimens would have produced a greater difference in the observed K_m between the flexure-bar and short-rod specimens.

Resistance Curve Interpretation

The variation in the apparent K_m for the various specimen configurations used is believed consistent with the concept of resistance to fracture rising with crack extension [*4*]. This concept represents the situation that as a crack propagates, more energy per unit crack area is required, perhaps due, for example, to aggregate interlock and pullout or to the increase of multiple fine cracks at the main crack tip. Figure 5 plots values of K_m from Table 3 versus crack area extended at the minimum Y, $\Delta A_m = (a_m - a_0)\, b_m/2$. Of course this calculated area is planar whereas the actual crack area is tortuous. While any variation in the K_m for the chevron-notched four-point flexure specimens (data sets 3 through 9 except 8) can not be distinguished within the precision of measurements, the dashed curve through all the data sets in Fig. 5 indicates a rising crack-growth resistance.

The resistance to fracture is more appropriately defined as the R-curve in Fig. 6. This curve represents the critical values of the stress-intensity factor for specimen configurations that have various extensions of crack area at criticality. [It should be noted that G can be also plotted versus ΔA]. The K-functions at the maximum P (solid line) or at a subcritical P (short-dashed line) were generated with the crack area extended for three selected specimen configurations (sets 8, 9, and 10). The crack area extended was calculated as follows

$$\begin{aligned} \Delta A &= (a - a_0)\, b/2, \quad a_0 \le a \le a_1 \\ &= (2a - a_1 - a_0)\, B/2, \quad a_1 \le a \le W \end{aligned} \tag{9}$$

For the purpose of clarification data sets 8 and 9 are represented separately in Figs. 6*a* and 6*b*. They are combined with data set 10 in Fig. 6*c*. Stable crack

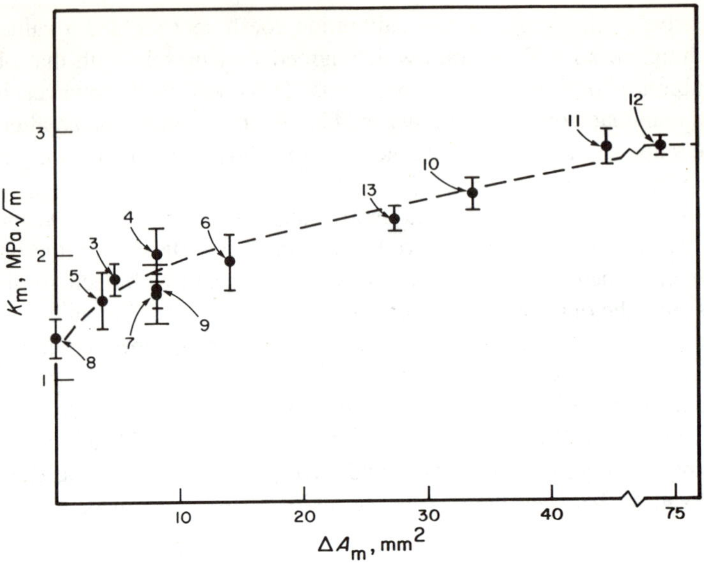

FIG. 5—*Mean values of* K_m *for the various specimen configurations of polymer concrete B versus crack extension area* ΔA_m *at the minimum value of* Y. *(The numbers refer to the data sets which are identified in Table 3. Error bars denote plus and minus one standard deviation.)*

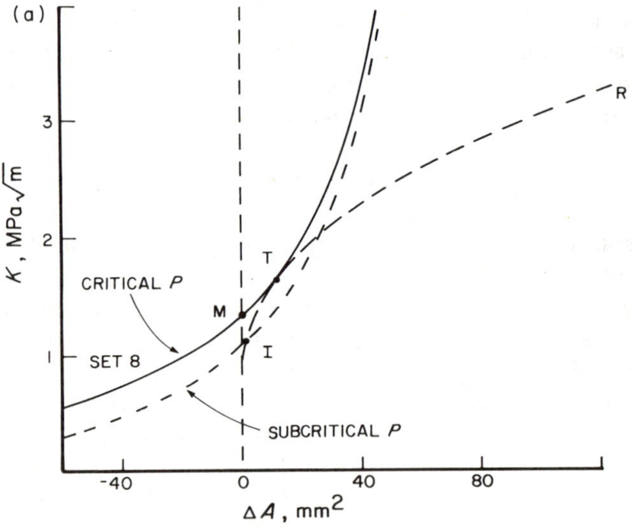

FIG. 6—*Postulated crack-growth resistance curve R for polymer concrete B versus crack extension area,* ΔA. *(Figures* (a) *and* (b) *show the applied* K*-functions that were generated for data sets 8 and 9, respectively; and Fig.* (c) *shows data sets 8, 9, and 10 together. The points I indicate subcritical crack extension. The points M refer to values at the minimum* Y *and the maximum* P *as defined in Table 3. The points T indicate where the slope of the R-curve equals the slope of the respective critical* K*-curves.)*

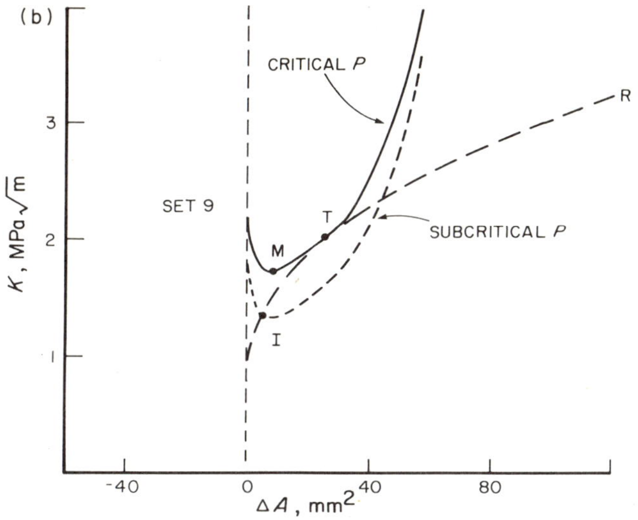

FIG. 6—*Continued.*

extension at a subcritical load is indicated by any point I. The points M indicate the K_m-values which were calculated from the minimum Y of the respective specimen configurations. The critical point on the K-curve for a given specimen configuration occurs when the slope of the K-curve, dK/dA, equals the slope of the R-curve, dK_R/dA. The points T indicate these tangential values. The R-curve in Fig. 6*c* is postulated as passing through these T points for the respective specimen configurations selected.

The question of a rising R-curve can be resolved, we believe, from fracture

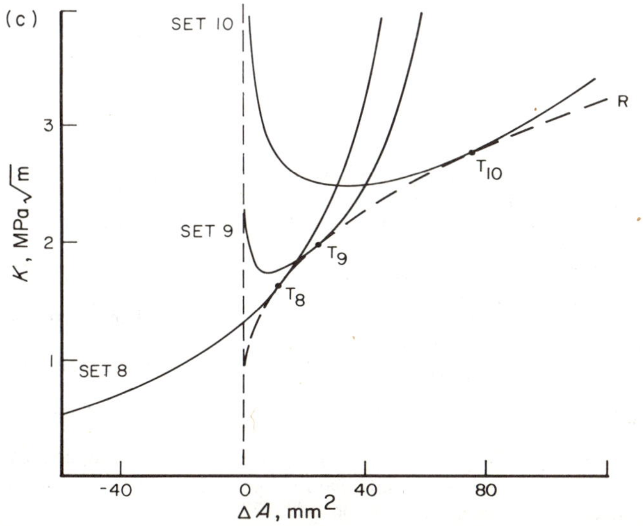

FIG. 6—*Continued.*

toughness measurements on various-sized specimens with the same chevron-notched geometry and test configuration. In this way any errors in the stress-intensity coefficients among various types of specimens will be avoided.

Conclusions

The width of the crack front which interacts with the varied microstructure of the polymer concrete materials has minor if any influence on evaluation of their fracture toughness. As the chevron-notched geometry of the same width, four-point flexure specimens was varied to produce a wide variation in the width of the crack front at minimum *Y*, the fracture toughness remained the same within the precision of measurements. Thus, in this regard there is no particular chevron-notched geometry that is to be preferred within the range of χ and a_0/W used in the present work.

The calculated amount of crack extension for all the various specimen configurations indicates a trend of fracture toughness which is consistent with a rising crack-growth resistance curve for the polymer concrete materials. The fracture toughness depends on specimen size and to a lesser extent on the chevron-notched geometry. Some error is introduced when fracture toughness is calculated from the minimum *Y*. The R-curve in Fig. 6*c* provides a better approximation of fracture toughness than the values of K_m in Table 3, which overestimate the values of K_R at the corresponding ΔA_m.

Acknowledgments

We are grateful to W. Reams, J. J. Fontana, and L. E. Kukacka of Brookhaven National Laboratory for preparing the polymer concrete materials.

This work was supported in part by the U.S. Department of Energy, Advanced Research and Technology Division, Fossil Energy Materials Program, under interagency agreement DE-A105-800R20679 with the National Bureau of Standards.

References

[1] Fontana, J. J., "Potential for Polymer Concrete to Conserve Alloys in Engineering Applications," BNL 31990, Brookhaven National Laboratory, Sept. 1982.

[2] Munz, D., Shannon, J. L., Jr., and Bubsey, R. T., *International Journal of Fracture*, Vol. 16, 1980, pp. R137–R141.

[3] Munz, D., Bubsey, R. T., and Shannon, J. L., Jr., *Journal of the American Ceramic Society*, Vol. 63, 1980, pp. 300–305.

[4] Broek, D., *Elementary Engineering Fracture Mechanics*, Noordhoff International Publishing, Leyden, The Netherlands, 1974, pp. 115–127.

[5] Velazco, G., Visalvanich, K., and Shah, S. P., *Cement and Concrete Research*, Vol. 10, 1980, pp. 41–51.

[6] Sugama, T., Kukacka, L. E., and Horn, W., *Cement and Concrete Research*, Vol. 9, 1979, pp. 461–471.

[7] Srawley, J. E., *International Journal of Fracture*, Vol. 12, 1976, pp. 475–476.

[8] Srawley, J. E. and Gross, B., *Cracks and Fracture*, *ASTM STP 601*, American Society for Testing Materials, Philadelphia, 1976, pp. 559–579.

[9] Bubsey, R. T., Munz, D., Pierce, W. S., and Shannon, J. L., Jr., *International Journal of Fracture,* Vol. 18, 1982, pp. 125–133.
[10] Shannon, J. L., Jr., Bubsey, R. T., Pierce, W. S., and Munz, D., *International Journal of Fracture,* Vol. 19, 1982, pp. R55–R58.
[11] Beech, J. F. and Ingraffea, A. R., *International Journal of Fracture,* Vol. 18, 1982, pp. 217–229.
[12] Barker, L. M., *Engineering Fracture Mechanics,* Vol. 17, 1983, pp. 289–312.
[13] Bluhm, J. I., *Engineering Fracture Mechanics,* Vol. 7, 1975, pp. 593–604.
[14] Shih, T. T., *Engineering Fracture Mechanics,* Vol. 14, 1981, pp. 821–837.
[15] Shih, T. T., *Journal of Testing and Evaluation,* Vol. 9, 1981, pp. 50–55.

J. J. Mecholsky[1] *and Lynn M. Barker*[1]

A Chevron-Notched Specimen for Fracture Toughness Measurements of Ceramic-Metal Interfaces

REFERENCE: Mecholsky, J. J. and Barker, L. M., **"A Chevron-Notched Specimen for Fracture Toughness Measurements of Ceramic-Metal Interfaces,"** *Chevron-Notched Specimens: Testing and Stress Analysis, ASTM STP 855,* J. H. Underwood, S. W. Freiman, and F. I. Baratta, Eds., American Society for Testing and Materials, Philadelphia, 1984, pp. 324–336.

ABSTRACT: A new double cantilever beam specimen with a chevron notch (short bar) has been developed to measure the fracture toughness, (or resistance), G_{Ic}, of ceramic-metal interfaces. The analysis of the specimen is based on previously developed compliance calibrations for homogeneous materials. Toughness measurements on a zinc-silicate glass ceramic-molybdenum system demonstrated that the bond at this interface is as tough (13 J/m^2) as the glass ceramic alone. Measurements on a lithium silicate glass ceramic-Hastelloy 276 metal system showed that the toughness of the interface depends on the degree of bonding ranging from 19 J/m^2 for weak bonding to 54 J/m^2 for strong bonding. This specimen is being suggested for use in the study of the effect of heat treatment and bonding chemistry on fracture of metal-ceramic interfaces.

KEY WORDS: fracture of metal-ceramic bonds, fracture toughness, chevron-notched specimen, ceramic-metal interface, fracture

Techniques for the measurement of crack growth resistance at the interface of dissimilar materials are crucial for the design and quality control of glass-to-metal, ceramic-to-metal, and ceramic-to-ceramic seals [*1*]. Early attempts at the measurements of the strength of the bond involved peel, tension, flexure, and shear tests [*2,3*]. The results of these tests, however, were disappointing because of a lack of reproducibility [*1*]. Part of the reason for this lack of success is that the measurements are dependent on the size of the initiating crack. Several techniques have been developed for measurement of fracture energy at thick-film interfaces [*4,5*] based on fracture mechanics principles. These techniques do not require a knowledge of the initial crack size. However, crack initiation is generally a necessary concern with these techniques. In addition to experimental techniques, there are many theoretical derivations for calculating the stress intensity factors of cracks along interfaces of dissimilar materials [*6–9*].

[1]Members of technical staff, Sandia National Laboratories, Albuquerque, N. Mex. 87185.

To help evaluate the appropriateness of particular theories and to measure the strength of interfacial bonds, it is important to have a reproducible, quantitative, experimental technique for measuring crack growth resistance at interfaces.

This paper presents an experimental technique for measuring the fracture toughness along the interface based on a double cantilever beam (DCB) type specimen utilizing a chevron notch. The advantage of this specimen over others is that it is relatively easy to make, the technique itself introduces its own sharp crack at the interface, and the test is relatively easy to perform.

Analysis

Stress-intensity factors are usually used to define the elastic stress field in the vicinity of a crack tip in a homogenous medium. Therefore, rather than to try to associate a critical stress-intensity factor, K_{Ic}, with the bond strength, it may be more germane to speak of the energy per unit crack surface area, G_{Ic}, which is required for a crack to advance in the bond plane. A consideration of the derivation of the fracture toughness equation for the short-bar fracture toughness specimen suggests a simple approach for measuring the G_{Ic} of the bond.

The derivation starts by assuming a specimen of the short-bar type in which the bond plane between the two materials coincides with the intended crack plane in the specimen (Fig. 1*a*). The specimen is loaded with a force F, causing the specimen mouth to flex open an amount x. During the time of interest, suppose that the crack advances by a small distance, Δl, creating the new crack area $b\Delta l$ shown cross hatched in Fig. 1*b*). If G_{Ic} is the work per unit crack surface area which is characteristic of the bond, the work required to advance the crack would be W

$$\Delta W = G_{Ic}\bar{b}\Delta l \quad (1)$$

If a load versus specimen mouth opening record is made of the test, it may look like that of Fig. 1*c*. The initial loading from O to E would correspond to the linear flexing of the specimen mouth before any crack growth initiation. At point E the crack begins to grow at the point of the chevron slot. Suppose that the crack has grown to the length l (Figs. 1*a* and *b*) by the time the load-displacement test record reaches point A. Then, if the load on the specimen were relaxed, and if the specimen response is linearly elastic, the unloading path would be along the straight line AO.

Now suppose that instead of unloading along AO, the mouth of specimen is further opened, causing the crack to advance the small additional distance Δl, and causing the load-displacement record to advance to point B. An unloading from B would proceed along the straight line BO, again assuming a linear elastic specimen response. Clearly, then, the additional irrecoverable work done on the specimen in advancing the crack the small distance Δl is equal to the cross-hatched area in Fig. 1*c*. This area is easily approximated as a triangle by defining

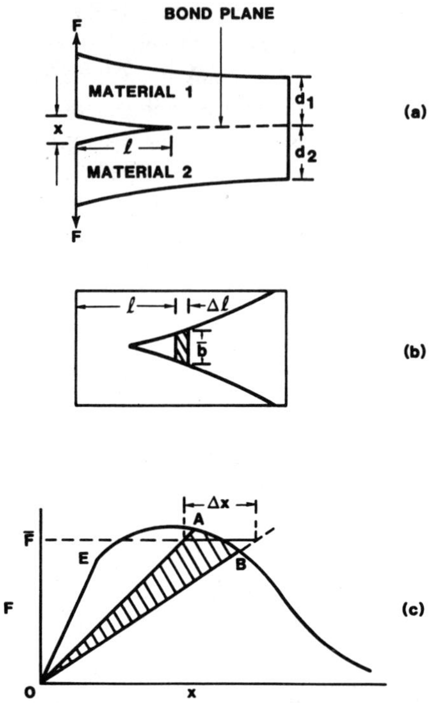

FIG. 1—*Crack extension due to load on chevron shaped crack* (a) *mouth opening displacement of two materials,* (b) *cross section at interface of two materials during propagation of crack,* (c) *load,* F, *versus mouth opening displacement,* x, *corresponding to* (a).

the average load, $\bar{F}$, between points A and B, and by letting Δx be the separation of the unloading paths at the load F. Thus, the irrecoverable work done on the specimen, which supplies the work to create the new crack surface area, is given by

$$\Delta W = {}^1\!/_2\, \bar{F}\Delta x = G_{Ic}\bar{b}\Delta l \qquad (2)$$

Now the change in the specimen compliance between points A and B is

$$\Delta c = \frac{\Delta x}{\bar{F}} \qquad (3)$$

If we use Eq 3 to eliminate Δx from Eq 2, then solve for G_{Ic} and take the limit as $\Delta l \rightarrow 0$, we find

$$G_{Ic} = \frac{F^2}{2b}\frac{dc}{dl} \qquad (4)$$

where $\bar{F} \rightarrow F$ and $\bar{b} \rightarrow b$ as $\Delta l \rightarrow 0$.

Note that in deriving Eq 4, we have made no assumptions which require the two halves of the specimen to be of the same material. However, we do assume that the crack remains coplaner. We have not required that the dimensions of the side comprising Material 1 should be the same as the dimensions of the side comprising Material 2 (Fig. 1*a*). Thus, Eq 4 is quite general, and one could use it directly in a test for the G_{Ic} of a bond between two materials. To do so, however, one would need to do a compliance calibration to evaluate dc/dl as a function of crack length for the particular specimen under investigation. It would be advantageous to somehow be able to use the published compliance calibrations on the short-rod and short-bar specimens [*10*].

The published compliance calibrations were done under the assumption that both halves of the specimen are composed of the same material. This allows the calibration to be given in the dimensionless form of A_o versus l/D, where A_o is defined as

$$A_o = \frac{1}{2b/D} \frac{d(cED)^{1/2}}{c(l/D)} \tag{5}$$

where

E = elastic modulus and
D = specimen breadth.

The published values of A_o ($\approx$23) are independent of the absolute size of the specimen, but a given scaled specimen geometry is required.

One way to modify Eq 4 to make use of the known compliance derivatives of Eq 5 is to divide the compliance of the specimen into two parts: c_1 attributable to the deflection of Material 1 (the top half of the specimen), and c_2 due to Material 2 (the bottom half—Fig. 1*a*). Thus

$$c = c_1 + c_2 \tag{6}$$

Now let the Material 1 side of the specimen be fabricated in the standard short-bar specimen configuration, and let the dimensions of the Material 2 side of the specimen be chosen such that $c_1 = c_2$. Since $c_1 = c_2$ is the same condition as in the calibrated specimen geometry, the Material 1 side (which has the standard dimensions) will then behave as though it were one half of a standard specimen. Also, c and dc/dl will be same as for a standard specimen composed entirely of Material 1. Thus, we multiply both sides of Eq 4 by E_1 (the modulus of Material 1), and use D_1 (the breadth of the Material 1 side)[2] to obtain

$$E_1 G_{Ic} = \frac{F^2}{D_1^3} \left[\frac{1}{2b/D_1} \frac{d(cE_1D_1)}{d(l/D_1)} \right] \tag{7}$$

Note that the term in brackets has only parameters associated with the standard

[2]Multiply and divide right side of Eq 4 by D_1^3.

side of the specimen, and it is therefore equal to A_o^2 (see Eq 5). Thus

$$G_{Ic} = \frac{A_o^2 F^2}{E_1 D_1^3} \tag{8}$$

Equation 8 can be used directly to evaluate G_{Ic}. Note that since the standard short-rod or short-bar fracture toughness test is based on the equation [*10*]

$$K_{IcSR} = \frac{A_o F}{D^{3/2}} \tag{9}$$

(where $D = D_1$ and K_{IcSR} is the critical stress intensity factor determined with the chevron-notched short-rod specimen) the bond toughness test can be run as though it were a standard K_{IcSR} test (except that one half of the specimen will be nonstandard in order to make $c_1 = c_2$). The value of G_{Ic} can be then found from

$$G_{Ic} = \frac{K_{IcSR}^2}{E_1} \tag{10}$$

where E_1 is the elastic modulus of the standard-dimensioned half of the specimen. It should be remembered that the value of K_{Ic} found from Eq 9 is not necessarily a meaningful critical stress intensity factor for the bond. However, the value of G_{Ic} from Eq 8 or 10 will be the actual toughness of the bond in terms of the work per unit area for a crack to separate the bonded materials.

We make the assumption that, to a good approximation, $E_1 I_1 = E_2 I_2$, where I_1 and I_2 are the moments of inertia of sides 1 and 2, respectively, will satisfy the criterion $c_1 = c_2$. The accuracy of this assumption is discussed in the appendix. Another implicit assumption in this paper is that we have coplaner crack propagation. With these assumptions, probably the easiest way to make the compliances of the two halves of the specimen equal is to keep all of the dimensions the same as the standard except for d_2 (Fig. 1*a*). If the value of d_2 is chosen such that

$$E_1 d_1^3 = E_2 d_2^3 \tag{11}$$

then according to elementary beam theory, neglecting shear, the compliances of the two "halves" are approximately equal.

Experimental Procedures

To experimentally verify the analysis, we tested two glass ceramic-metal systems. Glass ceramics were used because they exhibit the oxidation and reduction reactions necessary to form strong bonds with the metal. The two systems selected were zinc silicate glass-ceramic, bonded to molybdenum metal, and a lithium silicate glass-ceramic (S-glass) bonded to Hastelloy C-276 metal. Both of these systems are designed so that the thermal expansion coefficient of the glass ceramic and metal are matched for each. Thus, residual stresses are not developed at the interface on cooling. The compositions of the glass ceramics and the Hastelloy C-276 are given in Table 1. The heat treatment schedule for the zinc silicate-molybdenum system is 1050°C for 40 min and 760°C for 30 min, and for the S-glass Hastelloy system is 1000°C (15 min), 650°C (15 min), and 820°C (20 min), both in an argon atmosphere. The dimensions of the specimens were calculated based on the geometrical dimensions given in Fig. 2 with D equal to 12.7 mm, W = 19.05 mm, d_1 = 5.51 mm, and d_2 = 4.14 mm for the lithia-silica glass ceramic-Hastelloy C-276 combination and d_2 = 3.44 mm for the molybdenum-zinc silicate glass ceramic system. These values were based on the calculations of equal compliances (Eq 11); [E(S-glass) = 9.0×10^4 MPa; E(zinc-silicate = 7.9×10^4 MPa; E(Mo) = 32.4×10^4 MPa; E(C-276) = 20.5×10^4 MPa]. Each specimen was examined to determine the path of crack propagation. A valid test occurs when the crack propagates (macroscopically) coplaner, at least past the critical crack length, (l_c = 10.67 mm), and the value of G_{Ic} is less than the value of the glass ceramic.

Results

We show the results in Table 2 of chevron notch (short-bar) testing on the zinc silicate and the lithia silica glass ceramic (S-glass) systems with their respective metals. In the case of the zinc silicate system, the toughness values are approximately the same as that of the ceramic alone (13 ± 1 J/m^2). This is not really surprising since the zinc-silicate system is known to bond well to molybdenum [*11*], and we would expect the interface of the molybdenum-zinc silicate system to be as strong as the glass-ceramic itself. In the case of the S-glass system, values range from 19 J/m^2 to over 54 J/m^2 depending on the location of the failure. It is well known that heat treatment affects the bond of the lithia silica glass-Hastelloy C-276 system very dramatically [*12*] and that small temperature excursions and the presence of voids at the interface can make large differences in the adherence of the bond. We see from the ceramic fracture

TABLE 1—*Composition[a] of metal-ceramic materials in mole percentage.*

Oxide	S-Glass	Zinc Silicate	Metal	Hastelloy C-276
SiO_2	71.7	46.2	Ni	56.0
ZnO	0	32.3	Cr	15.5
B_2O_3	3.2	0	Fe	16.0
Al_2O_3	5.1	9.5	Co	2.5
P_2O_5	2.5	2.5	W	3.75
MO, M_2O[b]	17.5	9.6	V	0.35

[a]Compositions for glass ceramics are parent glasses.
[b]MO and M_2O are metal oxides.

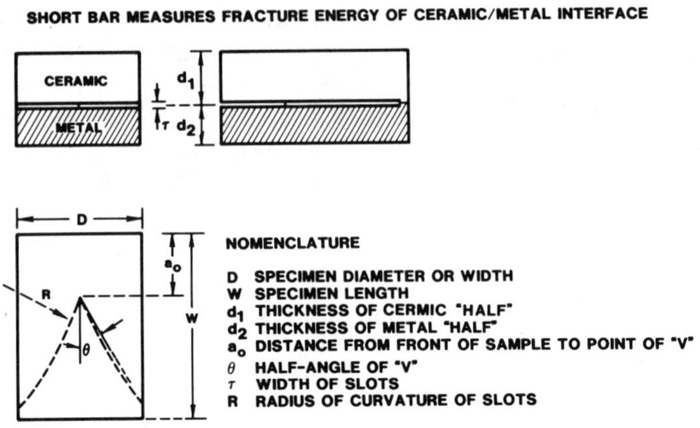

FIG. 2—*Schematic of short bar specimen geometry adapted for two dissimilar materials.*

surface in Fig. 3 that the values of fracture toughness correspond to different amounts of bonding. The 19 J/m^2 corresponds to failure right at the interface in the presence of the oxide associated with the metal and glass ceramic reaction. In the middle figure (33 J/m^2), there are islands of adhered material. The highest fracture energy was obtained when the failure went into the ceramic rather than the interface. In reality when the crack fails to follow the bond plane, there is no longer a valid test of the bond toughness, and the best we can say is that the fracture toughness was at least 54 J/m^2 (the fracture toughness of the ceramic). In fact, because of the sensitivity of the test, we can now determine the significance of heat treatments on the bonding chemistry and use the test to determine the effectiveness of the heat treatment on the bonds. This is one of the intended uses of this specimen.

The short-bar technique was compared to another type of DCB specimen: the constant moment DCB specimen (Fig. 4) developed for testing ceramics [*13*]. In this case, the elastic modulus times the moment of inertia were made equal in both arms of the specimen. This test is similar to that proposed by Becher and Newell [*4*] for thick films and was performed for the same systems given in the experimental procedure. Table 3 shows the relatively good agreement between the DCB specimen and the short-bar specimen in comparison tests using the S-glass ceramic system.

TABLE 2—*Fracture energy of glass ceramic-to-metal chevron-notched specimens.*

System	Failure Location	Toughness,[a] G_{Ic}, J/m^2 System	Toughness,[a] G_{Ic}, J/m^2 Glass Ceramic
Zinc silicate-molybdenum	interface failure	>13	13 ± 1
S-glass-Hastelloy C-276	interface failure	25 ± 1	54 ± 1
	ceramic failure	>54	54 ± 1

[a] ±95% confidence limits.

It must be realized that the fracture toughness in terms of the strain energy release rate, G_{Ic}, is an accurate measure of the resistance of the bond to fracture, that is, it is the energy to create the new surfaces. However, the fracture toughness in the form of the stress intensity factor is really a pseudocritical stress-intensity factor considering the fact that we must determine an effective modulus for the two systems. The value of K_{IcSR} depends on which material we call number one (see Eq 10). It is only K_{IcSR}^2/E_1, that is invariant. The strain energy release rate would be valid in any case.

Conclusion

We have demonstrated that the chevron-notched short-bar specimen can be used for ceramic-metal interface studies. It is a method to study the effect of interface reactions on fracture. We showed for the lithia-silica glass ceramic-Hastelloy C-276 metal system that chemical reactions at the interface are very critical to the fracture behavior during testing.

Acknowledgments

The authors would like to thank Juan Mata for performing many of the experimental tests and Clint Tuthill for producing the test specimens. The authors also thank Drs. J. R. Hellmann and L. S. Costin for reading the manuscript and making helpful, critical comments.

APPENDIX

We can estimate the validity of our assumption that the compliance of each "half" of the metal-ceramic specimens are equal, that is, $c_1 = c_2$, if $E_1I_1 = E_2I_2$. To do this, we examine the expression for the deflection of an end loaded cantilever beam [*14*]

$$y_o = \frac{Fl^3}{3EI} + \frac{\alpha Fl}{ZG} + \frac{kl^{n+1}F}{EI} \tag{12}$$

where

y_o = maximum deflection at the free end of the beam,
F = end load,
l = crack length,
I = moment of inertia ($I = bd^3/12$ for a rectangular section of width b and depth, d),
Z = cross-sectional area, G is shear modulus,
α = numerical factor close to unity [*15*] and,
k and n = constants.

The compliance can be calculated from the end deflection

$$c = \frac{y_o}{F} \tag{13}$$

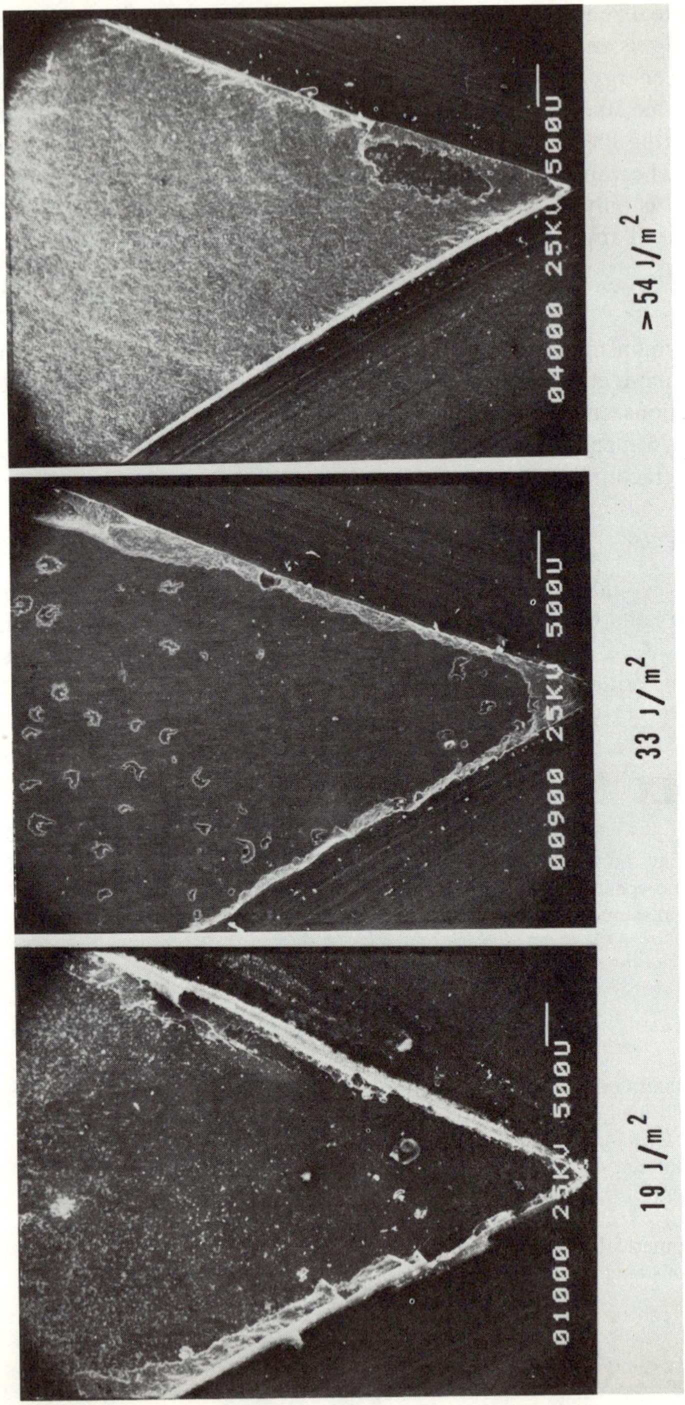

FIG. 3—*Fracture surfaces at the chevron crack on the S-glass ceramic half of the specimen. The critical crack size, that is, the crack location when the toughness measurement is made, is almost to the top of the picture in each.*

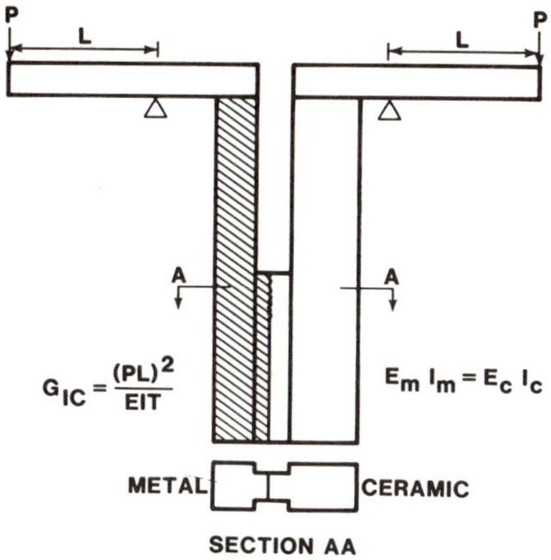

FIG. 4—*Schematic of the constant moment double cantilever specimen adapted to measure the fracture toughness,* G_{Ic}, *of ceramic-metal materials bonded together.*

So, from Eqs 12 and 13, we obtain

$$c = \frac{12l^3}{3Ebd^3} + \frac{\alpha l2(1 + \nu)}{Ebd} + \frac{12kl^{n+1}}{Ebd^{n+1}} \tag{14}$$

where the substitutions $G = E/2(1 + \nu)$ and $I = bd^3/12$ have been made; ν is Poisson's ratio and d^{n+1} is used for dimensional correctness [*15*]. The three terms in Eq 14 are due to flexure, shear, and end rotation, respectively. It can be shown [*14*] that shear effects can be neglected for crack lengths, $l > 3d/2$. Since we are primarily interested when the crack becomes unstable, that is, at the "critical" crack length, l_c, then $d < 2l_c/3$. Since for the standard short-bar specimen, the critical crack length is determined by the geometry, we know that $l_c = 10.7$ mm, and $d < 7$ mm. This is the case for our geometry ($d \leq 5.51$ mm). Thus, we can neglect shear effects; however, we cannot neglect end rotation. In order to evaluate the effect of end rotation on the compliance, we use Eq 14 while neglecting shear effects and substituting values for k and n obtained from Ref *15*, ($k = 0.66$ and $n \approx 1$)

$$c = \frac{4l^3}{Ebd^3} + \frac{12(0.66)l^2}{Ebd^2} \tag{15}$$

This expression can be used to determine the compliance for both sides of a metal-ceramic specimen. If we let the compliance in the ceramic be c_1 and the metal be c_2, then

TABLE 3—*Comparison of DCB and chevron-notched tests of interface toughness.*

	Toughness,[a] G_{Ic}, J/m²	
Failure Location	Chevron Notch	DCB
Interface	25 ± 5	15 ± 9
Into ceramic	54 ± 1	50 ± 9

[a]±95% confidence limits.

$$c_1 = \frac{l_c^3}{3E_1I_1}[1 + 2(d_1/l_c)] \tag{16a}$$

$$c_2 = \frac{l_c^3}{3E_2I_2}[1 + 2(d_2/l_c)] \tag{16b}$$

Since we are primarily interested at the point of instability, $l = l_c = 10.7$ mm. For the condition $c_1 = c_2$, Eqs 16*a* and 16*b* reduce to

$$\frac{E_2I_2}{E_1I_1} = \frac{1 + 2(d_2/l_c)}{1 + 2(d_1/l_c)} \tag{17}$$

However, we know for the standard size specimen side, that $d_1/l_c = 0.517$, so that

$$E_2I_2 \approx \frac{1 + 2(d_2/l_c)}{2} E_1I_1 \tag{18}$$

We could use Eq 18 with an iterative procedure to determine the value of d_2; however, if we can tolerate ~10% error in the compliance, then for $d_2/l_c \approx 0.39$, Eq 18 becomes

$$E_2I_2 = \frac{1 + 0.8}{2} E_1I_1 \tag{19}$$

or

$$E_2I_2 \approx E_1I_1 \tag{20}$$

We can calculate the fractional error (FE) obtained in making the approximation used in Eq 20 by evaluating the expression

$$FE = \frac{dc_I/dl - dc_A/dl}{dc_I/dl} \tag{21}$$

where dc_I/dl is the ideal compliance derivative assuming both halves have the same compliance, and that the compliance is that of a standard short-rod specimen, and dc_A/dl is the actual compliance derivative of the whole specimen consisting of one standard half and a nonstandard half

$$\frac{dc_A}{dl} = \frac{dc_1}{dl} + \frac{dc_2}{dl} \tag{22}$$

$$\frac{dc_I}{dl} = 2\frac{dc_1}{dl} \tag{23}$$

where one can obtain dc_1/dl and dc_2/dl from Eq 14

$$\frac{dc_1}{dl} = \frac{12l^2}{E_1bd_1^3} + \frac{2(1 + \nu_1)\alpha}{E_1bd_1} + \frac{16l}{E_1bd_1^2} \tag{24}$$

$$\frac{dc_2}{dl} = \frac{12l^2}{E_2bd_2^3} + \frac{2(1 + \nu_2)\alpha}{E_2bd_2} + \frac{16l}{E_2bd_2^2} \tag{25}$$

Substituting Eqs 24 and 25 into Eqs 22 and 23 we can evaluate Eq 21 (with $\nu \approx 0.3$, $\alpha \approx 1$, and $E_1 d_1^3 = E_2 d_2^3$)

$$FE = \frac{2.6(d_1^2 - d_2^2) + 16l(d_1 - d_2)}{24l^2 + 5.2d_1^2 + 32ld_1} \tag{26}$$

For the lithia-silica glass ceramic-Hastelloy C-276 combination, $d_1 = 5.51$ mm and $d_2 = 4.14$ mm, and the error is 5.6%. For the molybdenum-zinc-silicate glass ceramic system, $d_1 = 5.51$ mm and $d_2 = 3.44$ mm, and the error is 8.4%. This error can be corrected in the toughness value obtained through Eq 8, if desired. In Eq 8, the factor A_o^2 appears. Since E_1 and D_1 are constants, this value is

$$A_o^2 = \frac{E_1 D_1^3}{2b} \frac{dc}{dl} \tag{27}$$

The compliance calibration value of A_o used in the data analysis corresponds to the ideal compliance derivatives, that is, the A_o^2 that was used was $A_o^2 = A_I^2$

$$A_I^2 = \frac{E_1 D_1^3}{2b} \frac{dc_I}{dl} \tag{28}$$

However, to obtain the correct value of G_{Ic}, we need to use the value of A_o^2 corresponding to the actual specimens that were used, that is, $A_o^2 = A_A^2$

$$A_A^2 = \frac{E_1 D_1^3}{2b} \frac{dc_A}{dl} \tag{29}$$

To correct our data, we must multiply the toughnesses by

$$\frac{A_A^2}{A_I^2} = \frac{(dc_A/dl)}{(dc_I/dl)} = (1 - FE) \tag{30}$$

For the lithia-silica glass ceramic-Hastelloy C-276 combination, $(1 - FE) = 0.944$. For the molybdenum-zinc-silicate glass ceramic system, $(1 - FG) = 0.916$.

References

[1] Kohl, W. H., *Handbook of Materials and Techniques for Vacuum Devices,* Reinhold, New York, 1967.

[2] Hitch, T. T. in *Proceedings of the International Microelectronics Symposium,* International Society for Hybrid Microelectronics, Montgomery, Ala., 1971, pp. 7–7.1–4.

[3] Chapman, B. N., *Journal of Vacuum Science and Technology,* Vol. 11, 1974, p. 106.

[4] Becher, P. F. and Newell, W. L., *Journal of Materials Science,* Vol. 12, 1977, pp. 90–96.

[5] Elssner, G. and Pabst, R., *Proceedings,* British Ceramic Society, Vol. 25, 1975, p. 179.

[6] Erdogan, F., *Journal of Applied Mechanics,* Vol. 32, 1965, pp. 403–410.

[7] Theocaris, P. S. and Stassinakis, C. A., *Engineering Fracture Mechanics,* Vol. 14, 1981, pp. 363–372.

[8] Rice, J. R. and Sih, G. C., *Journal of Applied Mechanics,* Vol. 32, 1965, pp. 418–423.

[9] England, A. H., *Journal of Applied Mechanics,* Vol. 32, 1965, pp. 400–402.

[10] Barker, L. M., *Engineering Fracture Mechanics,* Vol. 17, 1983, pp. 289–312.

[11] Swearengen, J. C. and Eagan, R. J., *Journal of Materials Science,* Vol. 11, 1976, pp. 1857–1866.

[12] McCollister, H. L. and Reed, S. T., ''Glass Ceramic Seals to Inconel,'' Sandia National Laboratories, Document Number 55, 527 (patent pending).
[13] Freiman, S. W., Mulville, D. R., and Mast, P. W., *Journal of Materials Science,* Vol. 8, 1973, p. 1527.
[14] Gillis, P. P. and Gilman, J. J., *Journal of Applied Physics,* Vol. 35, No. 3, 1964, pp. 647–658.
[15] Wiederhorn, S. M., Shorb, A. M., and Moses, R. L., *Journal of Applied Physics,* Vol. 39, No. 3, 1965, pp. 1569–1572.

Summary

Summary

There were twenty-one presentations at the Symposium on Chevron-Notched Specimens: Testing and Stress Analysis. Of these, twenty appear as completed, reviewed manuscripts in this volume. At the symposium there was a division into two categories, analytical and experimental, with considerable overlap in some presentations. For this volume, three categories were chosen: stress analysis, test method development, and fracture toughness measurement. Nearly every paper included information in two of these areas, some all three. Regardless of the overlap, the categorization still helps those new to the topic of chevron-notched fracture testing. The basic geometry used for most of the testing and analysis is an edge-notched specimen loaded in tension with the deep, angled side grooves which join to make the V-shaped chevron-notch. Specimens with round cross section are commonly called short rod, with rectangular cross section, short bar.

Stress Analysis

The first of the six papers in this area is the most comprehensive and the only paper in the volume which is primarily a review of the overall topic. The author, J. C. Newman, heads the cooperative analysis program of the ASTM Task Group E24.01.04 on Chevron-Notched Specimens, and therefore is in good position to describe the development of the various specimens. He reviews the early stress intensity factor expressions based on empirical comparisons and experimental compliance and the more recent stress-intensity factor and displacement results from finite-element and boundary-element methods. He presents consensus results for stress-intensity factor and displacement and a discussion of the applicability of various specimens which will be useful in further work with chevron-notched specimens.

The paper by I. S. Raju and J. C. Newman gives results from three-dimensional finite-element analysis of various specimens. The authors present complete stress-intensity factor distributions along the crack front using a compliance method. Their stress-intensity factors and load-line displacements were up to 5% lower than reported experimental values.

The paper by A. R. Ingraffea, R. Perucchio, T. Y. Han, W. H. Gerstle, and Y. P. Huang describes three-dimensional finite- and boundary-element results. Both average and local variation values of stress-intensity factors along the crack front are given. Significant in their work are edge values of stress-intensity factor 20% higher than centerline values for an assumed straight crack front.

A. Mendleson and L. J. Ghosn present results from a three-dimensional boundary-element analysis. Load-line displacement and stress-intensity factors determined from both stress and compliance calculations were compared with the Raju and Newman results with close agreement.

The two remaining stress analysis papers used primarily experimental approaches. R. J. Sanford and R. Chona performed two-dimensional photoelastic experiments representing the midplane of a chevron-notched specimen. Numerical analysis of the photoelastic results using a ''local collocation'' around the crack tip gives the stress-intensity factor for the range of specimen geometry tested. The photoelastic results were also used to determine the size and shape of the near-field singular stress zone near the crack tip. The paper by I. Bar-On, F. R. Tuler, and I. Roman describes fracture toughness tests with various materials using both chevron-notched bend specimens and existing standard ASTM specimens. Analysis of these results gave experimental stress-intensity factors which compared well, in some cases, with results from a two-dimensional compliance analysis of a straight crack geometry.

Test Method Development

The seven papers in this section are all related to certain important variables and test procedures associated with fracture testing using chevron-notched specimens. The first, by L. M. Barker, describes systematic studies of several key test variables and procedures, including specimen size, elastic-plastic data analysis, and slot thickness and tip geometry. The paper describes the consistency of results in various metal alloys as related to the preceding and other test conditions. It also serves as a useful review of the general topic of chevron-notched testing.

The next three papers deal with fracture testing of hard, brittle materials, specifically glass and rock. R. T. Coyle and M. L. Buhl tested two glasses in a 30% relative humidity environment, and developed computer-assisted data collection procedures for measurement of crack velocity. The paper by A. R. Ingraffea, K. L. Gunsallus, J. F. Breech, and P. P. Nelson describes tests and test method development with limestone and granite. Chevron-notched results compare favorably with results from the conventional and more time consuming test methods. L. Chuck, E. R. Fuller, and S. W. Freiman describe chevron-notched bend testing of glass with humidity and loading rate as test variables. The authors focus on the experimental problems which they encountered, useful information for other prospective users of the test methods, information which too often is unreported.

The next two papers, both from the People's Republic of China, are comprehensive investigations of chevron-notched testing, including combined analysis and experiment. Thus, these papers provide a broad view of the topic, as well as a measure of progress of this topic in another country. Wu Shang-Xian concentrates on analytical compliance formulae for a wide range of chevron-notched geometries. These formulae are particularly useful for those who must use test specimens with unusual dimensions. The author compared fracture toughness measurements from chevron-notched and straight-notched standard specimens, with generally favorable results. The second paper, by Wang Chizhi, Yuan Maochan, and Chen Tzeguang, describes a compliance analysis for stress-intensity factor and an extensive series of tests with eight metallic materials, comparing chevron-notched and standard straight-notched results. A good comparison was noted when stable crack growth and limited plastic deformation were observed.

The last paper in this section, by J. L. Stokes and G. A. Hayes, describes an investigation of the use of acoustic emission with chevron-notched tests of four steels. Load versus deflection plots and load versus cumulative counts plots of the same chevron-notched test are directly compared.

Fracture Toughness Measurements

A primary purpose of these seven papers was to determine the fracture toughness of the particular materials in each of the investigations. In some cases, as discussed next, significant information on stress analysis and test method development was also included in the work. The first two papers describe fracture toughness tests of aluminum alloys. K. R. Brown gives data for seven alloys in various conditions, and points up conditions which affect comparisons between chevron-notched and standard fracture toughness measurements. Test conditions included in his work are toughness level, rising crack-growth resistance, and through-thickness material variation. J. Eschweiler, G. Marci, and D. G. Munz from West Germany, performed tests with one alloy, 7475-T7531, and a variety of test conditions, including specimen size, orientation, and different heats. The most significant difference between chevron-notched and standard toughness measurements was related to the overall toughness level, with higher fracture toughness leading to the larger difference between the two types of tests.

The next three papers involve tests of hard, brittle materials. J. L. Shannon and D. G. Munz describe tests of aluminum oxide with variations in specimen size, proportions, and chevron-notch angle. Differences in measured toughness are related primarily to differences in the amount of crack extension at maximum load. The rising crack growth resistance curve of the oxide is discussed as having important materials effect on measured toughness. The paper by J. R. Tingle, C. A. Shumaker, D. P. Jones, and R. A. Cutler describes toughness measurements of cemented tungsten carbides. Effects on toughness of the amount and distribution of tungsten and carbon were investigated. Also, up to 15% substitution of nickel for cobalt as the binder was found to have little effect. J. Hong and P. Schwarzkopf tested cemented tungsten carbide samples using nine alloys of various cobalt content and carbide particle size. Results from short-rod and four-point bend specimens were compared, including microstructural characterization using optical and electron micrography.

The paper by R. F. Krause and E. R. Fuller describes fracture toughness measurements of polymer concrete materials, which are polymerized mixtures of monomers, portland cement, and silica sand. Effects on toughness of various test conditions were considered, including chevron-notch angle, chevron-vertex position, width of specimen in the crack plane, and the material rising crack-growth resistance curve.

The paper by J. J. Mecholsky and L. M. Barker describes a chevron-notched specimen which was developed to measure the fracture toughness of ceramic-metal interfaces. Specimens were made with the chevron-notch plane aligned with the interface between a glass ceramic and molybdenum and a glass ceramic and Hastelloy 276. The toughness measured from such chevron-notched specimens is a direct measure of the bond strength between ceramic and metal.

What Is Ahead For Chevron-Notched Specimens?

There are several indications that chevron-notched specimens will be often used in the future. First, the basic idea of a self-initiating precrack is sound and useful. Because of the unavoidable complexity of the current standard fracture toughness tests, particularly involving precracking, the chevron-notch concept is attractive and will be used. A second indication of interest in chevron-notched specimens is the response to the symposium and this publication. Research and development work from a variety of perspectives was performed and reported. Finally, this body of work will certainly spur additional research, development, and testing with chevron-notched specimens.

A key requirement for continued technical development and productive use of chevron-notched specimens for fracture testing is a standardized test method. ASTM Task Group E24.01.04 on Chevron-Notched Test Methods is now preparing a draft standard method. It will be based upon the results of interlaboratory analyses and test programs, portions of which are included in this publication. Additional interlaboratory testing, and analysis if required, will be performed to validate the standard test method and demonstrate its precision and accuracy. Then the entire body of testing and analysis, plus any additional work, will be available to assess just which particular combinations of material, geometry, and test procedures, give reliable measures of fracture toughness. It is now clear that for some combinations of test conditions, chevron-notched specimens will provide virtually identical measures of fracture toughness as those obtained from the current ASTM standard methods. It is also clear that some chevron-notched test conditions will give different measures of fracture toughness than those from current standards.

John H. Underwood

Army Armament Research and Development Center, Watervliet, N.Y. 12189; symposium cochairman and coeditor.

Stephen W. Freiman

National Bureau of Standards, Washington, D.C. 20234; symposium cochairman and coeditor.

Francis I. Baratta

Army Materials and Mechanics Research Center, Watertown, Mass. 02172; symposium cochairman and coeditor.

Index